MECHANICS OF

SECONDARY
OIL RECOVERY

CHARLES ROBERT SMITH
Department of Petroleum Engineering
The University of Wyoming

ROBERT E. KRIEGER PUBLISHING COMPANY
MALABAR, FLORIDA

Original Edition 1966
Reprint Edition, 1975, 1983

Printed and Published by
ROBERT E. KRIEGER PUBLISHING COMPANY, INC.
KRIEGER DRIVE
MALABAR, FLORIDA 32950

Copyright © 1966 by
LITTON EDUCATIONAL PUBLISHING, INC.
Reprinted by Arrangement with
VAN NOSTRAND REINHOLD CO.

Printed in the United States of America

Library of Congress Cataloging in Publication Data

Smith, Charles Robert.
Mechanics of secondary oil recovery.

Reprint of the ed. published by Reinhold Pub.
Corp., New York.
Includes bibliographies.
1. Secondary recovery of oil. I. Title.
[TN871.S574 1975]   622'.33'82   74-32220
ISBN 0-88275-270-7

Preface

The mechanics of secondary oil recovery is a subject of more than passing interest to those concerned with the economical removal of valuable hydrocarbons stored in porous media. This book has been designed to serve both as a textbook for the student and as a handbook for the practicing engineer. Most of the content of this book has appeared in the technical literature of the subject within only the last fifteen years. Much of the information presented is only now being verified in the field. Some of the theory included will no doubt need revision, correction and replacement by better established interpretations and concepts.

A vast quantity of literature has appeared on the subject of secondary recovery, with only cursory treatments appearing in book form. It has been the immediate purpose of this work to formulate and correlate what appears to be known about the physical principles and facts underlying the mechanics of secondary recovery of oil. It is hoped that the information presented here will improve the understanding of the subject, with an attending improvement in the application of what is known to field cases. Significant progress along the line of field verification will probably make much of the book out of date in the future. However, the purposes of the work will be achieved just to the degree to which it will accelerate its own obsolescence.

The study of the mechanics of flow through porous media is unique among all applications of physics to the solution of engineering problems. The goal of the study is the prediction and interpretation of the detailed behavior of oil reservoirs under the different known methods of secondary recovery. Yet the behavior fields are irreversible transients which can only be observed or controlled at isolated points in the system. The behavior is controlled by three forces—potential gradients, gravity, and interfacial—acting on a three-phase system—oil, water and gas. No opportunity exists to apply the results directly from one field to another, since any major field experimentation or innovation results in an irreversible change in the system, which may not be exactly duplicated again. Such are the problems of the reservoir engineer!

iii

While a large part of this book has been drawn from the technical literature, the author has profited immensely from association with members of the industrial and academic community. The author has valued highly the comments of the many reviewers who have scrutinized the manuscript to insure accuracy and timeliness of the treatment. A partial list of the reviewers are: Dr. D. L. Stinson, Mr. R. A. Purvis, Dr. H. F. Silver, University of Wyoming; Dr. H. J. Ramey, Jr., Texas A & M University; Dr. Ben H. Caudle, University of Texas; Mr. R. D. Rinehart, Petroleum Engineering Associates; Mr. C. S. Land, Bureau of Mines; Mr. N. J. Clark, consultant; Mr. C. C. Wright, Oilwell Research, Inc.; Dr. Abdus Satter, Pan American Petroleum Corp.; Mr. F. I. Stalkup, The Atlantic Refining Company; Mr. L. K. Strange, Socony Mobil Oil Company, Inc. In addition, graduate students at the University of Wyoming have provided stimulating discussions on the general topic.

This book could not have been completed without the cheerful typing and retyping of most of the manuscript by Donna Thorne. DeeDee Anderson and Susan Clyde helped by correcting the grammar and proofreading the manuscript. Mr. Carl Lindberg prepared the illustrations. Last, but not least, the author is grateful to his wife, Beth, without whose help and understanding this work could not have been accomplished.

C. R. SMITH

Laramie, Wyoming
July, 1966

Dedicated to **Beth** and **Karen** and **Julie**

Contents

C H A P T E R 1

Introduction

Definition of Secondary Recovery

The glossary of the American Petroleum Institute[1] has defined secondary recovery as the oil, gas, or the combination of both, recovered by artificial flowing or pumping means, through the joint use of two or more well bores. Primary recovery has been defined as the oil, gas, or the combination of both, recovered by any method, either natural flow or artificial lift, through a single well bore. The difference between secondary and primary recovery is that secondary recovery involves the introduction of artificial forces, or energy, into a reservoir system, for the purpose of providing or supplementing the motive forces, inherent to the reservoir system. A secondary recovery program may be started at any time in the primary producing history of a given field, and is usually accomplished by the injection of fluids into the reservoir, through certain of the wells.

It could be noted that secondary recovery, as defined above, includes the so-called tertiary and quaternary recovery methods that have received attention in recent years. The above terms have been used to indicate the timing, or the relative numbering of the different producing mechanisms that have been used on a given field. An example would be the case in which a field was allowed to produce to an economic limit by primary means. Secondary recovery was accomplished by means of air injection, and subsequently by water injection (generally referred to as water flooding).

Water flooding in this case would then logically be termed a tertiary recovery operation.   If the definitions set out above were strictly adhered to, the water flooding of the reservoir would still be a secondary recovery technique. This apparent confusion should, however, be of little concern to the engineer interested in the maximum recovery of oil and gas from subsurface reservoirs, and the relative efficiencies of recovery methods.

Historical Background

Much has been written[2, 3] during the past fifty years, relative to the historical background of secondary recovery. In 1880, John F. Carll[4] reported that the back-pressure, exerted on an oil producing section by water seeping into the well bore from higher levels, severely restricted the production of oil from the well.   Crude packers made of seed bags were used to hold the water above the producing sand, and permit removal of oil by pump through tubing. It was also observed that lengthy periods of high water levels in wells where the seed bag packers had failed, resulted in increased production of oil from nearby wells.   Carll seemingly understood the apparent movements of fluids through the reservoir, which accounted for this increase.   His writings give some evidence that accidental, and perhaps intentional water flooding for an increase in oil recovery, was occurring in the Pithole City area of Pennsylvania, as early as 1865.   It is noteworthy that water flooding still accounts for the largest portion of secondary recovery oil.

The benefits of air or gas injection as a secondary recovery means were probably first recognized by James D. Dinsmoor[3], in 1888. Dinsmoor was working as a roustabout in a Pennsylvania field, when he noticed that oil production increased in nearby wells when a well containing both a higher pressure gas sand and a lower pressure oil sand were left temporarily with the two sands in communication. When the well was finally cased and the two sands isolated, production from the surrounding wells declined to the previous level.   Based on this observation, Dinsmoor purposely turned a gas sand into the oil sand on another well, in 1890 or 1891.   The oil production from nearby wells was almost doubled. This marks the humble beginning of the present day gas injection projects, which have accounted for many millions of barrels of secondary oil recovery.

It is interesting to note that the first patents, which can be construed to relate to the secondary recovery of oil, were filed as early as 1864, or only some five years after the completion of the Drake well. A rather extensive treatment of the early patent literature[5] is recorded in the testimony presented in the case of Petroleum Patents Company versus Walter Squires and Fredrick Squires. Most of the early patents[5] were concerned with means of stimulating individual wells by removing paraffin from the bottom of wells, with compressed gas or steam. However, the Richards patent[6] of 1884 states that compressed air, gas, or fluids may be injected as a suitable substitute for the gas, which has already escaped. He points out that maintenance of pressure on the injection well will force oil to the adjoining wells, where it could be recovered. His patent is apparently the first written record in which it is evident that the author clearly conceived that the injection of a number of different fluids down one well bore would have the effect of pushing oil to neighboring wells.

In 1917, Lewis[7] wrote at length concerning the results of injecting air and gas into shallow oil fields in Ohio. In addition, he describes some of the early water flooding operations in the Bradford Oil Field of Pennsylvania. Lewis stated that the Bradford Field was being water flooded, in part, around the turn of the century. It is certainly noteworthy that Lewis had correctly analyzed at this early date the effects of capillary pressure, viscosity of the reservoir fluid and the gases dissolved in the oil on the recovery of oil from underground reservoirs. It is apparent that he also had an appreciation for the heterogeneous character of oil sands. Considering the very sparse quantitative data with which he had to work, his conclusions are indeed significant.

Petroleum Use Rates of the United States and the World

In the increasingly complex industrial world of today, the balance of power between nations, their relative wealth, and the standard of living which their people enjoy can readily be related to their control and use of inanimate energy systems. Oil and natural gas supply a very large proportion of the energy. We need only contemplate the course of World War II to realize that victory or defeat depends upon whether fuels are available to power the ships, airplanes, trucks and other instruments of war. Then too, a country's use rate of inanimate energy is directly tied to the living standard which that country enjoys. The United States derives something in excess of 90 per cent of its energy from inanimate sources, whereas India at the present time uses less than 10 per cent from such sources. It is our use of the inanimate energy, derived primarily from hydrocarbon sources, which has made our country the power which it is today.

Latimer[8] reports that 75 per cent of the energy requirements of the United States are furnished by oil and gas at the present time, and that the use rate of the United States makes up roughly 40 per cent of the world's total consumption. The Office of Oil and Gas of the United States Department of the Interior reported that in 1962 the country had discovered a cumulative in-place total of 346 billion stock tank barrels of oil. Of this total some 68 billion had been produced and 32 billion barrels could be considered proven reserves (that is, can be produced by present technology). This would indicate that the United States has had to date access to 100 billion stock tank barrels, or 29 per cent recovery efficiency. Figure 1-1 shows the cumulative discoveries of in-place oil and the cumulative proven oil recovery. It is interesting to note that the recovery efficiency was 15 per cent in 1930, 21 per cent in 1945, and may reach 43 per cent by 1975. This increase in recovery efficiency in recent years can be largely attributed to the important role of secondary recovery. A major portion of future increases in recovery efficiency, by necessity, will be due to the increasing impact of secondary recovery.

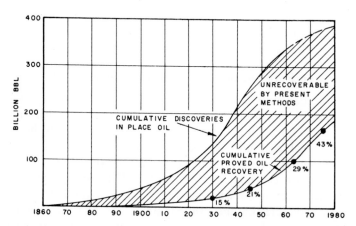

Figure 1-1. Graph showing recoverable and unrecoverable oil versus time. The actual and predicted recovery efficiencies are presented (from the Office of Oil and Gas, United States Department of the Interior).

The annual forecast of the Oil and Gas Journal for 1965[9], sets the proven producible oil reserves at 35,121 million stock tank barrels at the end of 1964, with an estimated ultimate recoverable oil reserve (including condensate) of 111,238 million stock tank barrels. Despite a production and consumption of 2,803 million stock tank barrels of domestic oil during 1964, the proven oil reserves increased by an apparent 528 million stock tank barrels. Figure 1-2 shows the trend of domestic crude production for the years 1954 to 1965. This publication[9] states that some 76 billion barrels of stock tank oil have been produced since the completion of the Drake well in 1859. Preliminary figures indicate that in addition to the 2.8 billion stock tank barrels of domestic oil consumed, an additional 1.2 million barrels per day of crude oil and 1.1 million barrels per day of finished petroleum products were imported, during the year 1964. To partially balance the imports of oil and finished products, the United States exported approximately 4 thousand barrels of crude per day, plus 194 thousand barrels of finished products per day. The domestic demand during the year 1964 was 11.1 million stock tank barrels of oil per day. Figure 1-3 presents a graphical picture of the domestic demand over the time interval, 1957 to 1965. It is reasonable to assume that demand will increase roughly 3 to 4 per cent, for at least the next several years [10].

The world-wide demand must also increase. Many of the emerging nations of the Far East and of Africa and South America will show petroleum use rate increases of more than the 3 to 4 per cent projected for the United States. This follows naturally from the ob-

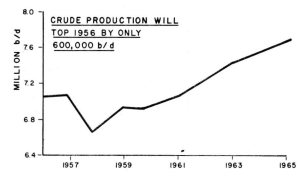

Figure 1-2. United States crude production rates (Oil and Gas Journal, Jan. 25, 1965).

servation that many of these nations have had very low use rates in the past, and will undoubtedly move rapidly toward the use of inanimate energy sources in the future. Even in the European countries, the use rate increase will exceed that of the United States, as people shift from bicycles, to motorcycles, to small cars, and of course to increased use of fuel oils and other energy sources for heating requirements.

We frequently hear that petroleum will very shortly be displaced by some of the more exotic energy sources, such as atomic energy. That some displacement will occur is inevitable. However, with an ever-increasing need for energy, it is almost certain that the demand for petroleum for the foreseeable future will continue to increase. Then too, petroleum is one of the most adaptable organic compounds for the synthesis of materials through the petrochemical industry.

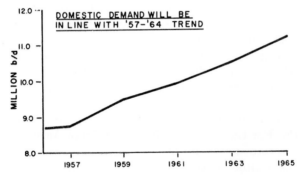

Figure 1-3. Trend of United States domestic demand (Oil and Gas Journal, Jan. 25, 1965).

Oil Reserves of the United States

What the ultimate oil reserves in the United States will be, both producible and in-place values, is of considerable interest to those who concern themselves with sources of energy for the nation. Hubbert[11] was recently quoted as saying that we have found half of our oil, and that the ultimate oil reserves should approximate 175 billion stock tank barrels of oil. He points out that it would be difficult to justify raising this number more than 50 billion barrels, or, to a very maximum, of 225 billion barrels. Apparently, it is intended that this will be the ultimate *recoverable* oil reserve, rather than the ultimate in-place oil reserves. Another source[9] sets the total in-place oil discovered to date at 346 billion barrels. If we were to take Hubbert at his word that roughly half of the oil has been found, then a case could be made for an ultimate in-place oil reserve of 700 billion stock tank barrels. Whether such a value is reasonable, only time will tell. Assuming that the 700 billion barrel value is not unreasonable, the next question is what fraction of this oil-in-place value can we reasonably expect to produce? This, of course, requires some crystal ball gazing, and a bit of optimism regarding the advances which the research and production segments of industry, government, and universities can provide toward the improved recovery of oil. This author finds it difficult to accept a value of less than 50 per cent total recovery, or some 350 billion barrels of ultimate oil recovery from the continental United States.

If we have been capable of recovering 29 per cent of the discovered oil to date (based on 68 billion produced, and 34 billion recoverable of a discovered total of 346 billion), there seems to be no good reason why the recovery factor cannot be increased to 50 per cent.

As a humorous footnote to this rather seriously toned discussion, it has been told that during World War II, when Russia was an Allied Power, one of their bureaucrats was inspecting our oil industry. He was informed that we were making good progress in reservoir control, and that we were now obtaining as much as 40 to 50 per cent of the original oil-in-place from fields which, in earlier years, would only have produced 15 to 20 per cent of the oil-in-place. A bit later, the Russians made the observation that such low recoveries were an illustration of capitalistic inefficiency, and stated that all of the oil was recovered from the Russian oil fields. When pressed for an explanation of this outstanding case of reservoir control, he stated that the wells were simply pumped until they quit producing oil!

The Cost of Producible Oil Reserves. It is a fact that the selling price for a barrel of crude from a given field has changed very little within recent memory. The cost of discovering oil has steadily increased, due to increased equipment cost and higher personnel costs. In addition, oil becomes more difficult to find as the more obvious structures are thoroughly explored. These factors constitute the main reasons for the recent great interest in secondary recovery. In this instance, the oil reserve is already known, and

it remains for the researcher, the engineer and the production personnel to devise methods to recover this oil at the lowest cost
possible. Many times, such secondary recovery "exploration" for
oil production can be done at a cost comparable with those for conventional exploration and, in an increasing number of instances, at
a lesser cost per recovered barrel of oil.

Due to the great divergence in accounting procedures, and in the
reporting form of exploration cost statistics, it is difficult to ar-
rive at an accurate estimate for the present cost of finding one
barrel of recoverable oil. It is certain though, that a cost in excess of $1.00 is experienced in the continental United States.
Bear in mind that a $2.50 possible selling price (an arbitrary value)
does not result in a net profit of $1.50 per barrel of recoverable
oil. Out of this $1.50 must be paid royalty, possible overrides,
taxes (local, state, and federal), wages, producing and treating
costs, plus the fact that recovery may be deferred for a rather long
period of time.

The observation could then be made that if secondary oil can be
"discovered" for less than $1.00 per barrel, then satisfactory
economics may result. Should a shortage of oil force the price of
crude upward, then a larger fraction of the oil presently unrecoverable due to high producing costs would immediately become attractive, and would be produced in our very competitive society.

References

1. Torrey, P. D., *Producers Monthly* (June, 1956).
2. "History of Petroleum Engineering," N.Y., Division of Production,
   A.P.I., 1961.
3. "Secondary Recovery of Oil in the United States," N.Y., Division of
   Production, A.P.I., 1950.
4. Carll, John F., "Second Geological Survey of Pennsylvania, 1875–
   79," III, 1880.
5. Petroleum Patents Co. *vs.* Walter Squires and Frederick Squires, doing business under the firm name of Squires Brothers, in Equity No.
   141, June, July, and October, 1920, U. S. District Court, South District of Ohio, Eastern District, Sater Judge.
6. Richards, William, U. S. Patent 308,522, issued in 1884.
7. Lewis, James O., *U. S. Bureau of Mines Bull.* **148**, 128 (1917).
8. Latimer, J. R., Jr., *Petrol. Engr.*, 62 (January, 1965).
9. Annual Forecast-Review, *Oil Gas J.*, 125 (January 25, 1965).
10. Hill, Kenneth E., AIME paper no. 1526-G, presented 35th Annual Fall
    Meeting, Society of Petroleum Engineers, Denver, October 2–5, 1960.
11. Hubbert, M. King, *Oil Gas J.*, 48 (January 21, 1964).

CHAPTER **2**

Determination of Oil-in-Place

The determination of the oil-in-place in a particular reservoir that is being considered as a candidate for one or more of the secondary recovery techniques is one of the factors of primary importance. It is intended here that secondary oil reserves be defined as that oil recoverable by means of the application of a particular technique designed to supplement the primary producing mechanism. Since many of the present day secondary recovery techniques are applied before the primary producing history of the reservoir is complete, actually the first part of the secondary recovery project may be better termed a pressure maintenance measure. Any pressure maintenance project, if successful, will recover more than the reserves producible by the primary producing mechanism.

The estimation of primary oil reserves cannot be reduced to a simple procedure, since several of the natural producing mechanisms are rate sensitive. The water drive, gas cap drive and combination drive producing mechanisms are good examples of rate-sensitive processes. The solution drive mechanism is the only producing mechanism that is truly rate-insensitive. The competitive and regulatory situation in a given area will usually control the actual primary oil recovery, which will ultimately be obtained as a result of the regulation of field-wide oil and gas producing rates.

Since several of the secondary recovery methods that will be described in the following chapters of this book are applicable to oil reservoirs that have not been producible until very recently, it is well to distinguish between oil reserves and oil deposits. Oil reserves refer to that oil which is recoverable by a known economic recovery method, but which has not yet been reduced to possession. Oil deposits refer to a location where oil exists, but which may or may not be presently recoverable by a known economic recovery method. Most of the tar sands which contain oil that will not flow under any reasonable pressure gradient would be classified as an oil deposit at this time, not as an oil reserve.

The importance of accurately determining the reserves that can be recoverable by primary producing means and that which can be recovered by the several applicable secondary recovery techniques cannot be overemphasized. Detailed design of the surface facilities and producing equipment is seldom able to influence the outcome of a host reservoir. On the other hand, the careful and detailed prediction of the performance to be expected will permit the judicious design of surface and producing facilities, to allow a maximizing of the economic benefits to be expected.

Determination of Primary Recovery

In those fields where a primary producing history is available as the result of the drilling and the producing of oil wells by natural or artificial oil lifting methods, the original oil-in-place may be determined by two means:

(1) by the use of material balance equations—provided that production history is known, and the properties of the producing fluids are known or can be developed, and

(2) by volumetric means—provided that information is available as to reservoir extent and thickness, reservoir porosity and water saturation, and the oil formation volume factor is known.

The primary oil recovery can be determined by three methods:

(1) material balance equations in conjunction with gas-oil ratio and saturation equations,

(2) decline curves when extensive production history is available, and

(3) by volumetric means if the residual oil saturation and the abandonment oil formation volume factor is known or can be estimated.

In addition, empirical procedures have been developed, which can result in a satisfactory determination of the producible oil reserves for special reservoir cases.

Each of the procedures of the estimation of oil recovery will be discussed individually in the following sections.

Material Balance Calculations

Although a number of procedures are available for the estimation of oil-in-place by volumetric means, it is wise to confirm these estimations by material balance calculations, if sufficient production data are available. A material balance, as the name implies, amounts to a balancing or inventory of the materials in a reservoir. Usually the material balance is written on a volumetric basis, although this is not the only basis upon which such an inventory of reservoir fluids may be made. Conceptually, the material balance equation will be seen to be a statement that the initial oil volume is equal to the oil volume remaining, plus the oil volume removed. It is more convenient, however, to make the balance on the gas present in the reservoir:

$$\text{Gas-in-place initially} = \text{Gas remaining} + \text{Gas produced} \quad (1)$$

Developing equation (1) results in the following material balance equation [1, 2, 3, 4, 5]:

$$N = \frac{N_p[B_o + B_g(R_c - R_s)] - B_w(W_e - W_p)}{mB_{oi}\left(\dfrac{B_g}{B_{gi}} - 1\right) + B_g(R_{si} - R_s) - (B_{oi} - B_o)} \quad (2)$$

where

N = original oil-in-place, STB

N_p = cumulative oil produced, STB

B_o = oil formation volume factor, volume at reservoir conditions per volume at standard conditions

B_g = gas formation volume factor, volume at reservoir conditions per volume at standard conditions

$\quad = \dfrac{14.7}{P}\dfrac{T_f}{520}z$

p = pressure, psia

z = gas deviation factor

R_c = net average cumulative gas-oil ratio, standard volume of gas per standard volume of stock tank oil

$\quad = G_p/(N_p \times 5.615)$

G_p = cumulative gas produced, scf

R_s = solution gas-oil ratio, standard volume of gas per standard volume of stock tank oil

B_w = water formation volume factor, volume at reservoir conditions per volume at standard conditions

W_e = cumulative water influx during the production of N_p stock tank barrels of stock tank oil, barrels

W_p = cumulative water produced during production of N_p stock tank barrels of stock tank oil, barrels

m = ratio of initial gas cap volume size to initial oil zone volume size, fraction

i = subscript indicating initial value or conditions

An insight into equation (2) may be gained by considering the physical significance of the groups of terms of which it is comprised. The set of terms, $N_p[B_o + B_g(R_c - R_s)]$, represents the reservoir volume of the cumulative oil and gas produced, while the total water which encroaches and is retained in the oil reservoir volume is represented by the terms $B_w(W_e - W_p)$. The expansion of the gas cap which occurs with the production of N_p barrels of stock tank oil is represented by the terms, $mB_{oi}(B_g/B_{gi} - 1)$. The reduction in the amount of gas initially present in stock tank barrels of oil, at reservoir conditions of pressure and temperature, is evidenced in the terms $B_g(R_{si} - R_s)$. The change in volume of the reservoir oil comprising one barrel of stock tank oil as the pressure in the system is lowered, is represented by the terms, $B_{oi} - B_o$.

It is evident from equation (2) that the volume of the original oil-in-place can be determined when certain geologic, laboratory and production data are available. A study of geologic data, in conjunction with production information, will usually indicate the type of producing mechanism which is operative. If the producing mechanism is not known, the problem will be complicated somewhat, since the producing mechanism must be preassumed and the oil-in-place calculated. If the assumption is not verified, as will be discussed later, the calculation must be repeated under a new assumption. There are essentially three unknowns in equation (2):

(1) the original oil-in-place (n),

(2) the cumulative water influx (W) and,

(3) the original size of the gas cap as compared to the oil zone size (m).

In general, it will be necessary to know the size of the gas cap (m), if the value of N is to be determined.

The reservoir fluid properties which must be known directly are B_o, R_s, B_g, and B_w. Since B_g is calculated from the natural gas law, the factors z and gas specific gravity, or gas composition, will also need to be known. Where reservoir pressures are moderate, B_w will approximate unity. Where the reservoir pressure is high, B_w should not be neglected. In this case, the specific gravity of the water (salt content) will need to be known so that the solubility of natural gas in water, with its resulting influence on B_w, may be determined[6,7]. In general, the reservoir fluid properties may be determined by laboratory studies, from information on similar fluids, or from published correlations. Amyx *et al.*[5], Burcik[8], Campbell[9], and Frick[10], have presented summaries of the available information on reservoir fluid properties.

The material balance equation presented in equation (2) is completely general. If a gas cap does not exist in a particular reservoir, then m is zero, and the gas cap term, $mB_{oi}(B_g/B_{gi} - 1)$, is zero. If no water drive exists, then W_e, the water encroachment, is zero, and is dropped from the equation. Above the bubble point of the reservoir fluid, the amount of gas in solution in the oil is constant and R_c, R_s and R_{si} are equal. This will result in a further simplification of the equation for a given reservoir case.

Material Balance—Solution Gas Drive Reservoirs. For the solution gas drive case, also called depletion drive, volumetric gas drive, or internal gas drive, no gas cap or water drive will be in evidence. Equation (2) then becomes:

$$N = \frac{N_p [B_o + B_g (R_c - R_s)] + B_w W_p}{B_g (R_{si} - R_s) - (B_{oi} - B_o)} \qquad (3)$$

If no water production occurs with the oil, the last term in the numerator also becomes zero. Strictly speaking, equation (3) is completely general, and applies both above and below the bubble-point pressure of the particular reservoir fluid. It should be noted that if a field has a primary producing history that indicates that solution gas is the driving mechanism, the initial pressure of the field will, in all probability, exceed the bubble-point pressure. If it were determined that the field was at bubble-point pressure at the time of discovery, this would constitute strong evidence that a gas cap existed. This follows from the simple reasoning that in this instance, the reservoir would be holding the maximum amount possible of gas in solution. It would be a considerable coincidence if there were exactly sufficient gas available to saturate the oil, but not enough to form a gas cap (even though small in size). When insufficient gas is present to saturate the oil at the prevailing initial reservoir pressure, the reservoir is said to be undersaturated. For this case, equation (3) may be simplified to the following form:

$$N = \frac{N_p B_o + W_p B_w}{B_o - B_{oi}} \tag{4}$$

Equations (3) and (4) may be used for the purpose of estimating N from oil and water production data and from information on the oil formation volume factor when the initial reservoir pressure is moderate (less than 2500 psia), the formation porosity is relatively high (larger than 20 per cent), and the water saturation is less than 30 per cent. When these conditions are not closely met, it will usually be necessary to include the effects of rock and water compressibility in the production of oil above the bubble-point pressure. It will be convenient to modify equation (4) so that the change in the oil formation volume factor, $B_o - B_{oi}$, will be represented in terms of the oil compressibility and the initial oil formation volume factor, B_{oi}.

By definition, the compressibility of oil is:

$$c_o = - \frac{1}{B_o} \frac{dB_o}{dp} \tag{5}$$

Separating variables and integrating between the original pressure and a lower pressure, larger than the bubble-point pressure, results in the following equation form:

$$B_o = B_{oi} e^{c_o(p_i - p)} \tag{6}$$

The expansion of a function of the form e^x is:

$$e^x = 1 + x + \frac{x^2}{2!} + \frac{x^3}{3!} + \dots \tag{7}$$

Since the exponent, $c_o(p_i - p)$, is much less than one, equation (6) may be rewritten, in view of the fact that exp $[c_o(p_i - p)]$ is closely approximated by $[1 + c_o(p_i - p)]$, in the following form:

$$B_o - B_{oi} = c_o B_{oi}(p_i - p) \tag{8}$$

Equation (4) then becomes:

$$N = \frac{N_p B_o + W_p B_w}{c_o B_{oi}(p_i - p)} \tag{9}$$

The effects of rock and water compressibility can be included in equation (9) by replacing the oil compressibility term, c_o, by an effective oil compressibility term, c_{oe}, defined by the following equation:

$$c_{oe} = c_o + \left(\frac{S_w}{1 - S_w}\right)c_w + \left(\frac{1 - \phi}{\phi(1 - S_w)}\right)c_f \qquad (10)$$

where

c_o, c_w, c_f = compressibilities of oil, water, and rock, respectively, vol/vol-psi
ϕ = porosity, fraction
S_w = water saturation, fraction

Equations (9) and (10) should be used with considerable caution for the calculation of the original oil-in-place from data available on fluids and field production, since small changes in the denominator can cause inordinately large changes in the calculated value of N. Above the bubble-point, it will usually be more satisfactory to calculate N on a volumetric basis. The utility of the material balance equation above the bubble-point can be shown by an example calculation:

EXAMPLE 1.
    Calculate the original oil-in-place for an undersaturated reservoir having an initial pressure of 5000 psia and an initial oil formation volume factor of 1.305. The following data applies at 3350 psia (p_b = 2750 psia):

$$B_o = 1.330 \qquad\qquad \phi = 10 \text{ per cent}$$
$$N_p = 1,510,000 \text{ STB} \qquad S_w = 21.6 \text{ per cent}$$
$$c_o = 1.5 \times 10^{-5} \text{ psi}^{-1}$$
$$c_f = 3 \times 10^{-7} \text{ psi}^{-1}$$
$$c_w = 3.5 \times 10^{-6} \text{ psi}^{-1}$$

From equation (10):

$$c_{oe} = (1.5 \times 10^{-5}) + \frac{(0.216)(3.5 \times 10^{-6})}{(1 - 0.216)} + \frac{(1 - 0.10)(3 \times 10^{-7})}{(0.10)(1 - 0.216)} =$$
$$1.94 \times 10^{-5} \text{ psi}^{-1}$$

It is evident that the contribution of the rock and water compressibilities will be roughly 10 per cent of the total equivalent oil compressibility. Equation (9) yields:

$$N = \frac{(1,510,000)(1.330)}{(19.4 \times 10^{-6})(1.305)(5000 - 3350)} = 48,100,000 \text{ STB}$$

Usually, production data used for the determination of the original oil-in-place will, in reservoirs producing by solution gas drive, be largely from producing history where field pressure is less than the bubble-point pressure. In this instance equation (3) applies. If the reservoir were initially under-saturated, an attempt should be made to estimate the amount of oil produced between the original pressure and the bubble-point pressure. This will require that a reasonable estimate be made for the original oil-in-place—usually by volumetric considerations. Equation (9) may then be rearranged to permit a calculation of the oil produced down to the bubble point of the reservoir:

$$N_p = \frac{N c_{oe} B_{oi}(p_i - p_b) - W_p B_w}{B_{ob}} \tag{11}$$

The cumulative oil value, N_p, used in equation (3), would then be equal to the total stock tank oil produced, less that produced, as indicated by equation (11). The concepts involved are best shown by an example calculation.

EXAMPLE 2.
   Calculate the original oil-in-place for the reservoir of example 1, where the following additional reservoir data are available at a reservoir pressure of 1500 psia:

$B_o = 1.250$ $z = 0.90$
$R_s = 375$ SCF/STB $T_f = 240\ °F$
$G_p = 3,732 \times 10^6$ SCF
$N_p = 6,436,000$ STB $B_g = \dfrac{(14.7)(700)(0.90)}{(1500)(520)} = 0.01187$ CF/SCF
$B_{ob} = 1.350$ (at 2750 psia)
                                 $R_{sb} = 500$ SCF/STB (at 2750 psia)

Assuming that the original oil-in-place approximates 48,300,000 STB, equation (11) can be used to determine how much of the 6,436,000 STB was produced, as the reservoir pressure was lowered from 5000 psia, to the bubble-point pressure of 2750 psia.

$$N_p = \frac{48,100,000\ (19.4 \times 10^{-6})(1.305)(5000 - 2750)}{1.350} = 2,040,000\ \text{STB}$$

This indicates that the oil produced, as the pressure was lowered from 2750 psia to 1500 psia, approximates:

$$N_p = 6,436,000 - 2,040,000 = 4,396,000\ \text{STB}$$

Equation (3) may now be used for the calculation of N, noting that the cumulative net producing gas-oil ratio will be calculated by dividing the total gas produced by the total oil produced, where all values are taken below the bubble point:

$$R_c = \frac{(3,732 \times 10^6) - 2,040,000\,(500)}{4,396,000} = 620 \text{ SCF/STB}$$

Then,

$$N = \frac{4,396,000 \left[1.250 + 0.01187 \left(\dfrac{620 - 375}{5.615} \right) \right]}{0.01187 \left(\dfrac{500 - 375}{5.615} \right) - (1.350 - 1.250)} = 47,100,000 \text{ STB}$$

The original oil-in-place would be 47,100,000 plus 2,040,000 STB. Strictly speaking, the calculation for the amount of oil produced between 5000 psia and the bubble-point pressure should be repeated, using an N value of 49,140,000 STB. This would result in some final adjustment to the indicated value for the original oil-in-place. The recalculation yields an original oil-in-place of 49,000,000 STB, an oil-in-place value at the bubble point of 46,900,000 STB and an oil production by fluid expansion of 2,075,000 STB.

A comparison of the answers in examples 1 and 2 shows some apparent disagreement as to the correct value for N. Since more data are available for the second calculation, the value of 49,000,000 would probably be the more accurate of the two values.

Since many of the displacement processes require that the oil saturation at the beginning of a project be sufficiently high that a frontal displacement occurs, a knowledge of its value will be necessary. The following saturation equation can be used for this purpose:

$$S_o = \frac{(1 - S_w)(N - N_p)(B_o)}{N B_{ob}} \tag{12}$$

The oil saturation, as a fraction of the total pore space, could then be readily calculated at a pressure of 1500 psia, for the data of example 2.

EXAMPLE 3.

Calculate the oil saturation at a reservoir pressure of 1500 psia for the reservoir data provided in example 2.

Equation (12) yields:

$$S_o = \frac{(1 - 0.216)(46,900,000 - 4,361,000)(1.250)}{46,900,000\,(1.350)} = 0.66 \text{ or } 66 \text{ per cent}$$

It should be noted that the N value used in the equation is that corresponding to oil-in-place at the bubble point. The oil-production below the bubble point amounts to 6,436,000 STB, minus 2,075,000 STB, or 4,361,000 STB.

Since secondary recovery will often be considered before the completion of the primary producing history, it will be useful to know the amount of primary oil to be recovered to the economic producing limit. This will provide a measure of the true recovery by primary producing means, and allow a calculation of the additional recovery, which would result from a secondary recovery project. The inclusion of part of the primary producing phase oil, as a justification for the installation of a secondary recovery project, can only properly be based upon an increase in the current producing rate, with a resulting benefit due to a shortened primary producing phase. Present income is of much greater interest than income which is deferred to a later time.

A number of authors [11, 12, 13, 14] have published techniques by which the primary recovery of a field, under solution gas drive, may be calculated. The calculation technique suggested by Pirson[14] requires the least amount of time if a desk calculation is attempted. The reader is referred to the original references for details in the calculation of primary recovery by solution gas drive. It is interesting to note that the solution gas drive type reservoirs provide some of the best candidates for secondary recovery, due to the usually high remaining oil saturations after the completion of the primary producing phase. Then too, due to the low recoveries from these types of reservoirs, secondary recovery (or pressure maintenance) operations are often instituted early in the primary producing history of such fields.

Material Balance—Water Drive Reservoirs. The calculation of the original oil-in-place in a water drive reservoir may also be accomplished using equation (2). If no gas cap exists, equation (2) may be rewritten in the following rearranged form:

$$N = \frac{N_p [B_o + B_g (R_c - R_s)] + B_w W_p}{D_i} - \frac{W_e B_w}{D_i} = N_i - \frac{W_e B_w}{D_i} \quad (13)$$

where

$$D_i = B_g (R_{si} - R_s) - (B_{oi} - B_o)$$

Study of equation (13) indicates that the first term on the right-hand side represents the contribution of solution gas drive (see equation (3)), while the second term indicates the contribution to the over-all driving mechanism that is supplied by water encroachment.

If the reservoir pressure exceeds the bubble-point pressure of the reservoir oil, equation (13) can be simplified, since R_c, R_s, and R_{si} will be equal:

$$N = \frac{N_p B_o + W_p B_w}{D_i} - \frac{W_e}{D_i} = N_i - \frac{W_e B_w}{D_i} \tag{14}$$

where

$$D_i = B_o - B_{oi} \simeq c_{oe} B_{oi} (p_i - p) \quad \text{[see equation (9)]}$$

The effective oil compressibility, c_{oe}, is defined by equation (10).

The normal procedure for determining the original oil-in-place in a water-driven oil reservoir, is to initially ignore the contribution of the water encroachment term, since W_e will seldom be available from measurements in the field. In special instances, a well drilled through the oil-water contact may be used to obtain data for a volumetric determination of the water encroachment rate. If data are available at two or more different times, the calculation of N, ignoring water encroachment, will result in increasing values for the original oil-in-place, as producing history increases. It is evident from either equation (13) or (14) that in this instance the water encroachment, which would serve to reduce the calculated value for N, has not been subtracted. Figure 2-1 provides a plot of the indicated original oil-in-place values, N_i, versus time. Extrapolation of the curve back to zero time, corresponding to zero water influx, results in the original oil-in-place, N. The primary objection to a plot, such as Figure 2-1, is that the plotted data points do not necessarily define a straight line. Extrapolating a curved line can result in significant errors.

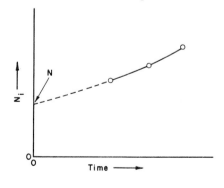

Figure 2-1. Plot of indicated oil-in-place versus producing time.

The calculation of N in a water drive field can be validly based upon expressions for water encroachment which, in turn, provide a means by which the future primary performance of the field can be based. Three methods have been widely used for the representation of water encroachment, W_e. The simplest approach is that due to Schilthuis [15]:

$$W_e = C_s \int_0^t \Delta p\, dt \simeq C_s \sum_0^t \Delta p\, \Delta t \qquad (15)$$

where

      C_s = proportionality constant, bbl/month-psi

      Δp = pressure difference between the aquifer and the reservoir, psi

      Δt = time over which the pressure difference, Δp, is applied, months

Equation (15) is strictly applicable only to those reservoirs that have high aquifer permeability or a close recharge source, so that water influx can be represented by steady-state methods. Since data are usually available at discrete intervals, the calculations will be in finite steps such that the pressure difference will be equal to the aquifer pressure minus the average reservoir pressure, over a given time interval.

Hurst [16] has also presented an equation to represent water influx, which modifies equation (15) for the case where the reservoir does not perform in a steady-state manner. The equation has the form:

$$W_e = C_h \int_0^t \frac{\Delta p\, dt}{\log t} \simeq C_h \sum_0^t \frac{\Delta p\, \Delta t}{\log t} \qquad (16)$$

where

      C_h = proportionality constant, bbl/month-psi

Equation (16) finds best application where the encroaching water has to come from greater and greater distances as field history lengthens. It should be noted that the time must be in months in this empirical equation. The log term creates problems otherwise, since, strictly speaking, the logged quantity should be dimensionless.

Where the size of the field warrants the approach, van Everdingen and Hurst[17] have described a method which can be used for water influx in finite or infinite aquifers for the unsteady-state case:

$$W_e = C_v \sum_0^t \Delta p \times Q_{(t)} \qquad (17)$$

where

C_v = water influx constant, bbl/psi

$$= 1.119 \, \phi \times c_{we} \times r_r^{\,2} \times h \times \frac{\theta}{360}$$

c_{we} = effective water compressibility, vol/vol-psi

$$= c_w + \frac{(1 - \phi) \, c_f}{\phi}$$

r_r = reservoir radius, feet

h = effective aquifer thickness, feet

θ = angle subtended by the reservoir circumference or the circumference over which water encroaches, degrees

Δp = pressure increment, psi

$Q_{(t)}$ = dimensionless water influx rate as a function of dimensionless time

Craft and Hawkins[18] have presented a particularly lucid treatment of the unsteady-state water influx problem.

The usefulness of the equations for water encroachment can be illustrated through the use of the simple form presented by Schilthuis. Equation (13) can be rewritten to include the Schilthuis water influx term:

$$N = N_i - \frac{C_s \sum_0^t \Delta p \, \Delta t}{D_i / B_w} \qquad (18)$$

If

$$N_i \text{ and } \left(\sum_0^t \Delta p \, \Delta t \right) / D_i$$

terms at the same times in the producing history of the field are plotted as shown in Figure 2-2, a straight line should result. The slope of the resulting straight line yields the steady-state water

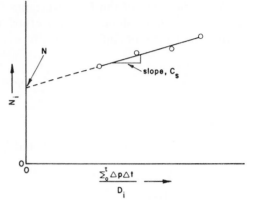

Figure 2-2. Plot to determine the original oil-in-place in a water-driven field.

influx constant, C_s, while an extrapolation to the ordinate results in a reliable value for N. It should be noted that if the reservoir pressure exceeds the bubble-point pressure, N_i and D_i are defined by equation (13), while if the pressure is less than the bubble-point pressure, equation (14) defines these factors.

The utility of the above method for determining the original oil-in-place, and providing a quantitative measure of the relative strength of the water drive, is illustrated in the following example.

EXAMPLE 4.

Determine the original oil-in-place in a field having the following production data and fluid properties:

Time, yr	Pressure, psia	N_p STB	W_p bbl	$G_p \times 10^{-6}$ SCF	B_o bbl/STB	R SCF/STB	B_g CF/SCF
0	2750*	0	0	0	1.350	500	0
2	2100	6,826,800	160,000	3,890	1.302	395	0.00829
4	1500	15,376,800	490,000	12,150	1.250	306	0.01187

*Initial pressure is the bubble-point pressure.

Since the data apply below the bubble point, equation (3) defines the values for N_i in equation (18). The value for N_i at 2100 psia, may be calculated as:

$$N_1 = \frac{6,826,800 \left[1.302 + 0.00829 \left(\dfrac{3,890 \times 10^6}{6,826,800(5.615)} - \dfrac{395}{5.615}\right)\right] + 1.02^*(160,000)}{0.00829 \left(\dfrac{500 - 395}{5.615}\right) - (1.350 - 1.302)}$$

$$= 101,200,000 \text{ STB}$$

*Estimated value for B_w.

A similar calculation would yield an indicated value for the oil-in-place, using production data at 1500 psia of 114,300,000 STB of oil (N_2). Water encroachment is indicated, since calculated N values are increasing. Equation (18), the Schilthuis water influx case, can be written using data corresponding to the 2100 psia reservoir pressure:

$$N = 101,200,000 - \frac{C_s \left(2750 - \dfrac{2750 + 2100}{2}\right)(2 \times 12)}{\left[0.00829 \left(\dfrac{500 - 395}{5.615}\right) - (1.350 - 1.302)\right]\bigg/1.02}$$

or,

$$N = 101,200,000 - 74,400\, C_s$$

The equivalent equation developed from the reservoir data at 1500 psia is:

$$N = 114,300,000 -$$

$$\frac{C_s \left[\left(2750 - \dfrac{2750 + 2100}{2}\right)(24) + \left(2750 - \dfrac{2100 + 1500}{2}\right)(24)\right]}{\left[0.01187 \left(\dfrac{500 - 306}{5.615}\right) - (1.350 - 1.250)\right]\bigg/1.02}$$

or,

$$N = 114,300,000 - 104,000\, C_s$$

The resulting two equations could be solved graphically, as in Figure 2-2, or as two simultaneous equations having two unknowns. The original oil-in-place, N, then is 68,200,000 STB, and the Schilthuis water influx constant, C_s, is 443 bbl/month-psi.

Example problem 4 shows the large influence which encroaching water can have on the calculated values of N where the water influx term is neglected. When this calculation approach is used, consideration should be given to the type of aquifer which is present. This would permit a decision to be made as to whether the Schilthuis, Hurst or van Everdingen-Hurst water influx terms should be employed.

Pirson[19] and others have published methods by which the primary recovery in water-drive fields may be determined. The reader is referred to these works for calculation details. Any secondary recovery project being considered for a field having at least a partial water drive should be based upon a careful determination of the remaining oil-in-place, from the information that primary recovery can provide. Failure to do this could result in an unsuccessful secondary recovery operation. Information in Chapters 5 and 7 is also directly applicable to the displacement of oil by natural water drive.

Material Balance—Gas Cap Drive Reservoirs. The material balance equation may also be used to determine the original oil-in-place when a gas cap is present. For this discussion, it will be assumed that the size of the gas cap relative to oil zone size, m, is known, and that no water influx is occurring. Equation (2) may be written in the following specialized form for this case:

$$N = \frac{N_p\left[B_o + B_g(R_c - R_s)\right] + B_w W_p}{mB_{ob}\left(\dfrac{B_g}{B_{gb}} - 1\right) + B_g(R_{sb} - R_s) - (B_{ob} - B_o)} \tag{14}$$

It should be noted that the subscript "i" has been replaced by a "b", since if a gas cap is present, the reservoir oil at initial conditions will be at the bubble-point. Equation (14) also only applies to those reservoirs where there is no active segregation of oil and gas as the field is produced, i.e., gas moving upward to a secondary gas cap, and oil moving downward to take the place of the gas. This condition will usually be satisfied if vertical permeability is less than approximately 200 millidarcys. If active segregation of the gas and oil is known to occur, the frontal displacement which occurs between the oil and gas zones must be considered. The reader is referred to the work of Pirson[20] for calculation procedures for this more complicated case.

Where active segregation of the fluids is not occurring, the original oil-in-place may be determined in the same manner as for the depletion drive case, outlined in the last section example calculation 2 of this chapter.

Material Balance—Combination Drive Reservoirs. The original oil-in-place may also be calculated with good accuracy where both a gas cap and water drive exist, if good production data and fluid property information are available. The size of the gas cap relative oil zone size must be known, or too many unknowns will exist for N to be determined accurately. Since, if a gas cap is present, the reservoir oil will be at bubble-point conditions initially, equation (2) may be written in the following specialized form:

$$N = \frac{N_p\left[B_o + B_g(R_c - R_s)\right] - B_w(W_e - W_p)}{mB_{ob}\left(\dfrac{B_g}{B_{gb}} - 1\right) + B_g(R_{sb} - R_s) - (B_{ob} - B_o)} \tag{15}$$

where the subscript "b" indicates bubble-point conditions existing in the oil zone. To be strictly applicable, no vertical segregation of oil and gas should occur if N is to be determined from production and fluid property information alone. This condition will usually be satisfied if the vertical permeability is less than approximately 200 millidarcys. Where water is encroaching, it will be convenient to alter equation (15) to the following form:

$$N = \frac{N_p [B_o + B_g (R_c - R_s)] + B_w W_p}{D_i} - \frac{B_w W_e}{D_i} \qquad (16)$$

where

$$D_i = mB_{ob} \left(\frac{B_g}{B_{gb}} - 1 \right) + B_g (R_{sb} - R_s) - (B_{ob} - B_o)$$

A consideration of the type of aquifer present will permit the proper choice of expression for W_e from equations (15), (16), or (17) of this chapter.

The determination of the original oil-in-place is done in the same fashion as that outlined in problem example 4 of this chapter. One additional group of terms is added to the development, $MB_{ob} (B_g/B_{gb} - 1)$, to account for the expansion of the gas cap if the reservoir pressure is lowered. For a determination of the primary oil recovery, the reader is referred to Pirson[20].

Volumetric Methods—Calculation of Oil Volumes

Perhaps one of the best understood methods for the estimation of primary and secondary oil recovery and original oil-in-place is the volumetric method. It is also one of the most abused techniques. In concept, the volumetric approach to oil reservoir or oil deposit volume determination is simple. Difficulty can arise, and often does, in the generation of values for the various factors which go into the determination of reserves by volumetric calculation means.

One would expect that the primary recovery obtained from a given field would be a function of the prevailing drive mechanism, the reservoir configuration, the fluids involved, the reservoir type and the prevailing reservoir pressure and temperature. If it is apparent that oil migration is not occurring across lease lines, it is also possible to calculate oil volumes on a lease basis. The basic equation for the determination of oil-in-place is an follows:

$$N = \frac{7758Ah\phi\,(1 - S_w)}{B_o} \tag{17}$$

where

> A = field or lease area, acres
> h = average oil zone thickness, feet
> B_o = oil formation volume factor, volume at reservoir conditions per volume at standard conditions
> S_w = water saturation as a fraction of the pore space

It is common practice to determine the acre-feet of oil reservoir rock, Ah, from a net oil pay isopachous map. Water saturation information is determined from information on a similar reservoir rock, from well logs, or from core saturation studies in the laboratory.

For a given field, the original oil-in-place, N, should be the same when calculated by either the volumetric equation, or by material balance. If the indicated volumetric result is less than that determined from the material balance equation, it is possible that present mapping of the field is conservative, and that an extension to the reservoir exists. This is a powerful means of oil exexploration when production information is available.

Equation (17) may also be modified to include a recovery factor so that the primary or secondary recovery oil reserves can be estimated. The form that the recovery factor takes will depend upon the producing mechanism of the reservoir.

Volumetric Methods—Solution Gas Drive Reservoirs. The oil recovered by primary producing means in a solution gas drive reservoir will be equal to the oil originally in place, minus the oil remaining in the reservoir at abandonment. The pressure at which the reservoir is abandoned will depend upon economic factors: the value of the oil, the operating costs and the oil producing rate. The oil recovered under primary producing operations would be:

$$N_{pt} = 7758Ah\phi \left(\frac{1 - S_w}{B_{oi}} - \frac{1 - S_w - S_{gr}}{B_{oab}} \right) \tag{18}$$

where

> N_{pt} = ultimate oil recovery by primary producing means, STB
> B_{oab} = oil formation volume factor at field abandonment pressure, bbl/STB
> S_{gr} = residual gas saturation in the reservoir at abandonment pressure as a fraction of pore space

The primary difficulty in the use of equation (18) is in the determination of a value for S_{gr}. As a first approximation, the S_{gr} value may be taken as 0.25, if the initial solution gas-oil ratio is 400 to 500 SCF/bbl, and the oil has a gravity between 30 and 40 °API. The S_{gr} would increase approximately 0.01 for every 3 °API increase in oil gravity. A decrease in oil gravity would result in a corresponding decrease in the S_{gr}. A 50 per cent decrease in the solution gas-oil ratio would decrease S_{gr} by 0.05. A doubling of the solution gas-oil ratio, or a complete lack of shaliness in a loosely cemented sandstone could increase S_{gr} by as much as 0.10. It is evident that S_{gr} can only be roughly estimated in this manner.

Wahl *et al.*[21] have prepared a series of charts from which the ultimate recovery, as a per cent of the residual oil originally in-place, may be determined, if the bubble-point pressure, the oil formation volume factor, and the solution gas-oil ratio are known. Figure 2-3 presents one of the charts for the case when S_w equals 30 per cent, and oil viscosity at reservoir temperature is 2 cp. The relative permeability relationships used are those represented by equation (26) of Chapter 3. The arrowed lines on the chart indicate the procedure for determining the ultimate recovery. For

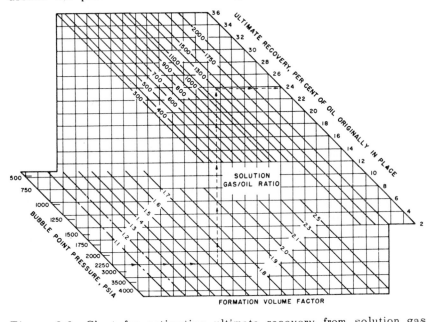

Figure 2-3. Chart for estimating ultimate recovery from solution gas drive reservoirs (from Wahl *et al.*[21])

the case where the bubble-point pressure is 2250 psia, the oil formation volume factor is 1.6, and the solution gas-oil ratio approximates 1300 SCF/bbl, the ultimate recovery is indicated to be 23.8 per cent, at an abandonment pressure of 14.7 psia. This may be converted to oil saturation at abandonment from the relationship that:

$$N_{pu} = \left[\frac{1 - S_w}{B_{ob}} - \frac{S_{oa}}{B_{oa}} \right] \div \left[\frac{1 - S_w}{B_{ob}} \right] \tag{19}$$

where, in this instance, N_{pu} represents the ultimate oil recovery down to atmospheric pressure from the bubble-point pressure as a per cent. The usefulness of this equation will be illustrated in the following example calculation:

EXAMPLE 5.

Determine the oil and gas saturation in a reservoir that has been produced down to atmospheric pressure for the conditions just previously described, and where the formation volume factor at atmospheric pressure is 1.06.

The ultimate recovery, N_{pu}, is 23.8 per cent from Figure 2-3. Substituting values into equation (19):

$$0.238 = \left[\frac{1 - 0.30}{1.60} - \frac{S_{oa}}{1.06} \right] \div \left[\frac{1 - 0.30}{1.60} \right]$$

or,

$$S_{oa} = 0.352$$

The gas saturation is:

$$S_{ga} = 1 - S_w - S_{oa} = 1 - 0.30 - 0.352 = 0.348 \text{ or } 34.8 \text{ per cent}$$

It should be recognized that reservoirs are seldom produced down even close to atmospheric pressure during primary operations. For this reason, the charts, of which Figure 2-3 is an example, will always yield an optimistic value for ultimate oil recovery. The main value of the charts is that of determining the upper limit for oil recovery by primary producing means. This value may be suitably reduced to reflect an abandonment pressure greater than 14.7 psia. This accomplished, the oil recovered by primary producing means could be obtained from the following equation:

$$N_{pt} = \frac{7758 A h \phi (1 - S_w) R}{B_{oi}} \tag{20}$$

where R is the recovery factor, as a fraction, of the original oil-in-place which can be recovery by solution gas drive. Notice that the recovery factor applies only to that oil recovered below the bubble-point pressure. If the reservoir pressure greatly exceeds the bubble-point pressure, additional oil recovery will be obtained.

An equation developed by combining equations (9) and (20) is useful in calculating the oil recovery between the discovery pressure and the bubble-point pressure in a solution gas drive reservoir:

$$N_p = \frac{7,758 A h \phi \, (p_i - p_b) c_{oe}}{B_{oi} [1 + c_{oe} \, (p_i - p_b)]} \tag{21}$$

where the effective oil compressibility, c_{oe}, includes oil, rock and water compressibilities, according to the relationships of equation (10). Where a reservoir has a primary producing history that includes behavior both above and below the bubble-point pressure, the total oil recovery would be that from equations (20) and (21).

Equation (20) can also be used for the estimation of total oil recovery under secondary recovery operations if the fraction, R, of the total original oil-in-place recovered, is known.

Volumetric Methods—Water Drive Reservoirs. In those reservoirs where oil recovery is effected by water encroachment, the factor influencing oil recovery is the efficiency of the replacement which occurs. Factors influencing this displacement are the dis-placing-displaced fluid mobility ratio, the reservoir dip, throughput rates and the heterogeneity of the reservoir rock. The recovery under water drive may be determined from the following equation:

$$N_p = 7758 A h \phi \left[\frac{1 - S_w}{B_{oi}} - \frac{S_{or}}{B_{oab}} \right] \tag{22}$$

where

S_{or} = oil remaining in the reservoir after water flooding, fraction of pore space

B_{oab} = oil formation volume factor at reservoir abandonment pressure, volume per volume at standard conditions

An approximation for S_{or} may be made from Arps's[22] treatment of the Craze and Buckley[23] data:

Reservoir Oil Viscosity, cp	Residual Oil Saturation, S_{or}
0.2	30
0.5	32
1.0	34.5
2.0	37
5.0	40.5
10.0	43.5
20.0	46.5

Since the trapped oil saturation, S_{or} is controlled by the pore open-sizes, a correction of the S_{or} trend correlated to oil viscosity, shown above, would be expected. The correction is shown as a deviation of S_{or} from the viscosity trend as a per cent:

Average Reservoir Permeability, md	Deviation of S_{or} from Viscosity Trend, %
50	+ 12
100	+ 9
200	+ 6
400	0
500	− 0.2
1,000	− 1
2,000	− 4.5
5,000	− 8.5

If relative permeability data are available, the frontal advance theory, described in Chapters 5 and 7, may be used to obtain a value for S_{or}, which will normally be more accurate than that provided from the correlations provided above.

Volumetric Methods—Gas Cap Drive Reservoir: Equation (22) may also be used for the case where a naturally expanding gas cap is providing the mechanism for the recovery of oil during the primary phase of an oil producing operation. As in the case of the water drive, values are needed for S_{or}. No correlations have been presented in the literature which relate S_{or} to permeability and oil and gas viscosity. As a first estimate, it will usually suffice to use values in the range of 10 to 20 per cent. To obtain a reliable value for S_{or}, it is usually necessary to make a frontal displacement calculation, as described in Chapters 5 and 9.

As in other facets of reservoir engineering, considerable care must be taken in recognizing the effect which possible gas counterflow, reservoir dip and gas-oil mobility ratios will have on the values for S_{or} which would actually be realized.

Decline Curves—Determination of Primary Oil Recovery

Decline curves are commonly used to determine both remaining primary oil reserves and the remaining productive life of a field or lease. The method requires that the wells be produced at capacity, so that the true decline in the reservoir's producing capability is observed in the performance data. Such an operating procedure would be attained in those states or countries where no proration exists, or where the wells have declined below the allowable producing rates where proration is applied. Usually, the oil producing rate will be chosen as the dependent variable; while producing time or cumulative oil production will be taken as the independent variable.

The two main types of decline curves are the rate-time and the rate-cumulative oil curves. The use of rate as the dependent variable is advantageous, since data for its determination will always be available from production statistics. Then too, the end point to which the resulting curve will be extrapolated can be estimated, since the producing rate at the economic limit can be readily calculated in most instances.

In the determination of the economic limit of an oil well, it is well to consider carefully actual reduction in operating expenditures, which the abandonment of one well might cause. It is usually unrealistic to include operational expenses beyond that necessary for the direct operation of the well itself.

The mathematical development of the rate-time and rate-cumulative curves has been discussed in detail in the literature[24]. Three types of decline curves exist: constant percentage decline, hyperbolic decline, and harmonic decline. In the constant percentage decline case, the drop in production per unit of time is a constant fraction of the production rate. In the hyperbolic decline case, the drop in production per unit of time, as a fraction of the producing rate, is proportional to the fractional power of the production rate, where this power ranges from zero to one. The harmonic decline case is a special case of the hyperbolic decline case, where the power is one and the drop in production rate per unit of time, as a fraction of the producing rate, is proportional to the production rate.

The equations for constant percentage decline for the rate-time and the rate-cumulative relationship, respectively, are:

$$q_t = q_i e^{-Dt} \tag{23}$$

$$Q_t = \frac{q_i - q_t}{D} \tag{24}$$

where

q_t = production rate at time t, bbl/month
q_i = production rate at beginning of production decline, bbl/month

D = decline as a fraction of the producing rate

$$= \frac{dq/dt}{q}$$

t = time, months

Q_t = cumulative oil production to time t, bbl

The equations applicable to the hyperbolic decline case are as follows:

$$q_t = q_i(1 + nD_i t)^{-1/n} \qquad (25)$$

$$Q_t = \frac{q_i^n}{(1-n)D_i}(q_i^{1-n} - q_t^{1-n}) \qquad (26)$$

The harmonic decline curve analysis procedure is defined by the equations:

$$q_t = q_i(1 + D_i t)^{-1} \qquad (27)$$

$$Q_t = \frac{q_i}{D_i} \ln(q_i/q_t) \qquad (28)$$

where

D_i = initial decline rate

$$= \frac{dq/dt}{q_i}$$

n = exponent

It is common practice to plot rate-time information applicable to a given well, lease or field, on semi-log paper. A straight line extrapolation of such a curve preassumes a constant percentage decline, as defined by equation (23). Usually, such an extrapolation will be pessimistic as to total primary recovery, since field curves of this type are normally observed to flatten as producing time increases. A horizontal shifting of the curve will often straighten the line, making an extrapolation to the economic limit more accurate than that of attempting to extend a curve.

Many rate-time curves for given oil fields will be found to be of the hyperbolic decline type, with the exponent n in equation (25) varying from zero to one. Most values for n will be in the range from 0.2 to 0.4. If producing rate-time data can be straightened by a horizontal shift of the data on log-log graph paper, the decline is of the hyperbolic type. Arps[24] has designed special graph paper where time can be plotted on a linear scale, and yet straight-line extrapolations to the economic limit can be made.

The equations for harmonic decline are included primarily for completeness, since rarely will a reservoir decline in this manner.

The utility of the decline curve analysis lies in the fact that often little other information will be available on many of the older oil fields. If a reasonable estimate can be made of the percentage of the oil-in-place produced by primary means, then the original oil-in-place value can be approximated. The difference between the original oil-in-place and the oil produced by primary means results in an estimation of the oil which remains in the reservoir. Information on porosity, thickness, and areal extent should then permit an estimation of the remaining oil saturation or remaining oil per acre-foot. Such information will be absolutely necessary if meaningful engineering analysis of a field is to be made before the choice and institution of a profitable secondary recovery technique can become a reality.

Empirical Methods for Estimation of Primary Oil Recovery

Where the methods previously described in this chapter cannot be used, or yield conflicting answers, it is sometimes possible to obtain meaningful answers by empirical methods. Guthrie and Greenberger[25] have made a statistical study of the Craze and Buckley[23] data on water drive fields, which resulted in the empirical relationship that:

$$N_{pt} = [0.2719 \log (k) + 0.25569 \, S_w + 0.1355 \log (\mu_o) - 15380 \, \phi - $$
$$0.00035 \, h + 0.11403] \left(7,758 A h \phi \, \frac{(1 - S_w)}{B_{oi}} \right) \quad (29)$$

In this expression, permeability (k) is in millidarcys, oil viscosity (μ_o) is in centipoises, and net pay thickness, (h), is in feet. Since equation (29) is a statistical analysis of the performance of a number of water drive fields, the effect of well spacing, producing rate, heterogeneity of the formation and pore-to-pore displacement efficiency have been included. The answer calculated for a given field should be used with judgment, since an exact mathematical modelling of a specific field would be indeed fortuitous.

References

1. Pirson, S. J., "Oil Reservoir Engineering," Chaps. 10, 12, 13, 14, New York, McGraw-Hill Book Co., Inc., 1958.

2. Craft, B. C., and Hawkins, M. F., "Applied Petroleum Reservoir Engineering," Chaps. 3, 4, 5, Englewood Cliffs, N. J., Prentice-Hall, Inc., 1959.
3. Frick, T. C., "Petroleum Production Handbook," Chaps. 34, 35, New York, McGraw-Hill Book Co., 1962.
4. Cole, F. W., "Reservoir Engineering Manual," Houston, Gulf Publishing Co., 1961.
5. Amyx, J. W., Bass, D. M., Jr., and Whiting, R. L., "Petroleum Reservoir Engineering," Chap. 8, New York, McGraw-Hill Book Co., Inc., 1960.
6. Dodson, C. R., and Standing, M. B., *Drlg. & Prod. Prac.*, *API*, 173 (1944).
7. Reference 2, p. 130.
8. Burcik, E. J., "Properties of Petroleum Reservoir Fluids," New York, John Wiley & Sons, Inc., 1957.
9. Campbell, J. M., "Oil Property Evaluation," p. 487, Englewood Cliffs, N. J., Prentice-Hall, Inc., 1959.
10. Reference 3, Chaps. 17 through 22.
11. Muskat, M., "Physical Principles of Oil Production," Chap. 10, New York, McGraw-Hill Book Co., Inc., 1949.
12. Tarner, J., *Oil Weekly*, 32 (June 12, 1944).
13. Tracy, G. W., *Trans. AIME* **204**, 243 (1955).
14. Pirson, S. J., "Oil Reservoir Engineering," Chap. 10, New York, McGraw-Hill Book Co., Inc., 1958.
15. Schilthuis, R. J., *Trans. AIME* **118**, 37 (1936).
16. Hurst, W., *Trans. AIME* **151**, 57 (1943).
17. van Everdingen, A. F., and Hurst, W., *Trans. AIME* **186**, 305 (1949).
18. Reference 2, Chap. 5.
19. Reference 1, Chap. 12.
20. Reference 1, Chap. 13.
21. Wahl, W. L., Mullins, L. D., and Elfrink, E. B., *Trans. AIME* **213**, 132 (1958).
22. Arps, J. J., *Trans. AIME* **204**, 182 (1956)
23. Craze, R. C., and Buckley, S. E., *Drill. & Prod. Prac.*, *API*, 144 (1945).
24. Arps, J. J., *Trans. AIME* **160**, 228 (1945).
25. Gutherie, R. K., and Greenberger, M. H., API paper 901-31-G, New Orleans, La. (March, 1955).

PROBLEMS

1. The following data are available on an oil field:

	At 2000 psia	At 1900 psia	At 1800 psia
Barrels of stock tank oil produced	250,000	1,850,000	3,550,000
Instantaneous GOR, SCF/STB	750	700	655
Gas in solution, SCF/STB	750	700	655
Oil formation volume factor, ft^3/STB	1.236	1.223	1.212
Gas formation volume factor, ft^3/SCF	0.00615	0.00651	0.00688

The oil formation volume factor is 1.228 at the discovery pressure of 2500 psia. Bubble-point pressure is believed to be 2000 psia.

(a) Could this field have a gas cap? Explain.

(b) What is the probable value of the original oil-in-place in stock tank barrels?

2. Derive the generalized material balance equation. Discuss the physical significance of each group of terms.

3. The following data are available on a 100-acre field:

Water saturation	26 per cent
Porosity	21.6 per cent
Initial pressure (bubble-point pressure)	1775 psia
Pay thickness	17 feet
Gas saturation at 1200 psia	5 per cent
Initial oil formation volume factor	1.35 bbl/STB
Oil formation volume factor at 1200 psia	1.275 bbl/STB
Formation temperature	154°F

Assuming a volumetric-type oil reservoir behavior, calculate:

(a) Barrels of water in the reservoir.

(b) Stock tank barrels of oil in the reservoir at 1200 psia reservoir pressure.

(c) Stock tank barrels of oil in the reservoir at 1775 psia reservoir pressure.

(d) The gas saturation at a reservoir pressure of 1775 psia.

(e) The standard cubic feet (60°F and 14.7 psia) of free gas in the reservoir when reservoir pressure is 1200 psia.

4. The Schilthuis, modified Hurst and van Everdingen-Hurst water influx terms are somewhat different in mathematical format. State briefly the assumptions necessary for the use of each. Give several examples of reservoir types to which each of the terms might be applicable.

5. The following information is available on a one-well oil reservoir:

Gas saturation at abandonment	35 per cent
Connate water saturation (irreducible)	25 per cent
Porosity	24 per cent
Initial oil formation volume factor	1.30 bbl/STB
Oil formation volume factor at abandonment	1.05 bbl/STB
Total oil production	117,900 STB

(a) Calculate the volume of the reservoir in acre-feet.

(b) If the average sand thickness is 17 feet, calculate the areal extent of the reservoir in acres.

Permeability Concepts

Permeability Concepts—Multiphase Systems

By definition, *absolute permeability* of a given porous material is the ability to pass a fluid through its interconnected pore and/or fracture network, provided that the fluid 100 per cent saturates the effective porosity. Here effective porosity is used to indicate that the pores are interconnected. The factors which influence effective porosity also directly affect the absolute permeability of the porous system, such as grain packing, petrofabric of the rock, grain size distribution, grain angularity and degree of cementation and consolidation. It is instructive to write the general form of the Darcy equation for flow in a linear horizontal system, in order to better understand the concept of permeability:

$$q = \frac{-Ak(\Delta p)}{\mu L} \qquad (1)$$

where

k = permeability, darcys
q = outlet flow rate, cc/sec
μ = fluid viscosity at temperature of the system, cp
L = system length, cm
A = system cross-sectional area, cm^2
Δp = pressure differential, atm

Darcy's law states that the velocity of a homogeneous fluid is proportional to the fluid's mobility, k/μ, and the pressure gradient, $\Delta p/L$. Flow must be laminar, since in turbulent flow, which occurs at higher fluid velocities, the flow rate does not increase in a linear fashion with the pressure gradient. In oil field usage, fluid velocities are seldom in the turbulent range, except very close to the sand face when producing or injection rates are high. The flow rate given by equation (1) is an apparent value, since rates inside the porous media would be equivalent to the apparent rate divided by the porosity, where the fluid 100 per cent saturates the porous media. Then too, Darcy's law does not describe the flow within a particular flow or pore channel; it is a statistical measure of the flow behavior within a large number of pore channels.

The constant of proportionality of the porous media, absolute permeability, has the units of darcys. To obtain a physical picture of this unit, a dimensional analysis can be made:

$$k = \left[\frac{L^3}{T} \cdot \frac{M}{LT} \cdot L \cdot \frac{1}{L^2} \cdot \frac{LT^2}{M} \right] = [L^2]$$

where

$$q = [L^3/T]$$

$$\mu = \frac{\text{shearing stress}}{\text{rate of shearing strain}} = \frac{F/A}{dv/dl} = \left[\frac{(ML/T^2)/L^2}{(L/T)/L} \right] = \left[\frac{M}{LT} \right]$$

$$L = [L]$$
$$A = [L^2]$$

$$\Delta p = \frac{\text{force}}{\text{unit area}} = \frac{[ML/T^2]}{[L^2]} = [M/LT^2]$$

where M, L, F and T refer to the units of mass, length, force and time, respectively. From the above analysis, it is seen that permeability has the units of $[L^2]$. This can be visualized as an area, presented at right angles to the direction of fluid flow, through which the flow occurs.

While a discussion of the techniques for the determination of absolute permeability will not be given here[1,2,3], it is well to note that a gas is usually used due to the facts that: (1) gas flow rates through a test core stabilize quickly, due to the low gas viscosity; (2) some liquids interact with the rock matrix, thereby negating results; (3) 100 per cent saturation of a dry core with gas is readily obtained. Although by definition, the absolute permeability of a rock is independent of the fluid used, the measured permeability is not totally independent of the mean pressure of the flowing fluid, if the fluid is a gas. This is especially true where a gas is flowing at relatively low pressures, since the mean free path of the gas molecules is relatively large, compared to the size of the pore openings. It is readily apparent that a high mean pressure of the flowing gas would result in a shorter mean free path of the molecules, and a resulting permeability which would closely approximate that where liquid was used. Klinkenberg[4] has developed an equation for determination of permeability at infinite pressure where a gas is flowing:

$$k_\infty = \frac{k_g}{1 + b/p_m} \qquad (2)$$

where

k_∞ = Klinkenberg permeability, darcys

k_g = measured gas permeability at p_m and q_m of the test, darcys

b = parameter showing influence of pore size and mean free path

p_m = mean pressure, $\dfrac{p_{inlet} + p_{outlet}}{2}$, atm

q_m = flow rate at the mean pressure of the system, cc/sec

Here, the Klinkenberg permeability can be considered as the equivalent liquid permeability, or the absolute permeability. The term b/p_m is a correction factor necessary for accurate permeability determinations, where low mean pressure prevails, and where tests are being performed on particularly tight reservoir rocks. Here, the correction factor may run as high as 50 to 100 per cent. With increasing permeability, b increases, and slope increases at a slower rate than k_∞, which results in a decrease in the percentage correction to k_g. Where the samples being tested have permeabilities exceeding perhaps 100 md, and where relatively high mean pressures are being used, the Klinkenberg correction to gas permeability is often validly ignored by testing laboratories doing routine core analysis. Figure 3-1 illustrates the Klinkenberg method.

By definition, the *effective permeability* to a given fluid at a saturation less than 100 per cent is the fluid conductivity of the porous media at a given saturation where other fluids are present.

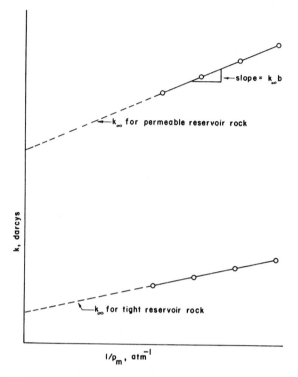

Figure 3-1. Klinkenberg method for determining equivalent liquid permeabilities.

This is measured in darcys or millidarcys, as in the case of absolute permeability; so has the same dimensional equivalent of L^2, or of cross-sectional area presented perpendicular to the direction of fluid flow. For the fluids gas, oil and water of petroleum engineering:

k_g = effective permeability to gas, darcys
k_o = effective permeability to oil, darcys
k_w = effective permeability to water, darcys

Figure 3-2 presents typical effective permeability relationships for an oil-water system in a water-wet porous media. Much can be learned about a given porous media by an examination of such permeability relationships. The curve shapes give direct evidence as to the wettability relationships prevailing. For instance, at an oil saturation or water saturation value of 50 per cent, it is apparent

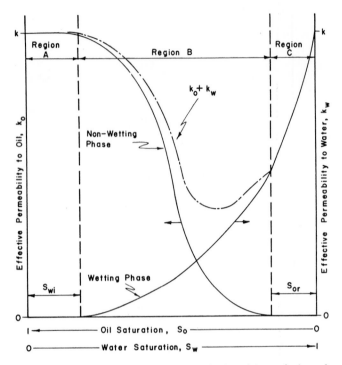

Figure 3-2. Typical graphical relationships of the effective permeabilities to oil and water in a water-wet porous media.

that the effective permeability to the wetting phase (water) is considerably less than that to the non-wetting phase. As the wetting phase preferentially adheres to the grain surfaces due to capillary forces etc., the effective cross-sectional area available to flow of the wetting phase is restricted, as compared to the effective cross-sectional area available to flow of the non-wetting phase. It could be observed that the more the wetting phase curve is displaced to right and downward in Figure 3-2, the more strongly water-wet the porous media would be.

The effective permeability relationships for water and oil, depicted by Figure 3-2, can be divided into three regions. Region A exhibits funicular saturation in the oil phase and pendular saturation in the water phase. Capillary forces are sufficiently strong to restrict the water saturation to sand grain contacts, with no interconnection of the resulting pendular rings which would permit the movement of the water phase. As the water saturation increases from zero to the irreducible value, the effective permeability to oil would be expected to decrease somewhat. The decrease will often be only slight, however, due to the fact that while the cross-sectional area for the flow of the funicular oil saturation is being reduced, the pore configuration is being simplified, due to water saturation in the grain contact positions. The analogy to the rough pipe-smooth pipe situation of fluid hydraulics could be made. Here, a smooth pipe of somewhat smaller cross-sectional area could move fluid at the same flow rate as a rough pipe having a larger cross-sectional area, for the same pressure drop across the system. With funicular saturations existing in the oil phase, a continuous oil phase exists, and a visual concept of effective permeability to oil is then possible.

Region B of Figure 3-2 shows the relationships existing when both the water and oil phases have funicular saturation configurations. It is evident that the sum of the k_o and k_w values will not equal the absolute permeability of the system. Again, the analogy to fluid mechanics and flow through pipes is useful. It can be readily shown that the flow rate through two small pipes having the same net cross-sectional area as one larger pipe, will be less than the flow possible through the larger pipe, with the same pressure drop. In hydraulics, the reason for this is tied directly to the Reynolds number, which contains the pipe diameter as a factor for the determination of the friction factor. Region C of the figure shows the effective permeability behavior when the oil saturation becomes insular or discontinuous and the water saturation remains funicular in configuration. Of course, at 100 per cent water saturation, the effective permeability to water becomes the absolute permeability of the system.

The typical behavior of effective permeabilities in an oil-water saturated porous system, where the oil is the wetting phase, can be mentally visualized by reversing the saturation relationships of Figure 3-2. Now the water becomes the non-wetting phase. It is readily apparent that in general, oil recoveries would be less from such a system since the water-oil ratios would be much higher at the same water saturation value than would be the case for a water-wet system. This would normally result in earlier abandonment of a field, with higher oil saturations remaining in the system in the oil-wet case for the same limiting producing water-oil ratios.

Figure 3-3 shows the typical graphical relationships between

effective permeabilities for gas and oil in a porous media. For the
same reasoning as applied to the oil-water case, the curve shape
in this instance would indicate that oil has now become the wetting phase in this system, while gas is the non-wetting phase.
This saturation arrangement is not difficult to visualize. It is
highly unlikely that the gas could ever exist as the wetting phase.
In most reservoirs of practical interest, all three phases would
actually be present. Where the flow rate of gas and oil are of
primary importance, as is the case for most reservoir calculations,
the assumption that the water present does not exceed its irreducible
value is necessary. With this assumption, the water can be considered as being immobile, serving merely to reduce the pore
space, and simplify the pore configuration. Note that Figure 3-3
has an abscissa which represents the hydrocarbon pore space.
The gas or oil saturations on a total pore space basis could be
readily determined by multiplying each saturation by $(1 - S_{wi})$ or,
that is, by the fraction that the hydrocarbon pore volume repre-

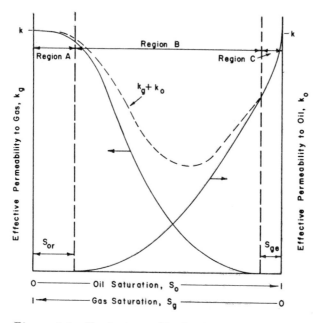

Figure 3-3. Typical graphical relationships for oil
and gas effective permeabilities where gas and oil
exist as separate and distinct phases.

sents of the total pore volume. Note that in the case represented by Figure 3-3, where only the hydrocarbon pore space has been accounted for, and where an irreducible water saturation does exist in the system, in Region A, the oil exists in a pendular or discontinuous state, while gas exists in a funicular or continuous filament state. In Region B, the gas and oil both exist in a funicular state. In Region C, the gas exists in a pendular state, while the oil exists in the funicular state. The equilibrium gas saturation, S_{ge}, represents the saturation at which the first continuous filament to the flow of gas, or that is, the first permeability to gas is attained. Similarly, the loss of permeability to the oil phase exists when the oil saturation is reduced to the residual value, S_{or}. For a given porous media of specific wettability and pore configuration containing given water, gas and oil of set compositions, the residual oil saturation is a set value and cannot be changed, due to the capillary forces involved.

The reader may misinterpret the aforegoing comments on multiphase flow as indicating that each individual pore contains oil, gas and water saturations in the percentages that the figures would indicate. This would only occur if each pore had exactly the same shape and wettability as all the other pores in a system. Such a system does not exist in nature, and cannot be formed in the laboratory. The saturations considered are therefore a statistical average representation of the entire system, and the "one pore basis" used for discussion purposes is considered to be representative of the entire system. In the case of gas-oil flow in a system, visual experiments[5] have shown that as the pressure is reduced to develop a gas phase, gas bubbles form at certain nucleation sites, and that effective permeability to gas occurs when a dendritic pattern, or interconnection of pores containing primarily a gas saturation, extends to the outlet of the system. In like fashion, a dendritic system is conducting oil through the system. While it is useful to consider the position of the various phases on a one pore basis, it must be recognized that the flow patterns and saturations which exist in the reservoir are considerably more complex. In the case of an oil-water system, the saturations through the system are perhaps more evenly distributed. Of course, the fact that pore configurations vary over a wide range, even in a so-called *homogeneous* system, would result in a wide range of oil-water saturations between pores. It will be evident in later discussion that this variation is tied to fluid mobility ratios.

In summary, it should be evident that the effective permeability values of k_o, k_g and k_w may vary from zero up to the absolute permeability, k:

$$0 \leq k_o, k_g, k_w \leq k$$

Relative permeability is a dimensionless quantity representing the ratio of effective permeability to absolute permeability, and is a fraction which varies between zero and one.

$$k_{ro} = \frac{k_o}{k} \qquad k_{rg} = \frac{k_g}{k} \qquad k_{rw} = \frac{k_w}{k}$$

and

$$0 \leq k_{ro}, k_{rg}, k_{rw} \leq 1$$

where k_{ro}, k_{rg} and k_{rw} equal the relative permeability to oil, gas and water, respectively.

Figures 3-2 and 3-3 could be altered to present relative permeability relationships as a function of the fluid saturations. The ordinates would merely be normalized by dividing the effective permeabilities by the absolute permeability. The curve shapes would remain the same. Figure 3-9 presents a petrophysical relationship between porosity and gas and oil relative permeabilities. Such a relationship would be expected, due to the fact that permeability has the units of length squared, $[L^2]$, and that, at least in the case of sandstone intergranular porosity, per cent porosity fraction should relate directly to the gas and oil relative permeabilities.

Determining Effective and Relative Permeabilities

Since most calculations of reservoir performance require either a direct or indirect use of Darcy's equation, it is necessary in many instances that absolute, effective and relative permeability values be determined. The sources of such data would be:

(1) From a study of production data on a similar field nearby, that has produced for a considerable period of time.

(2) From a study of production data on the subject field if such data are available.

(3) From published petrophysical considerations.

(4) From published generalized correlations.

(5) From detailed laboratory analysis on representative core samples.

(6) In some cases from well log analysis.

Permeabilities from Production Data.

The calculation of permeability values from production data presupposes that sufficient production from a given field has occurred so that meaningful data on field-wide pressures and quantities of fluids removed and/or injected are available. Certainly, this situation will commonly prevail where secondary recovery is being counted on to revive old fields. Where secondary recovery is being considered on a relatively new field, meaningful production data may not be available. Restricting this discussion to the relative permeability cases where two phases are flowing results in the two cases of gas-oil flow and water-oil flow. Gas-water flow will not be considered.

Gas-Oil Relative Permeability from Production Data. The determination of gas-oil relative permeability data from production data is of particular interest in those fields where the possibility of dispersed gas injection, pattern gas injection, or gas cap injection are being considered as a possible secondary recovery method. The details of such operations are discussed in Chapter 9. The equations necessary for determination of a "field" gas-oil relative permeability curve follow naturally from the definition for the instantaneous producing gas-oil ratio, or GOR:

$$\text{GOR} = \frac{\text{gas produced per day}}{\text{oil produced per day}}$$

$$= (\text{gas liberated from the produced oil per day} + \text{free flow gas produced per day}) \div \text{oil produced per day}$$

$$= R_s + \frac{7.07\, k_g\, h\,(\Delta p)}{B_g\, \mu_g\, \ln\,(r_e/r_w)} \bigg/ \frac{7.07\, k_o\, h\,(\Delta p)}{B_o\, \mu_o\, \ln\,(r_e/r_w)}$$

or,

$$R = R_s + \frac{B_o}{B_g} \cdot \frac{k_g}{k_o} \cdot \frac{\mu_o}{\mu_g} \tag{3}$$

where

R = instantaneous producing gas-oil ratio, SCF/CF of stock tank oil

R_s = solution gas-oil ratio, SCF/CF of stock tank oil

All other symbols are standard AIME nomenclature. By common usage, the instantaneous producing GOR, R, is often carried in the units of SCF/STB, which is readily obtained by noting that there are 5.615 CF/bbl. Equation (3) contains the instantaneous producing gas-oil ratio, R, which can be obtained from production data, plus the factors B_o, B_g, μ_o, and μ_g, which are primarily functions of pressure and can be determined from correlations or PVT data. The gas-oil permeability ratio, k_g/k_o, is a function of the total liquid saturations as existing in the reservoir. This can be expressed as:

$$S_l = S_w + S_o = S_w + \left[\frac{N - N_p}{N} \cdot \frac{B_o}{B_{ob}} \cdot (1 - S_w) \right] \qquad (4)$$

This equation is written directly by noting that the reservoir oil initially in place at the bubble-point is NB_{ob}, while the oil remaining after the production of N_p barrels of stock tank oil is $(N - N_p)B_o$. Of course, the total *fluid* saturation would be a sum of the total *liquid* saturation, S_l, and the gas saturation, S_g. In equation (4), the water saturation value, S_w, would be obtained from log and/or core analysis; the original oil-in-place term, N, would be approximated from volumetric considerations and/or material balance calculations. The cumulative stock tank oil produced, N_p, would be available from production data. The oil formation volume factors,

TABLE 3-1. Calculation of Gas-Oil Relative Permeability Values
from Production Data.

(1)	(2)	(3)	(4)	(5)	(6)	(7)	(8)	(9)
p, psia	N_p STB	R	$\dfrac{N - N_p}{N}$	B_o/B_{ob}	$S_l = S_w +$ $(1 - S_w) \cdot (4) \cdot (5)$	$F = \dfrac{B_o \mu_o}{B_g \mu_g}$	R_s	$k_g/k_o = \dfrac{R - R_s}{F}$
p_1	N_{p1}	R_{p1}	$\dfrac{N - N_{p1}}{N}$	B_{o1}/B_{ob}	etc.	$\dfrac{B_{o1} \mu_{o1}}{B_{g1} \mu_{g1}}$	R_{s1}	$(k_g/k_o) = \dfrac{R_{p1} R_{s1}}{F_1}$
p_2	N_{p2}	R_{p2}	$\dfrac{N - N_{p2}}{N}$	B_{o2}/B_{ob}	etc.	etc.	R_{s2}	etc.
p_3	N_{p3}	R_{p3}	$\dfrac{N - N_{p3}}{N}$	B_{o3}/B_{ob}	etc.	etc.	R_{s3}	etc.
↓	↓	↓	↓	↓	↓	↓	↓	↓

B_o and B_{ob}, would be available from correlations or PVT data, on a carefully tested bottom hole, or recombined reservoir fluid sample. Table 3-1 suggests a form which might be convenient for a calculation of the gas-oil relative permeability ratio, using equations (3) and (4). Note that the terms k_g/k_o and k_{rg}/k_{ro} are exactly equivalent.

Figure 3-4 is a plot of a typical gas-oil relative permeability curve, where considerable production was available to define the curve

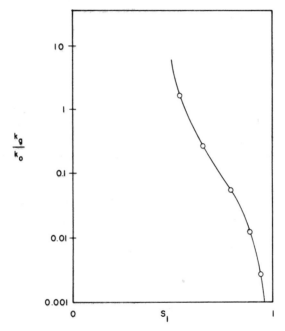

Figure 3-4. Typical gas-oil relative permeability curve.

down to low liquid saturation values. Such will not normally be the case. Most often, only sufficient production and field-wide pressure data will be available to define the shape of the curve over the high liquid saturation portion of the curve. This is, however, of considerable value, since it permits a check on the values which may have been derived from an empirical approach, or from laboratory determinations of the gas-oil relative permeability behavior.

The field-data approach to the determination of the relative permeability curve is of great value in establishing the validity of curves from other sources, within the limitations imposed by the quality of the production data. The proportionality constant, relative permeability, determined from production data, is a statistical treatment of the entire reservoir, and includes the effects of reservoir wettability, fluid interfacial tensions and other heterogeneities peculiar to the actual system, that cannot be accounted for by other methods.

Water-Oil Relative Permeability from Production Data. In a later section dealing with the immiscible displacement of oil by an immiscible displacing fluid (which could be gas or water), the need for data on the relative permeability between the displaced and displacing fluids will become apparent. Good water-oil relative permeability information from field production data is difficult to obtain, due to various ways by which water can be transported to the well for simultaneous production with the oil, i.e.:

(1) By coning of water from an underlying water section,

(2) By simultaneous production of alternating sand and shale sections, some of which have high mobile water saturations,

(3) Production of oil and water simultaneously from a section which has an extensive oil-water transition zone above an oil-water contact,

(4) Production from a thin producing section which has a mobile water saturation of uniform proportions through essentially the entire section,

(5) Unrecognized problems of water intrusion from over or underlying waterbearing sections, through a faulty cement job, or by vertical fracturing, either accidentally or intentionally.

For cases of practical importance, that is, where the resulting information would be useful to calculate k_w/k_o, only the case where there is simultaneous production of oil and water from a thin section, having a rather uniform water saturation in excess of the irreducible value would be of value. Here, the approach to the determination of an oil-water relative permeability would follow along much the same lines as the approach to the gas-oil relative permeability information from production data, as outlined in the preceding section. By definition:

$$\text{WOR} = \frac{\text{rate of water production}}{\text{rate of oil production}} = \frac{7.07 \, k_w \, h \, \Delta p}{B_w \mu_w \ln (r_e/r_w)} \bigg/ \frac{7.07 \, k_o \Delta p h}{B_o \mu_o \ln (r_e/r_w)}$$

or,

$$\text{WOR} = \frac{k_w}{k_o} \cdot \frac{\mu_o}{\mu_w} \cdot \frac{B_o}{B_w} \qquad (5)$$

Assuming:

$$(\Delta p)_o = (\Delta p)_w$$
$$(r_e)_o = (r_e)_w$$
$$h_o = h_w$$

From equation (5), it is apparent that the producing water-oil ratio is a direct function of the mobilities of the water and the oil (k_w/μ_w and k_o/μ_o), corrected to reservoir conditions by means of the oil and water formation volume factors, B_o and B_w, respectively. To use equation (5), the producing water-oil ratio would need to be known, corresponding to times where field-wide average reservoir pressure at datum elevations are also known. The oil viscosity, μ_o, could be obtained from a PVT analysis, or from published correlations by Beal[6, 7]. The oil formation volume factor, B_o, could be determined from the PVT analysis if one is available, from published information[8, 9], or possibly from data available on a similar crude produced in a nearby field. While there is no doubt that laboratory tests on the water produced would be beneficial, published correlations are sufficient in most cases since, in general, the values of water formation volume factor, B_w, and water viscosity, μ_w, vary over a rather narrow range. The volume of formation water is affected by temperature, pressure and gas in solution. Dodson and Standing[10] have investigated the PVT properties of formation water. If the accuracy of data available for the other factors in a calculation warrant it, Figures 3-5 and 3-6 may be used to determine values for B_w, as a function of reservoir pressure and temperature. The primary difficulty in the determination of meaningful water-oil relative permeability from field data is in the proper relating of it to liquid saturations. No one equation will adequately provide such a relationship. Benefit from such calculations will either support, or not support the reasonableness of water-oil relative permeability, as determined in the laboratory on small core samples, or as determined from petrophysical relationships to be developed at a later point in this chapter.

Caution should be used when utilizing generalized correlations since, for instance, use of charts prepared for paraffinic natural gases on a gas of radically different molecular structure would almost certainly result in grossly wrong answers.

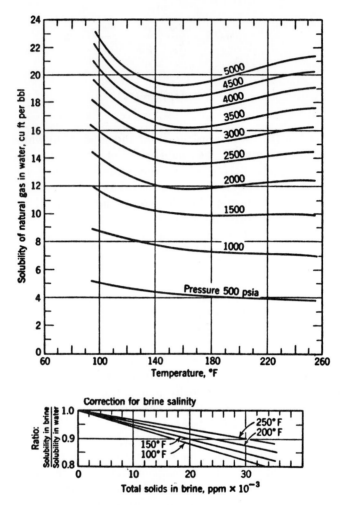

Figure 3-5. Solubility of gas in water (Dodson and Standing[10]).

Permeabilities from Petrophysical Considerations

The calculation of absolute, effective, or relative permeability values by means of petrophysical considerations is of considerable importance to the reservoir engineer, due to the fact that experimental measurement of the values is often difficult, or even impossible. Laboratory analysis preassumes that good representa-

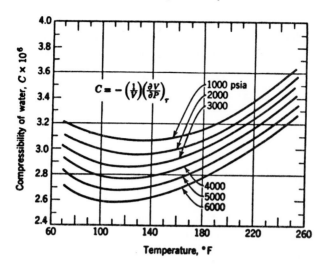

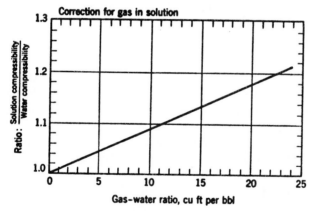

Figure 3-6. Effect of dissolved gas on the compressibility of water (Dodson and Standing[10]).

tive core samples are available. This may be true in large, recently drilled fields, but often will not be true in the older fields, where secondary recovery is being considered.

Absolute Permeability—Kozeny Equation. The classical approach to the calculation of absolute permeability originated with Kozeny[11], who treated the porous media as a bundle of capillary tubes of equal length, but of varied cross-sectional area. Numerous modifications of the original form of the Kozeny equation have been proposed. Carman[12] divided the Kozeny constant, C_k, into two parts; $K = K_o \tau^2$, where K_o was taken to be the shape factor, and τ the tortuosity. The resulting equation form is:

$$k = \frac{\phi^3 \times 0.987 \times 10^8}{K_o \tau^2 S_v^2 (1 - \phi)^2} = \frac{\phi \times 0.987 \times 10^8}{K_o \tau^2 S_\phi^2} =$$

$$\frac{\phi^3 \times 0.987 \times 10^8}{K_o \tau^2 S^2} \quad (6)$$

where

k = absolute permeability in darcys
K_o = shape factor
τ = tortuosity factor = L_e/L
L_e = length of fluid flow path through porous media
L = length of porous media
S_v = specific surface of the packing on a rock pore volume basis, cm^2/cm^3
S_ϕ = specific surface of the packing on a unit pore volume basis, cm^2/cm^3
S = specific surface of the packing per unit bulk volume, cm^2/cm^3

It is common practice to drop the unit conversion factor 0.987, due to the approximate nature of the Kozeny equation. The shape factor, K_o, varies from 2.0 to 3.0, with 2.5 often taken as a good average value. The tortuosity factor for unconsolidated sand is of the order of the square root of two. The Kozeny constant lies in the range from 4.5 to 5.5 for unconsolidated sand. Wyllie and Spangler[15] have published a number of Kozeny constants for consolidated sands, in the range from 17.8 to 34.5. Deviation from the values observed for unconsolidated sands is due to the tortuosity term.

Fair and Hatch[13] have treated the Kozeny constant, C_k, as being expressed by $2.5\tau^2$. To include the effect of varied grain shapes, a multiplying factor of t_s, the shape factor, was also included. It was experimentally determined that the shape factor in this instance should have values of 1.00, 1.02, 1.07, 1.17 and 1.27 for spherical, well-rounded, worn, sharp (subrounded) and angular grain shapes, respectively.

It is interesting to note that the Kozeny-Carman mathematical model disregards the presence of individual pores by grouping all pores in a single hydraulic radius term. This can be done with considerable precision if the pores are all identical, or if the range of sizes is small. Wyllie and Gregory[14] have demonstrated experimentally that the Kozeny-Carman equation is valid if the particle size is within certain limits. Since in natural consolidated sedimentary rock systems pore sizes vary greatly, the Kozeny-Carman model is no longer directly applicable. Wyllie and Spangler[15] have made an attempt to generalize the equation, taking into account the variation in pore sizes. Two good techniques presently exist that give data reflecting pore size distribution of a porous media, or at least the sizes that control the entry of fluids:

(1) Capillary pressure curves obtained by the mercury injection techniques, and

(2) Capillary pressure curves developed by the injection of a non-wetting phase into a porous media saturated with a wetting fluid.

Wyllie and Spangler have shown that the specific surface of a packing on a bulk volume basis, S, can be represented by $p_c \phi / \sigma$ when $\cos \theta$ is equal to one, where p_c, ϕ, and σ are capillary pressure, porosity and interfacial tension, respectively. Then, considering a "real" porous media, in which the sizes of pores differ, but where a number of the interconnecting media are composed of pores of uniform size, we could reason that the porosity of each "ideal" media will be less than the total porosity, ϕ, of the "real" medium. If an air-water capillary pressure curve is used, the porosity of each ideal medium will be $\phi \cdot \Delta S_w$, where ΔS_w is a small increment in wetting phase (water) saturation, over which the capillary pressure may be considered to be constant. Therefore, for an ideal medium:

$$S = p_c \phi \Delta S_w / \sigma. \tag{7}$$

where

capillary pressure, p_c, and interfacial tension, σ, have units of dynes/cm^2 and dynes/cm, respectively

Replacing S in equation (6) results in the following relationship:

$$k_i = \frac{\phi(\Delta S_w)\sigma^2 \times 10^8}{C_k \, p_c^2} \tag{8}$$

where

the factors $K_o \tau^2$ have been combined into a factor C_k, the Kozeny constant, and k_i is the permeability of the ideal system

The rate of flow per unit area through each ideal system is given by:

$$\Delta Q = \frac{(\phi \cdot \Delta S_w) \sigma^2}{C_k \, p_c^2} \times \frac{\Delta p}{\mu L} \times 10^8 \tag{9}$$

The total flow rate $Q = \Sigma \Delta Q$. Using Darcy's equation, Q is related to the permeability of the real medium by:

$$Q = \frac{\Delta p \cdot \phi \, \sigma^2 \times 10^8}{\mu L C_k} \sum \frac{\Delta S_w}{p_c^2} = \frac{k \cdot \Delta p}{\mu L} \tag{10}$$

Hence, the generalized Kozeny-Carman equation is given as:

$$k = \frac{\phi \, \sigma^2 \times 10^8}{C_k} \sum \frac{\Delta S_w}{p_c^2} = \frac{\phi \, \sigma^2 \times 10^8}{C_k} \int_0^1 \frac{dS_w}{p_c^2} \tag{11}$$

where k is the absolute permeability of the real medium in darcys.

Inspection of equation (11) indicates that the Kozeny-Carman constant, C_k, has been considered the same for all pore sizes. This may be justified by reasoning that the constant, which combines a tortuosity and shape factor, should be the same in each "ideal" media, so long as the shapes of the pores are not uniquely related to their sizes. This equation is primarily of interest for the determination of the relative permeabilities in porous media of gas, oil and water, as will be evident in a later section of this chapter, since the determination of meaningful interfacial tensions values at reservoir conditions of pressure and temperature is indeed difficult.

Absolute Permeability from Well Logs. As evidenced in the Kozeny equation, it would be expected that a correlation should exist between porosity, specific surface area and permeability. Certainly the specific surface of a reservoir rock should be a function of textural properties of the rock, the amount of cementing material present, the porosity type and the orientation of the grains in space. It could also be observed that the irreducible water saturation, S_{wi}, and the formation resistivity factor, F, should be tied very directly to the above-mentioned factors. The formation resistivity factor, F, is defined as:

$$F = R_o / R_w \tag{12}$$

where

R_o = the resistivity of a non-shaly formation 100 per cent saturated with water of resistivity R_w, resistivity units of $\dfrac{\text{ohm-meters}^2}{\text{meter}}$

Tixier[17] has presented a method for determining the formation permeability from true resistivity, R_t, in ohm-meters, from R_o and from the resistivity gradient in the transition zone above an oil-water contact, as evidenced on electric logs. Figures 3-7 and 3-8 permit a graphical solution for water salinities of 35,000 ppm and 150,000 ppm of the following equation:

$$\left[\frac{k}{20}\right]^{\frac{1}{2}} = \frac{2.3}{R_o(\rho_w - \rho_o)} \cdot \frac{\Delta R_t}{\Delta h} \tag{13}$$

where

ρ_w, ρ_o = specific gravity of water and oil, respectively
$\Delta R_t/\Delta h$ = resistivity gradient in ohm-meters per foot of depth, as determined from a deep investigation tool, lateral or focused logs and corrected for bore-hole effects

The approach must be used with considerable care, due to the limiting assumptions that capillary pressure, p_c, equals a constant divided by $k^{0.5}$, and that saturation exponent, n, in Archie's[18] equation:

$$S_w = \left[\frac{R_o}{R_t}\right]^{1/n} \tag{14}$$

being equal to 2.0.

The value of n lies between 1.5 and 3.0, with the more typical values generally being 1.8 to 2.0. Oil-wet rocks have been shown to have saturation exponents in the range of 3 or 4, by Fraser and Pirson[19]. If no other source of permeability information for a formation is available, the Tixier method will give an approximate value of permeability for water-wet, shale-free rocks, with an inter-granular-type porosity system.

Wyllie and Rose[20], through the use of petrophysical considerations, have modified the Tixier equation to provide a correlation between irreducible water saturation, permeability and formation

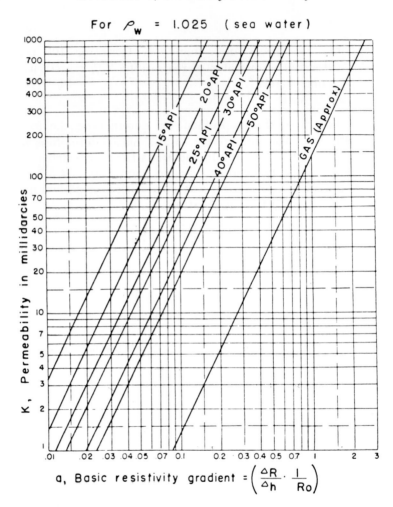

Figure 3-7. Permeability determination from resistivity gradient for water salinity of 35,000 ppm or sea water (Tixier, 1949, reprinted with permission of Schlumberger Well Surveying Corp.).

resistivity factor of the form:

$$S_{wi} = C\left[\frac{1}{k^{1/2}\,F^{0.67}}\right] + C' \qquad (15)$$

which can be written in the more general form:

$$S_{wi} = C\,\frac{\phi}{k^{1/2}} + C' \qquad (16)$$

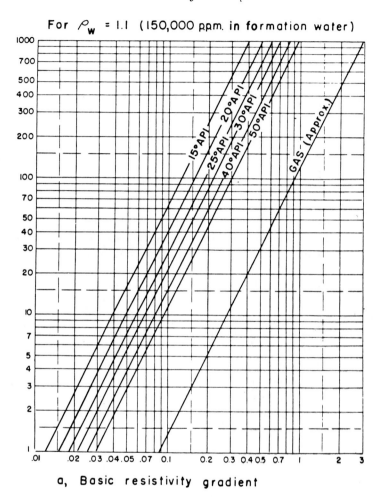

a, Basic resistivity gradient

Figure 3-8. Permeability determination from resistivity gradient for water salinity of 150,000 ppm (Tixier, 1949, reprinted with permission of Schlumberger Well Surveying Corp.).

where

C = constant to be determined having the dimensions of length.
C' = dimensionless constant which is related to per cent of bound water due to clay content of the reservoir rock.
$F = \phi^{-n}$, where $n = 1.5$

In a clean sand, the irreducible water saturation is zero when the interstitial water surface area is zero. Under these conditions,

C' would also be zero. Schlumberger[21] has modified equation (16) to an empirical form from field studies of porosity, irreducible water saturation and permeability. For oil sands containing medium gravity oils, the equation takes the form:

$$k^{1/2} = 250 \, \frac{\phi^3}{S_{wi}} \qquad (17)$$

The presence of heavier oils will result in pessimistic values for permeability. For dry gas sands, the equation becomes:

$$k^{1/2} = 79 \, \frac{\phi^3}{S_w} \qquad (18)$$

From equations (17) and (18), it is apparent that C is related to porosity. The above equations are useful primarily to give working ranges of permeability. It would then be possible to determine a permeability profile, where relative values would have meaning if a log(s) were available, which could provide a foot-by-foot indication of the porosity of a given section. Again, caution should be used in the application of these equations, due to the limiting assumptions made of a clean (shale or clay free) sand with a saturation exponent, n, of 2. Figure 3-9 provides a graphical solution of equations (17) and (18).

For a comprehensive treatment of petrophysical relationships underlying the aforegoing permeability treatment, the reader is referred to Pirson[22]. Another approach to the determination of absolute permeability from well logs has been proposed by Pirson[23]:

$$k = \left[\frac{850,000}{\text{API gravity}} - 3.5 \times \text{Depth in ft} \right] \frac{R_w^2}{FR_o R_{ti}} \qquad (19)$$

from which permeability is calculated directly in darcys. The factor R_{ti} is the true resistivity of the oil-bearing formation above the transition zone, or where an irreducible water saturation exists. The group of terms, $R_w^2/(R_o R_{ti} F)$, includes implicitly such fluid properties as fluid densities, surface tension and formation temperature, which in turn are instrumental in determining the relative wettability of the rock to oil or water. Pirson states that the equation should not be used for high gravity crudes, or for depths in excess of 6500 feet. Again, the method should be used with caution, but should provide order of magnitude values for the absolute permeability.

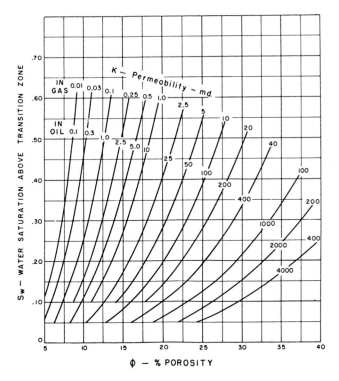

Figure 3-9. Empirical between irreducible water saturation, permeability, and porosity in oil and gas-bearing porous media (Schlumberger Well Surveying Corporation[21]).

A correlation based on Kozeny's equation has been developed by Sheffield[24]:

$$k = \frac{1}{F} \left(\frac{\phi}{1 - \phi} \right)^2 \frac{1}{S_{wi}}$$ (20)

Equation (20) was developed in part from a correlation of the factors for some well-known water-wet sands. The formation factor, F, may be determined from the Humble[25] equation, $F = 0.62\phi^{-2.15}$, or from Archie's formula, $F = \phi^{-m}$. The exponent m, the "cementation factor," is dependent upon the degree of cementation of the rock. Unconsolidated rocks and oolitic limestones have an m value of 1.3; very slightly cemented sands a value of 1.4 to 1.5; slightly cemented sands approximately 1.6 to 1.7; highly consolidated sands of 15 per cent porosity or less of the order of 1.8 to 1.9; and low porosity sands, intergranular porosity-type limestones and dolomites and chalks exhibit values of 2.0 to 2.2.

Effective and Relative Permeabilities. A more complete understanding of the petrophysical relationships controlling effective and relative permeabilities and their ratios in very recent years has resulted in the publication of a number of very good relationships. These developments now permit the calculation of permeability values that heretofore would have only been available from expensive and time-consuming laboratory analysis—if indeed samples were available for laboratory study. Then too, the selection of samples to properly characterize the reservoir under study can be a formidable chore in itself. It must be emphasized, however, that reservoirs that are markedly different from the more common types should be treated by laboratory analysis, since it is doubtful that meaningful relationships could be developed from the current knowledge of petrophysics.

As one would expect, the pore size distribution in a porous rock should have a bearing on the fundamental flow processes occurring where multiple fluid saturations exist. It is well known that pore size distribution and capillary pressures are directly related. This is the basis for the interest of a number of authors[26,27,28] in capillaric models as a foundation for an analytical approach to the understanding of multiple saturation fluid flow behavior. Without going into the mathematical arguments here, Burdine[29] and Wyllie and Spangler[30] have independently developed expressions for gas and oil relative permeabilities which can be expressed in the following form:

$$k_{ro} = \left[\frac{S_o}{1 - S_{wi}}\right]^2 \frac{\int_0^{S_o} dS_o/p_c^2}{\int_0^1 dS_o/p_c^2} \tag{21}$$

$$k_{rg} = \left[1 - \frac{S_o}{S_m - S_{wi}}\right]^2 \frac{\int_{S_o}^1 dS_o/p_c^2}{\int_0^1 dS_o/p_c^2} \tag{22}$$

where

S_m = lowest oil saturation at which the gas tortuosity is infinite.

Essentially the same equations have been developed by Wyllie and Gardner[31]. The expression $(S_o - S_{or})/(1 - S_{or})$ can be justified by an inspection of equation (11). The terms, $\phi \sigma^2/C_k$, in this equation for absolute permeability, do not entirely cancel when the ratio $k_{ro} = k_o/k$ is made as is required for the development of equation (21). It would be expected that the terms for total porosity, ϕ, and interfacial tension, σ, should cancel, but the Kozeny constant, C_k, incorporates grain shape and/or cementation factor and tortuosity, t. It is reasonable to assume that the factor $(S_o - S_{or})$ would incorporate some measure of a tortuosity coefficient. It could also be noted that the factor represents that fraction of the hydrocarbon pore space which contains movable hydrocarbons. It should then be, evident that Burdine's equation for gas and oil relative permeabilities can be evaluated if the saturation S_m and S_{or} are known, or can be determined, and if the ratios of integrals can be determined by integrating the areas under curves where a valid plot of the $1/p_c^2$ versus oil saturation on the porous media is available. Figure 3-10 is a plot of some typical $1/p_c^2$ versus oil saturation curves.

Corey[32] noted that for many oil-gas capillary pressure curves obtained on sedimentary porous material, the following was true to a close approximation:

$$1/p_c^2 = C \left[\frac{S_o}{1 - S_{wi}} \right] \qquad (23)$$

where C is a constant, and where S_o is larger than S_{or}.

The reciprocal of capillary pressure squared is zero when S_o is less than S_{or}. Due to the linear relationship evidenced by equation (23) and Figure 3-11, the integrals in equations (21) and (22) can be determined from the geometry of right triangles. It is then apparent that at an oil saturation corresponding to $(S_o)/(1 - S_{wi})$, the ratio of the integrals corresponding to the oil phase has a value of $S_o^2/(1 - S_{or})^2$. Similarly, the ratio of the integrals in the equation for relative permeability to gas is the complement of that to the oil phase, or $[1 - S_o^2/(1 - S_{wi})^2]$. Since only the ratio of areas under the $1/p_c^2$ curve of Figure 3-11 enters equations (24)

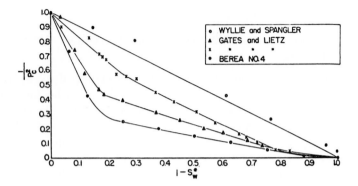

Figure 3-10. Examples of (capillary pressure)$^{-2}$ versus saturation curves with the data normalized (Wyllie and Gardner, 1958).

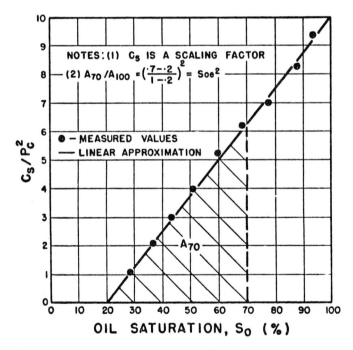

Figure 3-11. Plot of typical $1/P_c^2$ versus oil saturation function showing linear approximation of measured data values (Corey, 1954).

and (25), the calculated relative permeabilities are not sensitive to small changes in the actual shape of the capillary pressure curve (as evidenced in Figure 3-11). As a consequence, equations (21) and (22) may be written as:

$$k_{ro} = \left[\frac{S_o}{1 - S_{wi}} \right]^4 \tag{24}$$

$$k_{rg} = \left[1 - \frac{S_o}{S_m - S_{wi}} \right]^2 \cdot \left[1 - \left(\frac{S_o}{1 - S_{wi}} \right)^2 \right] \tag{25}$$

In equation (25), as a first approximation, S_m may be assumed to be unity. This should be very near to the physical case, since S_m is closely related to one, minus the critical gas saturation. S_m is a constant required to obtain the best straight line relation of $1/p_c^2$ to saturation, from a laboratory-determined k_{rg} curve. Deviation of S_m from $1 - S_{gc}$ means that $1/p_c^2$ is not a linear function of saturation. For intergranular-type rocks, the critical gas saturation is commonly less than 5 per cent. Due to the mathematical model upon which these relationships are based, these equations could be expected to have good application where the saturation changes are in the drainage direction, or where the wetting phase saturations are decreasing, rather than increasing. Corey has presented laboratory data which indicate excellent agreement with calculated k_{ro} and k_{rg} values for poorly consolidated sands, using equations (24) and (25). On consolidated sands, some deviation was evident in the region of low gas saturations. Deviations from the calculated values from laboratory data were apparent for sands cemented with dolomite and for poorly consolidated sands with stratifications. Measurement on many rock samples by Corey indicates that the equations are valid for reservoir materials not having extensive stratification, large solution channels, or large amounts of cementing material.

It should be readily apparent to the reader, as it was to Torcaso and Wyllie[33], that the gas-oil relative permeability ratio, k_g/k_o or k_{rg}/k_{ro}, could be determined within the bounds of the assumptions made in their derivation, by dividing equation (25) by equation (24).

Wahl *et al.*[34] have presented an equation for the k_{rg}/k_{ro} saturation relationship, based upon relative permeability data from field measurements on sandstone reservoirs. The equation is:

$$k_{rg}/k_{ro} = \xi(0.0435 - 0.4556\,\xi) \tag{26}$$

where

$$\xi = \frac{1 - S_{gc} - S_w - S_o}{S_o - C}$$

S_{gc} is the equilibrium, or critical gas saturation as a fraction of the total pore space, while C was represented as a constant equal to 0.25. Closer examination of the constant, C, would indicate that the residual oil saturation, S_{or}, is probably being represented.

The effect of stratification upon gas and oil relative permeability has been studied by Corey and Rathjens[35]. The authors point out that while the oil industry has been aware of the directional variation of permeability in porous media, the directional variation of relative permeability has been generally ignored. It is intuitively obvious that such a variation would be present. For uniform and isotropic cores, it was found that S_m, the lowest oil saturation at which the gas tortuosity is infinite, was essentially unity. Where cores had marked stratification perpendicular to the direction of flow, S_m exceeded one, where the stratification was parallel to the direction of flow, S_m was less than one. The authors found some indication that the oil relative permeabilities were less sensitive to stratification than were the gas relative permeabilities. The overall effect of stratification on reservoir performance could, however, only be ascertained by a complete study of the geologic setting in which the reservoir rock under study was laid down.

Torcaso and Wyllie have compared the equations for gas-oil relative permeability as presented by Wahl *et al.* and Corey, with the results as illustrated by Figure 3-12. The agreement between the two sets of curves is excellent. Deviation of the curves at high irreducible water saturations is perhaps due to the difficulty in obtaining good typical reservoir data upon which to base the equations. The excellent agreement between the equations justifies the mathematical model used by Corey, Burdine, Wyllie and Spangler, etc., since the Wahl *et al.* equation was based upon a correlation of field measurements. It is apparent that the Wahl *et al.* equation goes to infinite relative permeability values when the total liquid saturation is 0.25. The above-mentioned gas-oil relative permeability equations have application where reservoir

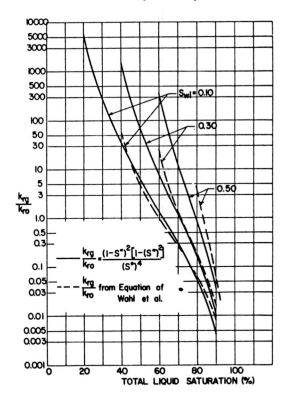

Figure 3-12. Comparison between k_{rg}/k_{ro} values calculated by the Corey and Wahl *et al.* techniques for three irreducible water saturation values (Torcaso and Wyllie[33]).

rocks contain an irreducible water saturation, and where the gas saturation increases at the expense of the oil saturation during the production process.

Wyllie and Gardner[31] have also presented an expression for the wetting phase relative permeability in the drainage direction, i.e., where the wetting phase saturation is decreasing:

$$k_{rw} = \left(\frac{S_w - S_{wi}}{1 - S_{wi}}\right)^2 \frac{\displaystyle\int_{S_{wi}}^{S_w} dS_w/p_c^2}{\displaystyle\int_{S_{wi}}^{1.0} dS_w/p_c^2} \qquad (27)$$

where: $(S_w - S_{wi})/(1 - S_{wi})$ is the fraction of the wetting phase saturation which is movable.

It could be noted that equation (21), for oil relative permeability, is a specialized case of equation (27). Inspection of Figure 3-11 justifies the form of the following equation:

$$\int_{S_{wi}}^{S_w} dS_w/p_c^2 + \int_{S_w}^{1.0} dS_w/p_c^2 = \int_{S_{wi}}^{1.0} dS_w/p_c^2 \qquad (28)$$

In view of the relationship which equation (28) represents, Wyllie and Gardner[31] have developed an additional relationship for k_{rw}:

$$k_{rw} = \frac{S_w - S_{wi}}{1 - S_{wi}} - k_{ro}\left[\frac{(S_w - S_{wi})/(1 - S_{wi})}{1 - (S_w - S_{wi})/(1 - S_{wi})}\right]^2 \qquad (29)$$

Based on the indicated linear relationship between $1/p_c^2$ and $(S_o)/(1 - S_{wi})$, equation (29) becomes:

$$k_{rw} = \left[\frac{S_w - S_{wi}}{1 - S_{wi}}\right]^4 \qquad (30)$$

The companion equation for oil relative permeability was presented as equation (25). It is well to note that the irreducible water saturation, S_{wi}, will only be so if $1/p_c^2$ is accurately proportional to $(S_w - S_{wi})/(1 - S_{wi})$. The major use of the equations as presented here would be the construction of relative permeability curves, by using whatever is considered to be the best value of S_{wi}. At this point it is well to reiterate that the preceding equations, based upon a linear relationship between the reciprocal of capillary pressure squared and effective saturation, apply for the *drainage* direction only, and have best application on intergranular type porous media, with a limited amount of grain cementing material present to complicate the pore configurations.

Naar and Henderson[36] describe a mathematical image patterned after Wyllie and Gardner's[31] and Marshall's[37] representations. This model, however, includes additional provisions, which allow the invading fluid to bypass and trap pores full of non-wetting fluid. An equation is derived for the hysteresis effect on a capillary pressure curve, giving the imbibition effective saturation in terms of the drainage effective saturation for a given capillary pressure:

$$(S_{we})_{\text{imbibition}} = (S_{we})_{\text{drainage}} - \tfrac{1}{2}(S_{we})^2_{\text{drainage}} \qquad (31)$$

where S_{we} is equal to $(S_w - S_{wi})/(1 - S_{wi})$ and the subscript w refers to the wetting phase.

Using equation (31) and equation (30) for the relative permeability to the wetting fluid on the drainage cycle, it is then possible to develop a meaningful relationship between relative permeability and saturation, for the imbibition of a wetting fluid into a porous media. Again, the resulting relationships would be valid for intergranular porous media systems, where the grain cementing material was present in limited quantities.

By using the concept represented by equation (31), the authors arrived at:

$$(S_{w, \text{imb.}})_{k_{ro} = 0} = 0.5 + \frac{S_{wi}}{2} \tag{32}$$

which gives a theoretical limit to the fractional recovery of oil-in-place by imbibition processes. The constant, 0.5, results from assumptions of random pore interconnections and of trapped non-wetting phase saturations. In addition, a relationship for the residual oil saturation where displacement is by imbibition was presented:

$$S_{or} = \tfrac{1}{2}(1 - S_{wi})\left(\frac{S_{oi}}{1 - S_{wi}}\right)^2 \tag{33}$$

where S_{oi} is the initial oil saturation, which would be equal to $(1 - S_{wi})$, if no gas saturation is present.

If a gas saturation is present, then the resulting residual oil saturation would be correspondingly reduced.

From petrophysical considerations, Pirson[22] has developed equations for the wetting and non-wetting phase relative permeabilities, for both the drainage and imbibition flow processes in porous media. Recent laboratory work[38] has resulted in some changes to the values of the exponents in the theoretically-derived formulas, but the form of the equations remain intact. The modified form of the equations where gas flows in the drainage direction, in rocks of intergranular-type porosity, in the presence of a flowing liquid are:

$$k_{rw} = S_{we}^{3/2} S_w^3 \tag{34}$$

and

$$k_{rg} = (1 - S_{we})\,[1 - S_{we}^{1/4} S_w^{1/2}]^{1/2} \tag{35}$$

Where there is simultaneous flow of oil and water in a water-wet intergranular-type porous media, the following modified equations for relative permeability to water and to oil can be used:

$$k_{rw} = S_w^4 \left(\frac{S_w - S_{wi}}{1 - S_{wi}} \right)^{1/2} \tag{36}$$

and

$$k_{ro} = \left[1 - \frac{S_w - S_{wi}}{1 - S_{wi} - S_{nwt}} \right]^2 \tag{37}$$

Equations (36) and (37) are valid when water flows in the imbibition direction. In the above equations, S_{we} is equal to $(S_w - S_{wi})$, the effective saturation to the wetting phase on a hydrocarbon pore space basis; S_{nwt} is the irreducible non-wetting phase saturation, in most instances the residual oil saturation. It should be noted that equations (34) and (35) are not limited to the case of a water-wet rock. If the system is predominately oil-wet, equation (34) is merely modified by replacing the wetting phase subscript, w, with the oil subscript, o. In this instance, the water saturation would have to be at the irreducible value. Essentially, the fluid phases in the oil-wet rocks are inverted, as compared to what existed in the water-wet rocks. It is well to note that the effective wetting phase saturation, S_{we}, for the oil-wet case would then be modified to the following form:

$$S_{oe} = \frac{S_o - S_{or}}{1 - S_{or}} \tag{38}$$

Pirson's equations should be used with considerable caution in the case of oil-wet reservoir rock systems, due to the fact that many rocks have a "dalmation" or "speckled" wettability—that is, only a fraction of the total rock surface is truly oil wet. In this instance, it is doubtful that any of the preceding equations for relative permeability have application except to give upper and lower limits, which would correspond to the water-wet and oil-wet rock cases, respectively. If representative core samples can be obtained, laboratory relative permeability relationships should be established. If sufficient field production data are available, it should be possible to partially define the controlling relative permeability curves, at least at the higher oil saturation values, by calculation techniques described previously in this chapter.

Wyllie and Gardner[31] have presented three-phase relative permeability relationships for water-wet systems, operating on the drainage cycle with respect to the water and the oil. In this instance, the gas saturation is increasing at the expense of the water and oil saturations. Where the formation can be represented as an unconsolidated sand with well-sorted grains, the following relationships apply:

$$k_{rg} = \frac{S_g^3}{(1 - S_{wi})^3} \tag{39}$$

$$k_{ro} = \frac{S_o^3}{(1 - S_{wi})^3} \tag{40}$$

$$k_{rw} = \left(\frac{S_w - S_{wi}}{1 - S_{wi}} \right)^3 \tag{41}$$

Where the reservoir rock is a cemented sandstone, oolitic limestone, or vugular rock, the equation forms are:

$$k_{rg} = \frac{S_g^2 \left[(1 - S_{wi})^2 - (S_w + S_o - S_{wi})^2 \right]}{(1 - S_{wi})^4} \tag{42}$$

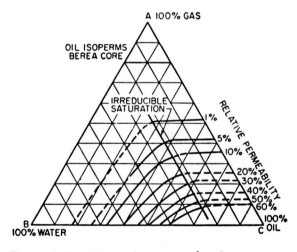

Figure 3-13. Lines of constant k_{ro} for a water-wet Berea sandstone core containing gas, oil and water—drainage cycle with respect to water and oil (Wyllie and Gardner[31]).

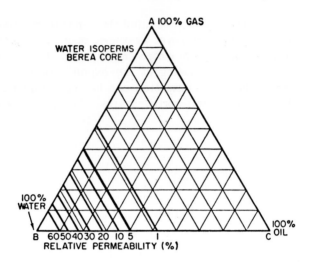

Figure 3-14. Lines of constant k_{rw} for a water-wet Berea sandstone core containing gas, oil and water—drainage cycle with respect to water and oil (Wyllie and Gardner[31]).

$$k_{ro} = \frac{S_o^3 (2S_w + S_o - 2S_{wi})}{(1 - S_{wi})^4} \tag{43}$$

$$k_{rw} = \left(\frac{S_w - S_{wi}}{1 - S_{wi}} \right)^4 \tag{44}$$

The six preceding equations may be modified for *totally* oil wet systems where the oil is the wetting phase, water the non-wetting phase and the gas non-wetting, with respect to both oil and water. In this case, the formulas apply if S_o is substituted for S_w, and vice versa. It should be noted that the above equations give *reasonable* values for three-phase relative permeabilities for the drainage cycle described. Due to the sensitivity of any three-phase flow case in porous media, the equations should be used with caution where an intermediate or "speckled" wettability might exist, or for a cycle of changing saturations which is not parallel in concept to the drainage cycle as described. Figures 3-13 and 3-14 present oil isoperms and water isoperms, respectively, for a water-wet Berea sandstone core containing oil, gas and water.

References

1. Pirson, S. J., "Oil Reservoir Engineering," p. 56, New York, McGraw-Hill Book Co., Inc., 1958.
2. Gatlin, C., "Petroleum Engineering: Drilling and Well Completions," p. 181, Englewood Cliffs, N. J., Prentice-Hall, Inc., 1960.
3. Amyx, J. W., Bass, D. M., Jr., and Whiting, R. L., "Petroleum Reservoir Engineering," p. 64, New York, McGraw-Hill Book Co., Inc., 1960.
4. Klinkenberg, L. J., *Drlg. & Prod. Prac., API,* 200 (1941).
5. Skidmore, Frederick A., "Bubble Nucleation and Gas Structure Growth in Synthetic Porous Media," M. Sc. Thesis, Petroleum Engineering Department, University of Texas, August, 1962.
6. Beal, C., *Trans. AIME* **165**, 94 (1946).
7. Burcik, E. J., "Properties of Reservoir Fluids," p. 126, New York, John Wiley & Sons, Inc., 1957.
8. Borden, G., and Rzasa, M. M., *Trans. AIME* **189**, 345 (1950).
9. Standing, M., *Oil and Gas J.,* 95 (May 17, 1947).
10. Dodson, C. R., and Standing, M. B., *Drlg. & Prod. Prac., API,* 173 (1944).
11. Kozeny, J., *S.-Ber. Wiener Akad. Abt. II a* **136**, 271 (1927).
12. Carman, P. C., *J. Agr. Sci.* **29**, 262 (1939).
13. Fair, G. M., and Hatch, L. P., *J. Am. Water Works Assoc.* **25**, 1551 (1933).
14. Wyllie, M. R. J., and Gregory, A. R., *Ind. Eng. Chem.* **47**, 1379 (1955).
15. Wyllie, M. R. J., and Spangler, M. B., *Bull. AAPG* **36**, 359 (1952).
16. Reference 1, p. 102.
17. Tixier, M. P., *Oil and Gas J.,* 113 (June 16, 1949).
18. Archie, G. E., *Trans. AIME* **146**, 54 (1942).
19. Fraser, C. D., and Pirson, S. J., *Petr. Engr.* **63**, 532 (July, 1961).
20. Wyllie, M. R. J., and Rose, W. D., *Trans. AIME* **201**, 43 (1954).
21. "Log Interpretation Charts," p. E-3, Houston, Schlumberger Well Surveying Corp., 1962.
22. Reference 1, Chap. 3.
23. Pirson, S. J., "Handbook of Well Log Analysis," p. 266, Englewood Cliffs, N. J., Prentice-Hall, Inc., 1963.
24. Sheffield, M., Unpublished Master's Thesis, The University of Texas, 1956.
25. Winsauer, W. O., Shearin, H. M., Masson, P. H., and Williams, M., *AAPG Bull.* **36**, 2 (Feb., 1952).
26. Rose, W. D., and Bruce, W. A., *Trans. AIME* **186**, 127 (1949).
27. Purcell, W. R., *Trans. AIME* **186**, 39 (1949).
28. Gates, J. I., and Lietz, W. T., *Drlg. & Prod. Prac., API,* 285 (1950).
29. Burdine, N. T., *Trans. AIME* **198**, 71 (1953).
30. Wyllie, M. R. J., and Spangler, M. R., *AAPG Bull.* **36**, 359 (1952).
31. Wyllie, M. R. J., and Gardner, G. H. F., *World Oil,* 121 (March, 1958), 210 (April, 1958).
32. Corey, A. T., *Producers Monthly,* 38 (November, 1954).

33. Torcaso, M. A., and Wyllie, M. R. J., *Trans. AIME* **213**, 436 (1958).
34. Wahl, W. L., Mullins, L. D., and Elfrink, E. B., *Trans. AIME* **213**, 132 (1958).
35. Corey, A. T., and Rathjens, C. H., *Trans. AIME* **207**, 358 (1956).
36. Naar, J., and Henderson, J. H., *Soc. Petr. Engrs. Jour. AIME*, 61 (June, 1961).
37. Marshall, T. J., *J. Soil Sci.* **9**, No. 1 (1958).
38. Pirson, S. J., Boatman, E. M., and Nettle, R. L., *J. Petrol. Technol.*, 564 (May, 1964).

PROBLEMS

1. From the following past performance history of a field, it is desired to establish a valid k_g/k_o curve:

Field Pressure, psia	B_o	B_g	$N_p \times 10^{-3}$ BSTO	R_s, SCF/ ft^3	Prod. GOR, SCF/STB	Oil Viscosity, cp
2050	1.262		0	96.25	535	
1690*	1.265	0.0064	not avail.	96.25	535	1.18
1480	1.238	0.00745	1785	86.50	825	1.20
1425	1.232	0.00777	2180	84.00	920	1.22
1050	1.186	0.01120	4400	66.00	1750	1.31

*Bubble-point pressure.

Assume that no gas cap exists, and that gas viscosity remains constant at 0.02 cp as pressure declines. Use an equilibrium gas saturation of 5 per cent in drawing the k_g/k_o curve. The irreducible water saturation is 25 per cent and the oil compressibility is 6×10^{-5} psi^{-1}.

2. Determine the surface area in cm^2/cm^3 and ft^2/ft^3 exposed to flow in an unconsolidated sand pack having a porosity of 34 per cent and a permeability of 1.5 darcys. The sand grains appear to be subrounded under a microscope. What range of surface area values could be calculated if the above sand were consolidated but still had the same porosity?

3. Prepare a graph of irreducible water saturation versus porosity with permeability as a parameter similar to that presented by Figure 3-9, through use of equation (20) and the Humble formula for formation factor. Compare the resulting relationships to those presented in Figure 3-9, and logically explain any differences in curve shapes.

4. Prepare gas-oil relative permeability curves using the equations of Corey and Wahl *et al.* for the following conditions:

> Irreducible water saturation   23 per cent
> Critical gas saturation   3 per cent
> Residual oil saturation   20 per cent

(a) Give reasons for the difference between the two curves. The curves should be plotted as the log of the gas-oil relative permeability versus total liquid saturation.

5. Prepare plots of oil and water relative permeability as a function of water saturation by the methods of Wyllie and Gardner, and Pirson, for the drainage and imbibition directions in a water-wet porous media. Compare the resulting plots and explain any differences in the curve shapes which result. The conditions of the problem are as follows:

Irreducible water saturation	21 per cent
Initial oil saturation	79 per cent
Residual oil saturation	22 per cent

The displacement occurs above the bubble-point pressure of the oil.

6. Develop ternary diagrams of oil and water isoperms for the three-phase relative permeability of gas, oil and water, where the following conditions are present:

Irreducible water saturation	18 per cent
Initial oil saturation	82 per cent
Residual oil saturation	16 per cent

The reservoir rock is a cemented sandstone and is water-wet.

Flood Patterns
and Coverage

A number of the calculation techniques that may be used to predict the production and injection characteristics of a given displacement process, whether water displacing oil, gas displacing
oil, miscible fluid displacement, etc., assume a linear system.
Most flow geometries of interest to secondary recovery of oil are
not of the linear type. Still, many times the linear calculation approach to reservoir performance may be validly used, if the concept of areal sweep efficiency is included. In this application,
the areal sweep efficiency will be that per cent of the porous media
contacted by the injected fluid at a given amount or number of pore
volumes of the displacing fluid injected. The linear system calculations, on a linear non-stratified system, will yield the displacement efficiency at a given throughput of the displacing phase,
or at a given displacing phase produced fraction at the outlet end.
The multiplication of the displacement efficiency by the areal
sweep efficiency results in the overall sweep efficiency of a single
homogeneous porous media of constant pay thickness. If the system is stratified or varies in pay thickness, it will also be necessary to multiply by an additional vertical displacement efficiency
factor. A number of the performance calculation techniques include an analytical accounting for the effect of stratification.

The primary interest of the following section will center on flood
patterns and the flood efficiencies which can be obtained at given
amounts of displacing phase injection, for a range of mobility

ratios. By definition, the mobility ratio is:

$$M = \frac{\text{Mobility of displacing fluid}}{\text{Mobility of displaced fluid}} = \frac{\lambda_1}{\lambda_2} \tag{1}$$

where λ is mobility which is the ratio of permeability to viscosity, k/μ. If M has a value of one, then the mobilities of the displaced and displacing phases are identical. Where the mobility ratio is less than one, the mobility ratio is said to be favorable, and the areal displacement efficiency is usually high; when the mobility ratio is high, the cause is often due to low mobility of the displaced phase. A case in point would be a viscous oil. Here, the injection of gas to displace the oil would result in low recoveries due to a high mobility ratio, resulting from high gas mobility.

The Basic Flooding Networks

Many of the older fields that are now being flooded for the secondary recovery of oil were developed on an irregular well spacing. However, in more recent years, a better understanding of reservoir behavior, and the rights of offset operators to recover a fair share of the oil in place, have resulted in relatively uniform well spacings and drilling patterns. This means that at the time of secondary recovery planning, the field will usually already be completely developed on a regular pattern basis. Since new wells are expensive, usually the presently completed wells will be used, with possibly some additional infill drilling of producing or injection wells, if such is warranted. In very recent years, operators have been instituting pressure maintenance (here used as being secondary recovery started early during the primary producing phase) very soon after field discovery. In this instance, it is possible to choose a development pattern which is of advantage in maximizing oil recovery in both the primary and secondary phases of oil production.

Were it possible, the ideal flood advance would be as a plane. If gravitational forces were acting to segregate the fluids, whether miscible or immiscible, then the plane should be horizontal, in order to achieve something approaching 100 per cent total coverage. This is possible in some reservoirs over at least a part of the secondary recovery or repressuring program. A vertical sweep from

one end of the field or system to the other end would also result
in excellent total coverage. This is not possible due to configuration of many fields, due to differing fluid densities which result
in either slumping or overriding of the injection fluid and also to
the fact that the injected fluid cannot be injected over an entire
vertical plane. The only way that this could be accomplished
would be with injection into a line of wells of very close spacing
along that plane. Economics would normally prohibit such a system.

One of the ways that something approaching a vertical planar
advance could be attained would be by the use of a direct line-
drive, where the production wells directly offset the injection
wells, as shown in Figure 4-1(a). The further the injection wells

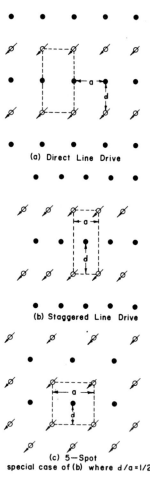

(a) Direct Line Drive

(b) Staggered Line Drive

(c) 5—Spot
special case of (b) where d/a = 1/2

Figure 4-1. Input and output well configurations for typical line flooding networks (after
Muskat and Wyckoff[1]).

are from the producing wells, the more nearly the case of a vertical front would be approximated. It is obvious that the horizontal
sweep efficiencies of the direct line drive well network would
vary from zero to nearly 100 per cent. Muskat[2] has expressed this
aspect of well spacing as the ratio d/a, where d is the distance
between the lines of producing and injection wells, and a is the
distance or spacing between wells in lines in the regular pattern
(see Figure 4-1). Figure 4-2 presents areal sweep efficiencies for
a variety of practical d/a ratios.

In a line flood, there are really "two degrees of freedom." The
d/a ratio can be changed or the wells may be staggered, thus
modifying the geometry of the network. We will only concern our-
selves with a shifting so that the injection and producing wells as
rows are moved in such a manner that the members of the alternate
rows are displaced one-half the inter-well distance. Figure 4-1(b)
illustrates this condition. A study of Figure 4-2 shows that the
effect of this "staggering" of wells is to significantly increase
the breakthrough efficiency, as compared to the direct line drive,
where d/a ratios are less than 2.0. At larger d/a ratios, the advantage of the staggered line drive is not so marked. Such a consideration of sweep efficiencies is of great importance if it is
possible to select the well configuration, due to the new drilling
of a field, or to the redrilling of an old abandoned field. Figure
4-2 also shows the differences between the sweep efficiencies for
the staggered line drive well network for a range of d/a values, as

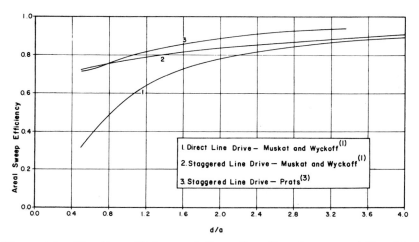

Figure 4-2. Flooding efficiency of direct-line (I) and staggered-line drive
(II and III) well networks as a function of d/a.

determined by different authors. The relationship (curve II) developed by Muskat and Wyckoff[1] assumed that "the streamline of highest average velocity is the line of centers between input and output wells." This would be true when the d/a ratio is 0.5 (i.e., the five-spot well pattern). The fact that the line of centers cannot coincide with the breakthrough streamline at other d/a ratios is evident, due to radial flow requirements in the immediate vicinity of the producing and injection wells. Prats[3] has presented (curve III) the results of calculations where the breakthrough sweep efficiency and pattern geometry have been determined by using the proper shape of the breakthrough streamline.

Also of considerable interest is the effect of various d/a ratios in the staggered and direct line drive well configurations upon the conductivity of the flooding network. Muskat[1] has investigated this aspect of the problem, with the results shown in Figure 4-3. Conductivity of the network has been taken as $q/\Delta p$, or the production rate per well per unit pressure differential. For the staggered and direct line drive cases, this figure shows that the flood conductivity decreases as the network element becomes elongated, as would be expected, due to lengthened fluid flow paths for comparable drainage areas. It is also apparent from Figure 4-3 that the staggering of the wells has had very little effect upon system

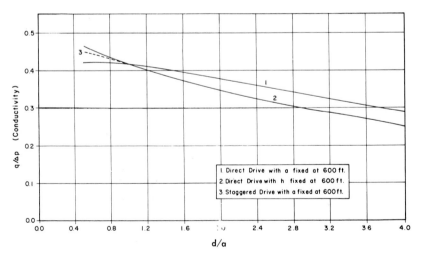

Figure 4-3. Conductivity of direct-line drive and staggered-line drive well networks, as a function of d/a (after Muskat and Wyckoff[1]).

conductivity, except for lower values of d/a, where the conductivity is lowered somewhat by the staggering.

Figure 4-4 shows the effect of the lengthening of the flow paths in a two-well element of a staggered line drive well network. It should be apparent that the shortest flow paths for a given well drainage area and for the staggered line drive well network, would occur when the d/a ratio is 0.5, and would, of course, have the greatest conductivity. Prats[3] has recently improved Muskat[1] values for the sweep efficiency of the staggered line drive. Muskat's development had assumed that the streamline of highest velocity would be on the line of centers between the injection and producing wells. This could only be true when d/a equaled one-half

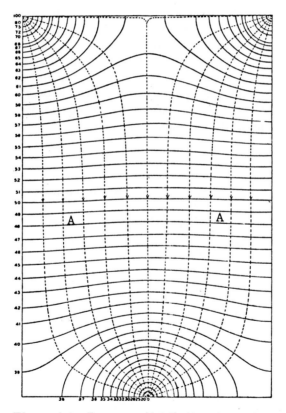

Figure 4-4. Pressure distribution in a staggered-line drive well network (after Muskat and Wyckoff[1]).

(the five-spot well pattern). The streamline marked *A* on Figure 4-4 clearly shows that the line must leave and enter the injection and producing wells at 45 degrees to the horizontal. Figure 4-2 presents the exact sweep efficiency values for various *d/a* ratios (curve III) obtained through use of rigorous complex potential theory. Figure 4-4 illustrates the fact that the five-spot well network is a special case of the staggered line drive, where the *d/a* ratio is 0.5. At the present time, this spacing is more widely used than any other spacing. This is due largely to regular well spacings that have been required, or at least used in many states.

Figure 4-5 shows the results of a potentiometric model study of the five-spot network for a horizontal system, a fluid mobility ratio of one, and steady-state conditions. The isopotential lines, the streamlines and two flood fronts are shown. It has been verified by a variety of modeling techniques [1,4,5,6] that the sweep efficiency for the five-spot for the above conditions is about 72 per cent. Notice that for a large number of wells, the number of injection wells equals the number of producing wells for the direct-line drive, the five-spot and staggered-line drive networks. The ratio of producing wells to injection wells becomes important where problems of injectivity or productivity exist.

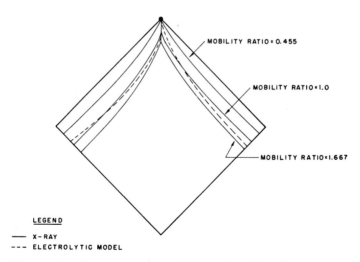

Figure 4-5. Potentiometric model study of the five-spot network, showing the flood fronts, the stream lines and the isopotential lines (after Frick[7], 41-25).

There are two common cases of well patterns where the ratio of injection wells to producing wells is not one. Figure 4-6 presents the diagrammatic representation of the seven-spot well network, while Figure 4-7 is a representation of the nine-spot well system. Inspection of Figure 4-6 will show that the ratio of injection to producing wells is two. This pattern has merit if well injectivity is low, but would, of course, entail considerable expense to convert wells, and layout the extensive injection system that would be required. Not often will an already developed field have this pattern of wells. Actually, the normal seven-spot could also be termed a four-spot, where the well pattern is also shown on Fig-

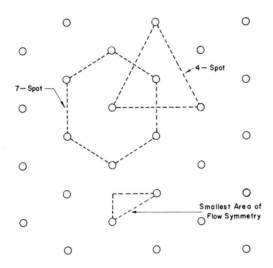

Figure 4-6. Diagrammatic representation of a seven-spot and four-spot well network.

ure 4-6. It is conventional to name a repetitive injection pattern by the number of injection wells that surround each producer. Thus, the so-called inverted seven-spot is actually the four-spot flooding network. If model studies are to be done to determine the sweep efficiencies for a confined network, the smallest area which can be modelled is that shown in the diagram. The smallest area of flow symmetry may always be chosen along streamlines, at stagnation points or points of zero flow and along boundaries across which there is no flow, due to pattern symmetry.

The four-spot flooding network has also been used occasionally.

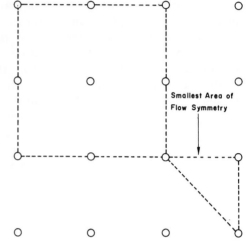

Figure 4-7. Diagrammatic representation of a nine-spot well network.

In this instance, the ratio of injection to producing wells is one-half. In Figure 4-6, the roles of the producing and injection wells would be merely reversed. Such a pattern would be useful where the displacing fluid is readily accepted by the formation. The conversion of a lesser number of wells to an injection status would then be a benefit, as compared to the five-spot system. Note though that the seven-spot cannot be formed from regular well spacings by conversion of producing wells to injection wells. Infill drilling of three wells for each original well is required. This results in 1.33 injection wells for each well of the original regular spacing. When the mobility ratio of the driving and driven fluids is unity, the sweep efficiency approximates 74.5 per cent for both and the normal and "inverted" well systems. Figures 4-8 and 4-9 present sketches of three positions of the flood fronts for the normal and "inverted" seven-spot flooding patterns, as observed with electrolytic models.

Figure 4-7 presents the well placement for the nine-spot flooding element. In this instance, the ratio of injection wells to producing wells is three—a system which has merit if the formation permeability is low, or if injection rates are low due to swelling clays or problems of a similar nature. Actually, the "inverted" nine-spot is the more common of the nine-spot configurations since, in this case, the producing wells outnumber the injection wells by a factor of three. Such a well configuration has found rather wide application where displacing fluid injectivity is high, and where

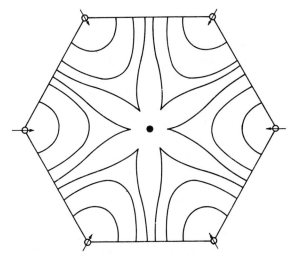

Figure 4-8. Sketch of three flood fronts observed in a normal seven-spot flood element as obtained with electrolytic models (from Wyckoff *et al.*[4]).

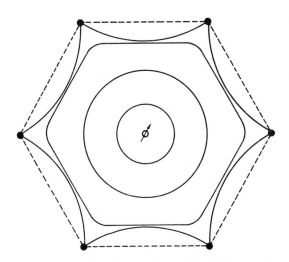

Figure 4-9. Sketch of three flood fronts observed in an inverted seven-spot flood element as obtained with electrolytic models (From Wyckoff *et al.*[4]).

considerable savings in injection lines and well conversions can
be effected, since fewer injection wells are required. When producing rates and injection rates are balanced, i.e., so that steady-
state prevails in the reservoir, and all the producing wells produce
at constant rates, and the injection wells accept the displacing
fluid at constant rates (three times the individual producing well
rates), the sweep efficiency at breakthrough is 74 per cent for a
fluid mobility ratio of one. When the side wells are shut in at
breakthrough, and the producing rates doubled at the corner wells,
the overall sweep efficiency when the displacing fluid reaches the
corner wells approximates 80 per cent. Where the producing wells
have sufficient producing capability to institute such a scheme, an
additional advantage of handling less produced water up to breakthrough at all producing wells is also obtained. On many flooding
projects, however, the shutting-in of producing wells and sub-
stitition of allowables to corner wells would not be possible, due
to low oil well producing capabilities. More will be said about the
influence of different well producing rates for various mobility
ratios later in this chapter.

Injection Beyond Breakthrough and the Effect
of Fluid Mobility Ratio

In those flooding patterns of other than linear geometry, and
especially where mobility ratios are somewhat adverse, the oil
recovery, due to the continued injection of displacing fluid past
breakthrough of the displacing fluid at the producing wells, can
sometimes exceed the production obtained up to the time of breakthrough. A number of authors have treated this particular problem.
Since most of the investigations have utilized potentiometric models, electrolytic models, immiscible and miscible liquid-liquid
displacements in porous plate models, fluid mapper systems, or
analytical approaches, the mobility ratio represented is for only
two fluids flowing. The effect of an initial gas saturation, or a
mobile interstitial water saturation is usually not modelled. If
there is no flowable gas present in the system, this is equivalent
to saying that no oil bank is formed.

The Five-Spot Well Network

Many authors have studied the five-spot well network because
of its widespread use in secondary recovery operations. The ef-

fect of fluid mobility ratio on an areal sweep efficiency on breakthrough for an enclosed five-spot well pattern has been studied by
Aronofsky and Ramey[8] with an electrolytic tank; by Cheek and
Menzie[9], by fluid mapper models; by Dyes, Caudle and Erickson[10]
and Caudle, Erickson and Slobod[11], by means of porous plate models and X-ray shadowgraph techniques; by Fay and Prats[5] and by
Sheldon and Dougherty[12], by means of analytical methods; and by
Bradley, Heller, and Odeh[13], with potentiometric models—to
mention but a few. Table 4-1 provides a summary of most of the
published areal sweep efficiencies for an enclosed five-spot well
pattern and mobility ratios ranging from 0.1 to infinite values.
Table 4-1 illustrates the lack of complete agreement on areal
sweep efficiency values that have been obtained by the different modeling techniques.

TABLE 4-1. Effect of Mobility Ratio on Areal Sweep Efficiency at
Breakthrough Enclosed Five-Spot Well Pattern
(taken from reference 13)

Areal Sweep Efficiency—Per cent

Mobility Ratio M	Potentiometric Analyzer		Fluid Mapper[9]	X-Ray Shadowgraph[10,11]	Analytical Calculation
	Conductive Cloth-Uskon	Electrolytic Tank[8]			
infinite	62.6[14]	62.5			
10		64.5	52.7	51.0	
4	66.4	65.8	62.0	54.0	45.0[3]
2	68.8	68.0	68.0	60.4	
1	71.6	70.0	71.7	69.8	71.5
0.25	82.2	88.5	78.0	87.0	
0.1		94.5	82.0	100.0	

Figure 4-10 is a graphical comparison of the published areal sweep
efficiencies due to Craig, Geffen and Morse[15], Dyes[16] and Nobles
and Janzen[17], for the enclosed five-spot network over a range of
fluid mobility ratios. All of these approaches to sweep efficiency
values assume a constant thickness homogeneous formation, with
an infinite array of wells so that the streamlines will not be distorted outside of, or within the enclosed pattern.

Dyes, Caudle and Erickson[10] and Caudle and Witte[18] have used
the X-ray shadowgraph model, developed by Slobod and Caudle[6],
as a technique by which the influence of fluid mobility ratio on
sweep efficiency on horizontal systems of various well geometries
could be determined. The models used were cut from one-quarter

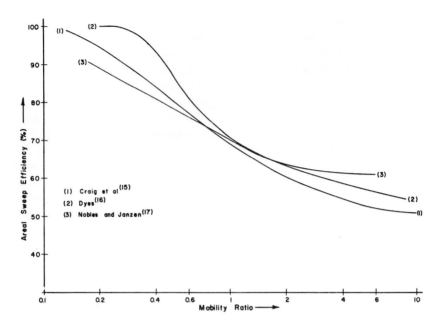

Figure 4-10. Areal sweep efficiency at breakthrough of the displacing phase as a function of mobility ratio for a confined five-spot well pattern.

inch thick alundum plates, or from uniform packings of Ottawa sand, consolidated with an epoxy resin, producing and injection wells affixed, and the surfaces sealed with a ceramic glaze or an epoxy resin.   The saturation gradients common to immiscible fluid displacements were avoided by the use of miscible fluids, usually naphtha and mineral oil.   Since no capillary forces or interfacial tensions are present, the mobility ratio becomes merely a ratio of the viscosities of the two miscible fluids.   To the injected fluid is added an X-ray absorbing compound, such as iodobenzene or potassium iodide.   The injected fluid front position at set time intervals is obtained by taking radiographs or X-ray pictures.

Figure 4-11 is a schematic drawing of the apparatus (for the constant pressure and variable rate case) for studying sweep-out patterns.   The constant rate case with variable injection pressure is modelled by replacing the constant air pressure with a constant volume pump.   The authors of the technique term the resulting developed negatives shadow-graphs.   Figure 4-12 is plot of the horizontal area swept versus the fluid mobility ratio, with the pro-

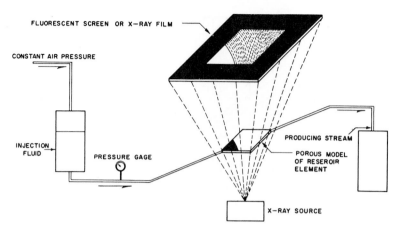

Figure 4-11. Apparatus for studying sweepout patterns and conductance ratios (Caudle and Witte [18]).

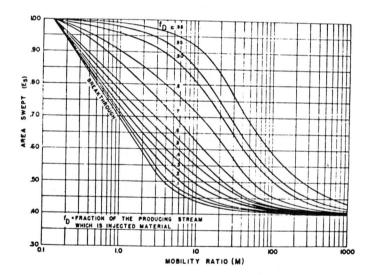

Figure 4-12. Effect of mobility ratio on areal sweep efficiencies for the five-spot pattern (Caudle and Witte [18]).

ducing water cut (reservoir conditions) as a parameter for an enclosed five-spot pattern. Figure 4-13 presents the data obtained from the shadowgraphs in a form where displacement volume injected is the parameter. The displaceable volume injected, V_D, is the product of the pore volume of the unit pattern and the displace-

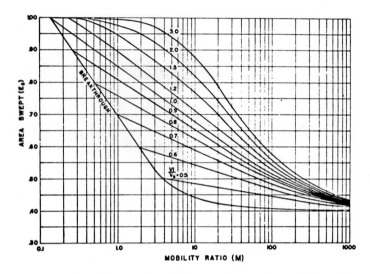

Figure 4-13. Effect of mobility ratio on the displaceable volumes injected for the five-spot pattern (Caudle and Witte [18]).

ment efficiency (the change in oil saturation attributable to the flood). Due to the fingering of the displacement front which occurs at adverse mobility ratios, the information provided in Figures 4-12 and 4-13 for mobilities larger than two, is only approximate. This too, is one of the main reasons for the lack of agreement between the different modelling techniques as to the sweep efficiencies at adverse mobility ratios, as clearly shown by the data of Table 4-1. The utility of data provided by Figures 4-12 and 4-13 is best shown by an example problem:

EXAMPLE 1:
   Water flood recoveries from a five-spot well pattern.

Data: Well spacing = 40 acres μ_o = 5 cp
       Pay thickness = 15 ft μ_w = 0.8 cp
       Porosity = 20 per cent k_{ro} = 0.75
       S_{oi} = 70 per cent k_{rw} = 0.25
       S_{or} = 30 per cent   Injection rate = 200 reservoir
                                                          B/D

SOLUTION:

$$V_D = 7{,}758 \times 40 \times 15 \times 0.20 \times (0.70 - 0.30)$$
$$= 372{,}400 \text{ reservoir barrels}$$
$$\text{Mobility Ratio} = \frac{0.25 \times 5}{0.75 \times 0.8} = 2.08$$

Figure 4.12 yields an areal sweep efficiency of 0.58 at breakthrough to the producing well. The oil recovery to this point, if the oil formation volume factor is 1.25, will be:

$$N_p = \frac{0.58 \times 372{,}400}{1.25} = 168{,}200 \text{ STB}$$

Since the injection rate has been constant at 200 barrels per day, the time to breakthrough is:

$$t_B = \frac{372{,}400 \times 0.58}{200} = 1080 \text{ days}$$

The performance of the five-spot past breakthrough can be generated at values of the displacement volume, from which the total oil recovery and cumulative time are readily determined, i.e., at a displaceable pore volume injected of 1.5:

$$N_p = \frac{0.89 \times 372{,}400}{1.25} = 265{,}000 \text{ STB}$$

$$t = \frac{372{,}400 \times 1.5}{200} = 2795 \text{ days}$$

From the water flood calculation technique set out in Example 1, it is evident that Figures 4-12 and 4-13 will have best application to systems where the mobility ratio is favorable, since, in this case, the displacement would be piston-like, with the increase in oil recovery past breakthrough being due only to the increase in areal coverage. This could, of course, be taken partially into account by a judicious choice of the residual oil saturation value used in the calculation of the displacement volume. Actually, if the "pore-to-pore" sweep efficiency is going to change markedly during the after-breakthrough phase of the performance, then a modification may be necessary. The Buckley-Leverett theory for immiscible displacement, or the viscous fingering theory, provides a means by which this change in pore-to-pore sweep efficiency can be satisfactorily predicted. Where the saturations behind the front are changing with continued injection past breakthrough, the relative permeability of the displacing and displaced phases are changing—in which case the mobility ratio, necessary for a proper use of Figures 4-12 and 4-13, is changing also.

It is evident that if the displacing fluid is to be injected at a constant rate, as was the case in Example 1, the injection pressure will vary during the production performance of the five-spot well pattern flood. Certainly, if the water viscosity is low and the oil viscosity somewhat higher, the pressure drop through the flooded out zone per given distance will be less than the pressure drop experienced through the same distance in the zone where only oil flows. Then too, the length of the flow path ahead of the displacement front and behind it will change as the flooding proceeds. It is convenient to term this change the conductance ratio, γ, which is equivalent to:

$$\gamma = (q/\Delta p)/(q/\Delta p)_i \tag{2}$$

where the subscript i refers to initial conditions, q is the reservoir flow rate at either the producing or injection well in this steady-state system, and Δp is the pressure drop from the injection to the producing well. Caudle and Witte[18] and Prats *et al.*[19] have published experimental and calculated values of conductance ratio for the five-spot well network. Figures 4-14 and 4-15 are a plot of conductance ratio as a function of mobility ratio and the areal sweep efficiency, and as a function of mobility ratio and the displaceable pore volume, V_i/V_D. V_i is the cumulative volume injected, while V_D is the displaceable, or floodable pore volume of the system.

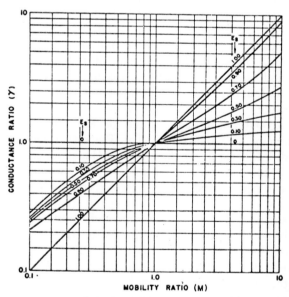

Figure 4-14. Conductance ratio as a function of mobility ratio and the areal sweep efficiency (E_s) for the five-spot well pattern (Caudle and Witte[18]).

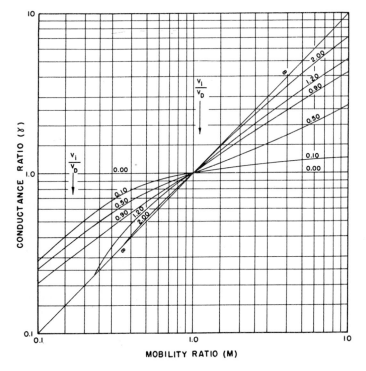

Figure 4-15. Conductance ratio as a function of mobility ratio, and the displaceable pore volumes injected (V_i/V_d) for the five-spot well pattern (Caudle and Witte [18]).

The usefulness of Figures 4-14 and 4-15 in determining the pressure and injectivity behavior of the five-spot well system can best be shown by an example problem:

EXAMPLE 1 (con't):
Pressure and injectivity behavior of a five-spot well system.

Additional Data: Initial Δp = 1000 psia
Initial q = 200 B/D

SOLUTION:
Two solutions are possible, that of constant rate and variable injection pressure or variable rate and constant injection pressure.

(1) Constant rate and variable injection pressure:
At breakthrough, the areal sweep efficiency is 0.58 and from example problem 1, the mobility ratio is 2.08. Figure 4-14 provides a value of γ equal to 1.2. Then from equation (2):

$$\Delta p = (q/\gamma)/(q/\Delta p)_i = \frac{1000}{1.2} = 830 \text{ psia}$$

At a displaceable pore volume injected of 1.5, the corresponding pressure drop, Δp, is:

$$\Delta p = \frac{1000}{1.46} = 685 \text{ psia}$$

(2) Variable rate and constant injection pressure:
At breakthrough, the injection rate would amount to:

$$q = \gamma \cdot \Delta p \cdot (q/\Delta p)_i = 1.2 \times 200 = 240 \ B/D$$

At 1.5 displacement volumes, the injection rate would be:

$$q = 1.46 \times 200 = 296 \ B/D$$

From problem example 1, it is apparent that an injectivity or pressure behavior for a five-spot can be estimated within the assumptions which are inherent in the experimental data itself, i.e., constant pay thickness, a "piston-like" displacement, homogeneous reservoir and steady-state operations. In prediction of a five-spot well pattern performance under fluid injection, either miscible or immiscible, the pressure drop that would result, and conversely the injection pressure required at the beginning of injection, would not be known. Sometimes experience in reservoirs of a similar type with similar fluids would permit good estimates to be made. Where this is not possible, an equation developed by Muskat[20] may be used:

$$\left(\frac{q}{\Delta p}\right)_{M=1} = \frac{hk_o}{\mu_o} \frac{3.535}{\ln (d/r_w) - 0.619} \tag{3}$$

where d is the distance between the producing and injection wells in feet, and the field units used are darcys, psia, cp, and B/D. If a reasonable estimate is available for oil permeability from laboratory studies, from pressure build-up or productivity tests or from petrophysical considerations, then either the injection rate q can be calculated, if Δp can be stated or, conversely, Δp may be determined for a given injection rate. This approach does assume a mobility ratio of one—the case prevailing when injection is first begun.

Aronofsky and Ramey[8] have studied "nose advance" in a five-spot well network and related it to the mobility ratio and the conductance ratio. Here, nose advance is defined as the percentile ratio between the injection well and the interface position along a straight line, joining the production and injection wells and the total distance between the wells, d. Of course, the nose advance

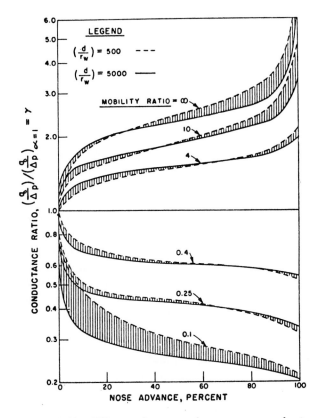

Figure 4-16. Effect of nose advance on conductance ratio for various mobility ratios and d/r_w values of 500 and 5000 in an enclosed five-spot well
pattern (Aronofsky and Ramey[8]).

at breakthrough would be 100 per cent. Figure 4-16 shows the resulting relationships for ratios of d/r_w, equal to 500 and 5000.

Figure 4-16 provides a means by which the injection rate, (pressure drop known) or the pressure drop (injection rate known) may
be determined. At breakthrough, the nose advance is 100 per cent.
For a given value of mobility ratio and d/r_w, the conductance
ratio, γ, can be read from the graph directly. It is then a simple
matter to calculate the injection rate or the pressure drop from
equation (2). For estimated positions of the nose advance, further
values for the conductance ratio may also be developed, and corresponding pressure drops or injection rates determined. Note that

the pressure drop so determined is the pressure drop across the formation. The injection pressure at the pump would be equal to:

$$p_i = \Delta p_{\text{surf. lines}} + \Delta p_{\text{injection well}} + \Delta p_{\text{reservoir}} +$$
$$\Delta p_{\text{pump submergence}} \pm \Delta p_{\text{surface elevations}} -$$
$$\Delta p_{\text{injection well fluid head}} \quad (4)$$

where the Δp's are the pressure decrements corresponding to a particular part of the system. The pressure drop through the surface lines may be readily estimated from the well known Hazen-Williams or similar type equation, and the pressure drop in a well bore can be calculated by methods suggested by Craft, Holden and Graves[21]. The average pressure head (pump submergence) in the producing well will depend upon the type of lifting equipment used, and how close to being "pumped off" the well is kept. In those producing wells that are flowing, the bottom hole pressure, now corresponding to the submergence pressure, will be considerably larger. If gas lifting techniques are being used, calculation techniques for bottom hole pressure have been published[21]. Many times it will suffice to ignore the differences in pressure, due to changes in surface elevation, within the producing property boundaries. There are instances, though, where the differences in elevation are such that the pressure change due to surface elevation is sizeable.

Sheldon and Dougherty[12] have used numerical techniques to compute the simultaneous dynamic behavior of multiple fluid-fluid interfaces in two dimensions, and for five-spot systems in particular. Since the technique is an analytical one, the mobility ratio can be specified and some of the pitfalls avoided that might result from experimental data taken from displacement processes, using miscible fluids in porous media. No doubt, much of the disagreement between the recoveries observed or calculated by published techniques is due to the problems of fingering or saturation instabilities, when the mobility ratios are adverse. Most of the analytical methods avoid the problems of fingering, either rightly or wrongly, due to the mathematical techniques utilized. The mathematical arguments of Sheldon and Dougherty will not be reproduced here. Figure 4-17 is plot of oil recovery versus the pore volumes of displacing fluid injected (miscible fluids), with mobility ratio as a parameter for the five-spot well pattern. The experimental results plotted on Figure 4-17 are from work by Paulsell[22],

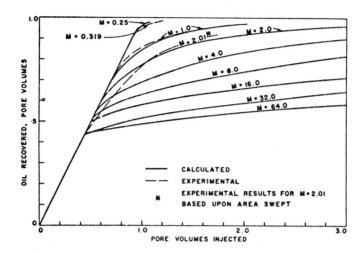

Figure 4-17. Plot of oil recovered as a function of pore volumes of displacing fluid (miscible) injected with mobility ratio as a parameter for the five-spot well network (after Sheldon and Dougherty[12]).

where miscible fluids were used. Recovery curves computed from Paulsell's data are in fairly good agreement with the calculated curves for mobility ratios of 0.25 and 1.0. In addition, Paulsell's data for recovery when mobility ratio equaled 2.01 agrees quite well with the results of Sheldon and Dougherty, with the exception that breakthrough occurred at 43 per cent oil recovery, as compared to 60 per cent by the analytical approach. At high mobility ratios the results published by Dyes et al.[10], Caudle and Witte[18], and perhaps by Sheldon, Dougherty[12], would not necessarily be correct, and would most probably error on the side of indicating too favorable of values for oil recovery at breakthrough of the displacing phase. This could be explained by observing that the areal sweep efficiencies would differ, depending upon whether the efficiencies were based upon (1) the pore volume injected, (2) area measured (i.e., as done in the X-ray shadowgraph technique), (3) mathematics where viscous fingering could not occur. Habermann[23] has compared experimental results based upon area measured and pore volume injected, in order to obtain an indication of the size of the mixing zone. Figure 4-18 shows that the mixing zone may result in a difference of more than 0.1 pore volume, at high mobility ratios.

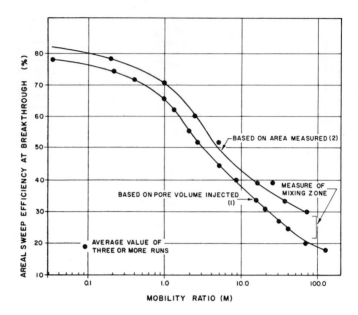

Figure 4-18. Breakthrough sweep efficiencies, and a measure of the mixing zone for a range of mobility ratios, five-spot well network (after Habermann[23]).

Figure 4-19 presents the displacement fronts for a variety of fluid mobility ratios which have been observed in laboratory models where miscible displacing and displaced fluids were used. It is not difficult to see why agreement between various published sweep efficiencies will be very good for favorable mobility ratios, but show poor agreement for floods where fluids having adverse mobility ratios are used. Then too, the proper sizing of a laboratory model in order to duplicate the fingering and resulting sweep efficiencies which would occur in the field case is not well understood at the present time.

Other Well Networks

A sizeable quantity of information has been published for the line-drive well network—for the staggered and direct line-drive configurations. Dyes, Caudle and Erickson[10] have presented graphical results showing the effect of mobility ratio on the areal sweep efficiency for a range of displacing phase fractional produc-

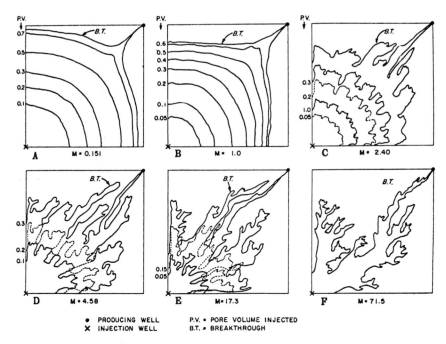

Figure 4-19. Displacement fronts for different mobility ratios and injected pore volumes until breakthrough, quarter of a five-spot well network (after Habermann[23]).

ing rates, or displaceable pore volumes, V_i/V_p, for the direct line-drive (square pattern) and the staggered line-drive. The displaceable volume is the product of the pore volume of the unit pattern, and the displacement efficiency achieved in the swept region. Figures 4-20 through 4-23 show the relationships for the direct and staggered line-drive well configurations. The discussion of similar graphical relationships for the five-spot well network was presented in the preceding section, where a simplified performance calculation has been outlined. The same technique applies for the direct and staggered line-drive networks. It should be noted that at adverse mobility ratios, such performance calculations could be considerably in error, since large amounts of oil production would come from the subordinate phase or, that is, from behind the displacement front. Figures 4-20 through 4-23 are developed where there is no saturation gradient, either ahead of, or behind the front (i.e., constant mobility ratio), and where no further oil

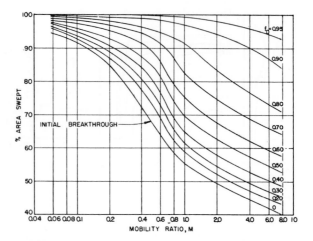

Figure 4-20. Effect of mobility ratio on areal sweep efficiencies with the fraction of the displacing phase produced as a parameter for the direct line drive well network (square pattern) (after Dyes, Caudle and Erickson[10]).

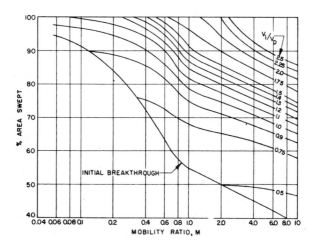

Figure 4-21. Effect of mobility ratio on areal sweep efficiencies with displaceable volumes as a parameter for the direct line drive well network (square pattern) (after Dyes, Caudle and Erickson[10]).

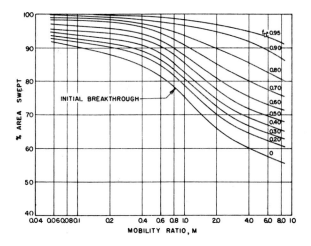

Figure 4-22. Effect of mobility ratio on areal sweep efficiencies with the fraction of the displacing phase produced as a parameter for the staggered line drive well network (after Dyes, Caudle and Erickson[10]).

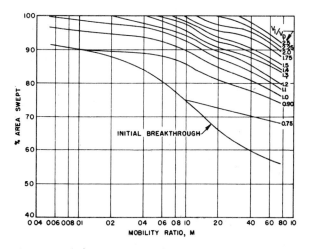

Figure 4-23. Effect of mobility ratio on areal sweep efficiencies with displaceable volume as a parameter for the staggered line drive well network (after Dyes, Caudle and Erickson[10]).

production is realized from the areas behind the displacement front. Then too, the models used represented a steady-state injection operation where only two phases are flowing, the oil and the displacing phase. This corresponds to flooding operations above the bubble point, or where the gas saturations are less than the critical value where the first permeability to gas would occur.

The relationships depicted by Figures 4-22 and 4-23 are for the case where the ratio of the distance between injection and producing lines, to the distance between either producers, or injection wells was unity, i.e., d/a equal to one. If the d/a ratio were to have a value between 1 and 0.5 (the five-spot pattern), interpolation between the areal sweep efficiencies given by Figures 4-19 and 4-20, and Figures 4-12 and 4-13, would permit displacement values to be obtained for other staggered line-drive well configurations.

Morel-Seytoux[24] has used complex potential theory and numerical techniques as presented by Sheldon and Dougherty[12], to show the influence of fluid mobility ratio on oil recovery and upon producing water-oil ratio for the direct-line drive well network, where d/a is equal to 0.5. Figures 4-24 and 4-25 present the results of this analytic treatment. It should be noted that no provision is provided for the inclusion of the residual oil saturation, or the irreducible water saturation on this curve. In actuality, the oil

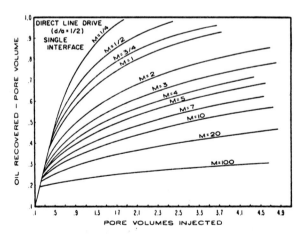

Figure 4-24. Influence of mobility ratio on oil recovery for the direct line drive well network system (after Morel-Seytoux[24]).

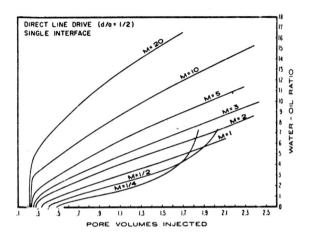

Figure 4-25. Influence of mobility ratio on producing water-oil ratio for the direct line drive
well network system (after Morel-Seytoux[24]).

recovery is the oil recovered as a fraction of the *recoverable* hydrocarbon pore volume, $1 - S_{wi} - S_{or}$. Morel-Seytoux has also
studied the influence of mobility ratio on injectivity, with injectiv-
ity defined as $\dfrac{q\mu_o}{kh(\Delta p)}$, where q is the injection flow rate in bar-
rels/day, μ_o is the oil viscosity in poises, k is the formation permeability in darcies, h is the formation thickness in feet and Δp
is the pressure drop between the injector and producer in psi.
Figure 4-26 presents these relationships, and provides a convenient means for estimating either the injection pressures (q, μ_o,
k and h known), or the injection rates (μ_o, k, h, and Δp known) as
the pore volumes injected, or history of the injection project goes
forward. Equation (4) permits a calculation of the injection pressure at the pump so that sizing of equipment for the injection facilities may be conveniently made. It is readily apparent from
Figure 4-26 that the injectivity would increase sharply at breakthrough of injected fluid at the producing well for mobility ratios
larger than one. This is occasioned by the fact that the injected
fluid is more mobile than the displaced fluid. Of course, where
the injected fluid is less mobile than the produced fluid, injectivity decreases quite rapidly during radial movement of the injected
fluid away from the well, less rapidly as the displacement front

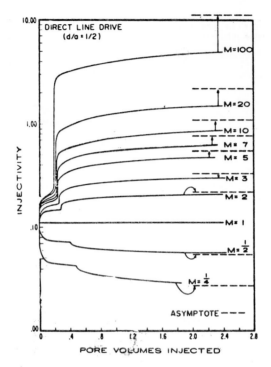

Figure 4-26. Influence of mobility ratio on injectivity as the injection history proceeds for the direct line drive well network system (after Morel-Seytoux[24]).

starts to cusp toward the producing well, and then takes a finite decrease in injectivity at breakthrough to the producing well. This type of injectivity behavior has been frequently observed during field secondary recovery operations in all flooding well networks, in addition to the direct line-drive well network case.

Recently, Kimbler, Caudle and Cooper[25] have published information on the areal sweepout behavior in an inverted nine-spot injection pattern. Figure 4-27 shows the system studied, and locates the positions of the "corner wells" and the "side wells." The results presented are strictly applicable to an infinite array of wells in a homogeneous horizontal reservoir, where two-phase flow is occurring (i.e., no flowable gas is present), and no oil bank is formed. The results presented make use of a producing rate ratio, which is defined as the total producing rate of a corner

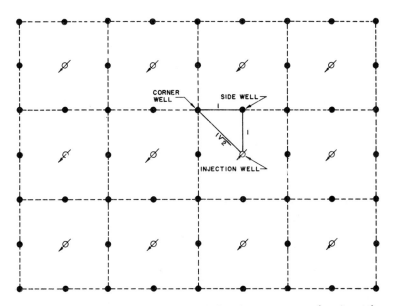

Figure 4-27. Inverted nine-spot injection system showing the position of the side wells and corner wells (after Kimbler *et al.*[25]).

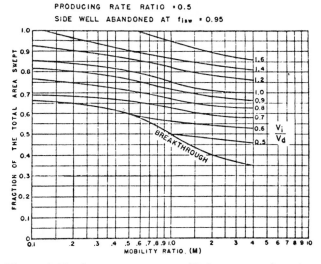

Figure 4-28. Sweepout pattern efficiency as a function of mobility ratio for the nine-spot pattern at various displaceable volumes (V_i/V_d) injected (after Kimbler *et al.*[25]).

well divided by the total producing rate of a side well. Since steady-state flow conditions apply in this system, the amounts of withdrawals at the corner and side wells will be exactly balanced by the input to the injection wells. It is the ratio of the producing rates of the corner wells and side wells which affects the break-through recovery. Figures 4-28 through 4-37 present the sweepout pattern efficiencies as a function of mobility ratio at various displaceable pore volumes injected (V_i/V_D), at various side well producing cuts (f_{isw}) and at various corner well producing cuts (f_{icw}), for producing rate ratios of 0.5, 1.0 and 5.0. Nearly all of the runs were made for a mobility ratio range from 0.1 to 4.0—a useful range for many of the present day water floods, where oils are relatively mobile. Then too, the data presented by the graphs are strictly applicable only to those cases where the mobility behind the displacement front is constant. This would be a reasonable assumption for the field case when the mobility ratio is favorable, but could result in a large error if any attempt were made to extrapolate the graphs for information at mobility ratios larger than four. The results may, of course, be used for other types of dis-

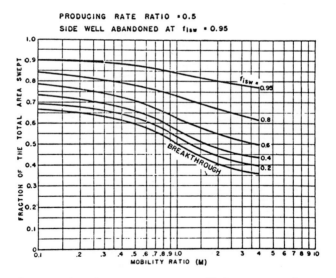

Figure 4-29. Sweepout pattern efficiency as a function of mobility ratio for the nine-spot pattern at various side well producing cuts (f_{isw}) (after Kimbler et al.[25]).

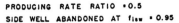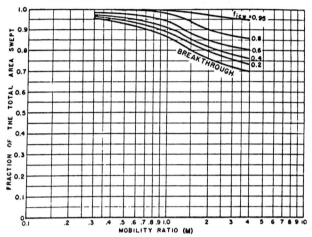

Figure 4-30. Sweepout pattern efficiency as a function of mobility ratio for the nine-spot pattern at various corner well producing cuts (f_{icw}) (after Kimbler *et al.*[25]).

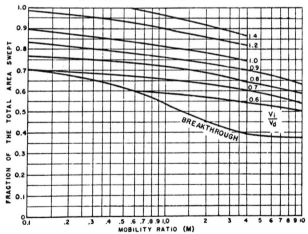

Figure 4-31. Sweepout pattern efficiency as a function of mobility ratio for the nine-spot pattern at various displaceable volumes (V_i/V_d) injected (after Kimbler *et al.*[25]).

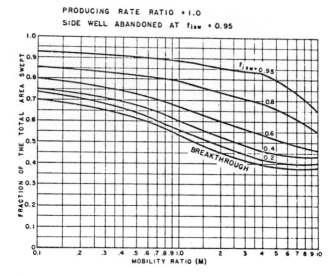

Figure 4-32. Sweepout pattern efficiency as a function of mobility ratio for the nine-spot pattern at various side well producing cuts (f_{isw}) (after Kimbler et al.[25]).

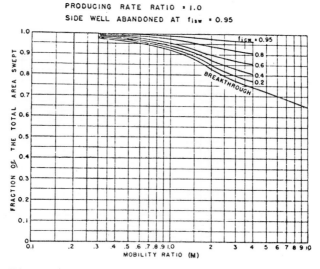

Figure 4-33. Sweepout pattern efficiency as a function of mobility ratio for the nine-spot pattern at various corner well producing cuts (f_{icw}) (after Kimbler et al.[25]).

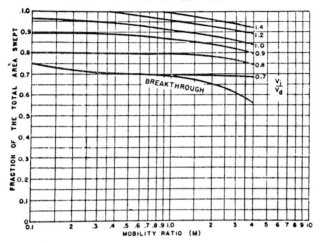

Figure 4-34. Sweepout pattern efficiency as a function of mobility ratio for the nine-spot pattern at various displaceable volumes (V_i/V_d) injected (after Kimbler *et al.*[25]).

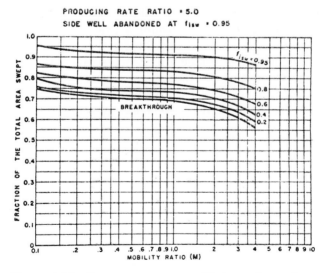

Figure 4-35. Sweepout pattern efficiency as a function of mobility ratio for the nine-spot pattern at various side well producing cuts (f_{isw}) (after Kimbler *et al.*[25]).

PRODUCING RATE RATIO =5.0

SIDE WELL ABANDONED AT $f_{i_{sw}}$ = 0.95

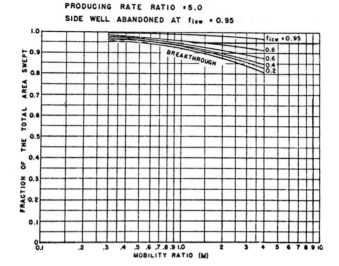

Figure 4-36. Sweepout pattern efficiency as a function of mobility ratio for the nine-spot pattern at various corner well producing cuts (f_{icw}) (after Kimbler et al.[25]).

placement than where water is displacing oil. The X-ray shadowgraph technique, used in the model studies by Kimbler *et al.*[25], has been described by Slobod and Caudle[6].

The information presented in Figures 4-27 through 4-36 is strictly applicable only for those cases where the mobility behind the flood front is constant. This restriction is usually satisfied so long as the mobility ratio does not greatly exceed a value of

Figure 4-37. Reservoir model used to study the performances of an unconfined five-spot (after Caudle and Loncaric[26]).

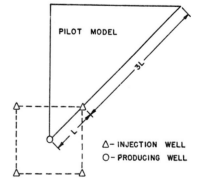

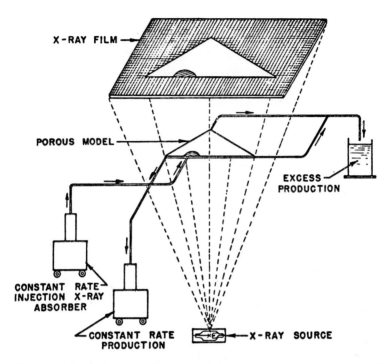

Figure 4-38. Schematic diagram of the laboratory equipment used to study the performance of an unconfined five-spot pilot flood (after Caudle and Loncaric[26]).

one. For this reason, the data presented apply to many water flood or miscible gas-drive processes. The graphical relationships are given in reservoir units so that stock tank barrels of oil could be obtained by means of an oil formation volume factor. If the connate water and residual oil saturations can be estimated, and the cumulative injected fluid volume (in reservoir barrels) is known at a given time, the fractional displaceable volume (V_i/V_D) can be determined. The mobility ratio is determined by calculating the ratio of the displaced fluid mobility behind the front and the displaced fluid mobility ahead of the front. The figures may then be used, consistent with the producing rate ratio which exists in the field problem, or to that required in the flood design calculations. The proper plot of fraction of area swept (E_s) versus mobility ratio, with V_i/V_D as a parameter, is used to determine whether or not the well of interest has broken through. The sweep efficiency at breakthrough, or at any subsequent value of V_i/V_D, is taken directly from the plot. Reference to the corresponding plot, for either the side or corner well and the proper producing rate ratio, permits the determination of the producing cut in reservoir barrels for the well of interest. Repetition of these steps at vari-.ous throughput values will allow the production histories of the side and corner wells to be generated, and the overall producing history of the pattern and the field determined. Of course, the limiting assumptions of constant formation thickness, homogeneous formation properties and piston-like displacement, with no production occurring from the sweep area apply, and should be recognized if the method is used to predict performance in the nine-spot well injection configuration.

The Pilot Flood—Theory and Interpretation of Results

Pilot flooding is a commonly used means of evaluating a proposed secondary recovery project before large quantities of capital have been invested, where basic questions exist which bear on whether flooding would be successful on a field-wide basis as projected. A field-wide flood normally consists of an array of nearly identical well patterns. In such an extensive well pattern,

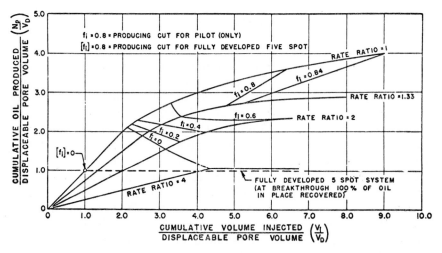

Figure 4-39. Production histories of pilot floods at various rate ratios and of a confined five-spot flood. Mobility ratio = 0.1. (after Caudle and Loncaric[26]).

each basic flooding unit, the five-spot, can be visualized as a repetition of "confined" floods. On the other hand, a pilot flood involving one or more well patterns is essentially, or at least partially, "unconfined." The result is that some of the oil produced from a pilot flood may come from outside the perimeter of the area, and some of the displaced oil may completely leave the pilot area. Similarly, only a portion of the fluid injected is useful in moving oil to the producing wells, while the rest of the injected fluid escapes to the surrounding reservoir. In addition, the recovery realized from a pilot flood can be greatly influenced by the operating conditions, i.e., mostly by the bottom-hole pressures at the producing and injection wells.

Caudle and Loncaric[26] have studied oil recovery in five-spot pilot floods by means of laboratory models constructed from round-grained sand, consolidated with an epoxy resin. The X-ray shadowgraph technique was used to determine the sweepout pattern efficiencies[6]. The mobility ratios studied were approximately 0.1, 0.3, 1, 3 and 10. Figure 4-37 presents the single five-spot pilot area studied, while Figure 4-38 shows the schematic layout of the laboratory equipment. Miscible oils were used in the model to avoid the effects of capillary forces. The authors found that the model dimensions of L and $3L$ (see Figure 4-37) were sufficiently large so that the distance to the outside edge of the model had

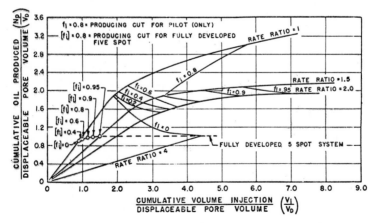

Figure 4-40. Production histories of pilot floods at various rate ratios and of a confined five-spot flood. Mobility ratio = 0.29. (after Caudle and Loncaric[26]).

little influence upon fluid distributions in the vicinity of the pilot
area.   Since it is common to have producing rates in a five-spot
which are considerably less than the displacing fluid injection
rates, the laboratory flow model has been constructed so that part
of the injected fluid may be withdrawn (i.e., escapes) from the
model without being taken from the producing well.   Notice too,
that the results obtained apply strictly to a single five-spot well
configuration where the injectors make a´square pattern around a
single producing well.   The results should have good application
where the comparable field case has little free gas, uniform permeability in all directions and constant formation thickness in the
pilot flooding area.   Figures 4-39 through 4-43 are production histories of the pilot floods at various rate ratios (q_i/q_p) and of a
confined five-spot unit where the fluid mobility ratios are 0.1,
0.29, 1.0, 3.7, and 10.

Notice that all the data taken from or used on the graphs should
be in reservoir units.   The displaceable pore volume, V_D, is the
displaceable oil volume in the confined five-spot area, i.e., the
reservoir oil in place less the residual oil—all at reservoir conditions of reservoir temperature and pressure.   The cumulative oil

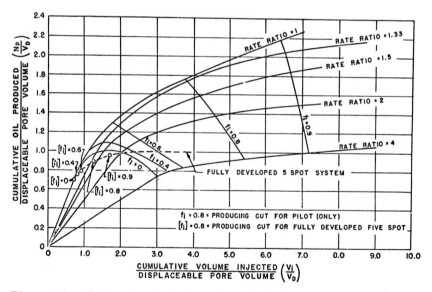

Figure 4-41. Production histories of pilot floods at various rate ratios
and of a confined five-spot flood. Mobility ratio = 1.0. (after Caudle and
Loncaric[26]).

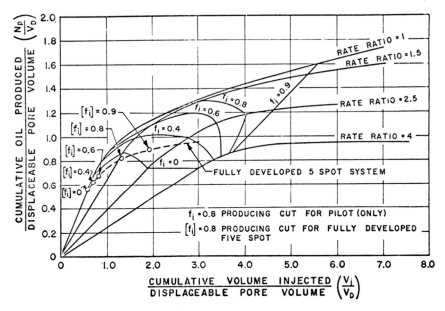

Figure 4-42. Production histories of pilot floods at various rate ratios and of a confined five-spot flood. Mobility ratio = 3.7. (after Caudle and Loncaric[26]).

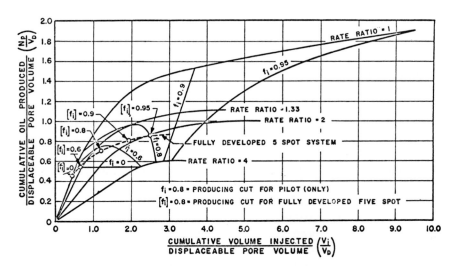

Figure 4-43. Production histories of pilot floods at various rate ratios and of a confined five-spot flood. Mobility ratio = 10.0. (after Caudle and Loncaric[26]).

produced, N_p, is that at reservoir conditions also. The fraction of the displacing phase produced is represented as f_i. The rate ratio has been arbitrarily defined as the injection rate at all four injection wells of the pilot, divided by the production rate at the single producing well of the pilot.

The problem normally encountered in reservoir engineering will be how to correct the results from a five-spot pilot flood to that which would have occurred had the flood been totally confined— as would occur during actual flooding operations. Or possibly, it might be necessary to develop beforehand the anticipated performance of a pilot flood before actually going to the expense of doing the pilot flood.

To illustrate the calculation procedure, let us assume that a pilot flood has had a rate ratio of 1.5 and a mobility ratio of 1.0. Then from Figure 4-41, the correction factor for the ordinate at breakthrough to the producing well would be the ratio 0.70/0.88, while that for the abscissa would be 0.70/1.40. Then if pilot flood performance curves are available, a number of such correction factors would permit the generation of the predicted performance of a totally confined five-spot. If the actual flood was on a forty-acre basis while the pilot flood was of a ten-acre size, the cumulative oil produced and cumulative water injected values, after using the correction factor, would merely be multiplied by four.

Where a field has been depleted by primary means and a high gas saturation exists in conjunction with low reservoir pressure, the method of Caudle and Loncaric does not have good application due to the difficulty in specifying the proper rate ratio, and also due to the fact that a banking of oil will probably have occurred. Dalton *et al.*[27] have presented the results of laboratory work where the effect of injection and producing well bottom-hole pressures and an oil bank have been included. Three pilot flooding well networks were studied in addition to the normal five-spot pilot studied by Caudle and Loncaric. These patterns were: a group of four inverted five-spots, six inverted five-spots and a single inverted five-spot. An inverted five-spot has producing wells at the four corners with a single injection well at the center of the pilot flood area. The results are applicable to fields having isotropic, homogeneous formations of constant pay thickness, a uniformly distributed free gas saturation at the start of the pilot flood and having a fluid mobility ratio throughout the flood of unity. Note that the results would have only qualitative value if the fluid mobility ratio being considered is much different from one.

Dalton *et al.* have observed that a significant feature of an unconfined five-spot is the competition of the surrounding reservoir and pilot producing wells for the injected fluid and the displaced oil. If the reservoir pressure is low, it is not possible for the producing wells to provide a sizeable pressure sink which would attract large quantities of this oil and displacing fluid. On the other hand, if reservoir pressure is high and the pressure at the producing wells of the pilot low, the better the displaced and displacing fluids are confined to the pilot area. The influence of pressure in pilot flooding operations has been defined by Dalton *et al.* as:

$$\pi = \frac{p_s - p_{wp}}{p_{wi} - p_s} \tag{5}$$

where π is a dimensionless pressure parameter, p_s is the static reservoir pressure, p_{wp} is the pressure at the producing well(s) and p_{wi} is the pressure at the injection well(s). Actually, the

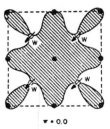

$\pi = 0.0$

Figure 4-44. Illustration of area supplying oil to the producing wells of a pilot area for two values of the π ratio (after Dalton *et al.*[27]).

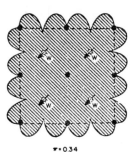

$\pi = 0.34$

FOUR INVERTED 5-SPOTS

π factor represents a scaling coefficient necessary to adapt the laboratory results to the field situation. Then too, the initial fluid saturations, the field geometry, the well distributions, the well penetrations, and the ratio of well bore size to well spacing should be the same in both the field and model cases, if the resulting flow rates are to be directly comparable.

Figure 4-44 is an illustration of the area supplying oil to the producing wells of the pilot area, for four inverted five-spots and two π ratio values. When the static reservoir pressure is nearly zero, the π factor is nearly zero also. As a result, the area supplying oil to the producing wells is low too, since there is no sizeable pressure sink at the producing wells to attract the displaced oil. At sizeable π ratios, the area supplying oil to the producing wells can substantially exceed that contained by the basic pilot area.

Dalton *et al.*[27] used both potentiometric models and two-phase (water-oil) flow model tests. A gas saturation of 15 per cent was used throughout the tests, with the assumption that 8 per cent of

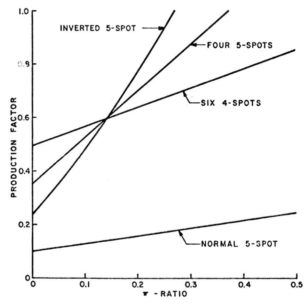

Figure 4-45. Effect of π-ratios upon the production factor for various pilot flooding well networks (after Dalton *et al.*[27]).

the pore volume of gas would be displaced ahead of the oil bank, with the remaining 7 pore per cent being compressed and dissolved within the leading edge of the oil bank. This type of behavior is known to occur in many water floods. In addition, a "production factor" was defined as the ratio of cumulative liquid production to cumulative injection, while "areal recovery factor" was defined as the area which supplies oil to the basic pilot area. Figures 4-45 and 4-46 show the effect of operating conditions (π ratios) on the

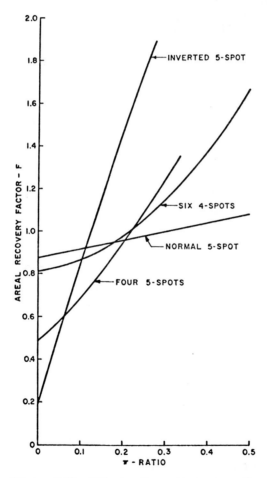

Figure 4-46. Effect of π-ratios upon the production factor for various pilot flooding well networks (after Dalton *et al.*[27]).

"production factor" and the "areal recovery factor" for various basic pilot flooding well networks. All the graphical results were computed for the case when total liquid production from the producing wells of the pilot area totalled 1.5 pore volumes (of the pilot area).

Notice from Figure 4-45 that at a π ratio of zero, the production factor equals 10 per cent in the case of the normal five-spot—that is for four injection wells surrounding one producing well. This means that the producing well captures only 10 per cent by volume of the total fluid injected to the five-spot. This means that in a normal five-spot pilot, large quantities of liquid injection will need to be injected in order to obtain liquid production from the producing well, totaling 1.5 times the pore volume of the pilot area. On the other hand, Figure 4-46 and Figure 4-47 show that the normal five-spot pilot experiences relatively small changes in the production factor and the areal recovery factor, for large changes in the π ratio. Note that the inverted five-spot, one in-

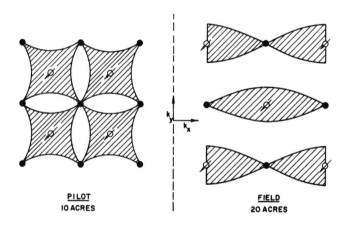

PILOT	FIELD
10 ACRES	20 ACRES

5—SPOT SWEEP EFFICIENCIES

k_y/k_x	PILOT	FIELD
1	72	72
3	77	43
10	90	12
100	100	1.0

Figure 4-47. Five-spot flood network (confined) sweep efficiencies for varied degrees of permeability anisotropy for a fluid mobility ratio of one (after Landrum and Crawford[31]).

jecting and four producing wells, is much more sensitive to the operating conditions, i.e., the reservoir static pressure and the producing and injection well bottom-hole pressures. The finding by Dalton *et al.* that the normal five-spot is relatively insensitive to pressures is, in part, a vindication of the work by Caudle and Loncaric[26], where the factor was not considered. This would indicate the results of Caudle and Loncaric, though not developed for the case where substantial gas saturation exists, may be applicable without concern for the operating conditions as outlined by Dalton *et al.*

Practical application of the laboratory results of Dalton *et al.* is achieved by adjusting the recovery curve of a field pilot flood, conducted at a set π ratio, to the confined pattern recovery curve by dividing the pilot recoveries and other liquid recovery values by the appropriate areal recovery factor. Notice that the correction has been accomplished using only the areal recovery factor obtained when the pilot producing wells have produced 1.5 times the pore volume of the pilot area. This approximation will usually be somewhat incorrect at the time of breakthrough of the displacing fluid in the confined pattern. Of course, the mobility ratio in the field case will need to be of the order of unity.

Fischer, Rosenbaum and Matthews[28] have noted that when the response of a pilot is earlier than that predicted for a homogeneous reservoir, i.e., by one of the foregoing methods, the behavior observed is a measure of the reservoir inhomogeneities. Should the response be much later than predicted, then formation permeability near the producing wells is lower than the average formation permeability, or there are thief sands. If response is early, then perhaps the formation is somewhat stratified, or directional permeability may be a problem.

Directional Permeabilities—Effect on Areal Sweep Efficiencies and Productivities

It is well known that vertical and lateral permeabilities can vary considerably within a given reservoir rock. If this change is random, usually it will suffice to average the measured permeability values and treat the system as if the permeability were constant. Where there are sizeable regions where there are large changes in the lateral permeability, i.e., where a directional permeability trend is apparent, considerable care should be exercised in the selection and directional orientation of the basic flooding patterns. Recognition of the directional permeability can result in the selection of patterns which will result in excellent areal sweep efficiencies. On the other hand, poor placement of the injection wells can result in fast breakthroughs of the displacing fluid at the producing wells, and very poor areal sweep efficiencies.

Assuming that the directional permeabilities are different but uniform, Muskat[29] has shown that Darcy's law may be written for a homogeneous but anisotropic medium as:

$$v_x = \frac{-k_x}{\mu} \frac{\partial p}{\partial x}, \quad v_y = \frac{-k_y}{\mu} \frac{\partial p}{\partial y}, \quad v_z = \frac{-k_z}{\mu} \left(\frac{\partial p}{\partial z} - \rho g \right) \quad (6)$$

where

$\quad k_x, k_y, k_z$ = directional permeabilities for the x, y, z coordinate
$\qquad$ system
$\quad \rho$ = reservoir fluid density
$\quad g$ = acceleration of gravity
and where z is positive downward.
Applying the equation of continuity for incompressible fluids, the pressure at a point, $p(x, y, z)$, may be written as:

$$k_x \frac{\partial^2 p}{\partial x^2} + k_y \frac{\partial^2 p}{\partial y^2} + k_z \frac{\partial^2 p}{\partial z^2} = 0 \quad (7)$$

Equation (7) may be reduced to the Laplace equation by the following substitutions:

$$\bar{x} = \frac{x}{\sqrt{k_x}}, \quad \bar{y} = \frac{y}{\sqrt{k_y}}, \quad \bar{z} = \frac{z}{\sqrt{k_z}} \quad (8)$$

where the coordinate system becomes $(\bar{x}, \bar{y}, \bar{z})$. The Laplace equation then results:

$$\frac{\partial^2 p}{\partial \bar{x}^2} + \frac{\partial^2 p}{\partial \bar{y}^2} + \frac{\partial^2 p}{\partial \bar{z}^2} = 0 \quad (9)$$

It is then evident that the effect of permeability anisotropy can be represented by an adjustment of the coordinate system as suggested by equation (8). Of course, the coordinates of the pressure at $p(\bar{x}, \bar{y}, \bar{z})$ in a radial system would be altered to the coordinates of the same pressure at $p(x, y, z)$ by multiplying $\bar{x}, \bar{y}, \bar{z}$ by the respective factors $\sqrt{k_x}, \sqrt{k_y},$ and $\sqrt{k_z}$.

The utility of the above outlined concept can be shown by considering a five-spot well pattern where the permeability in the y direction is four times that in the x direction. It should then be evident that the pattern geometry could be changed so that breakthrough would occur at all the producing wells simultaneously by shortening the distance to the well in the x-direction by a factor of 2. Such a procedure has been suggested by Johnson and Hughes[30]. The position of the flood front at any particular time could be positioned by proper scaling of the fronts for the homogeneous (with respect to permeability) five-spot flood to the pattern, with the adjusted producing well positions.

Usually though, the position of the wells is predetermined by drilling before the consideration of secondary recovery of oil, and before there is any real information available on directional trends

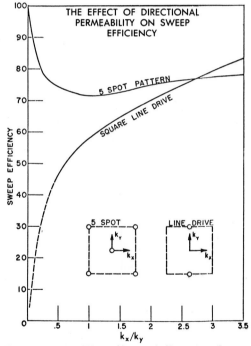

Figure 4-48. The effect of directional permeability on sweep efficiencies for varied degrees of permeability anisotropy for a fluid mobility ratio of one (after Landrum and Crawford[31]).

of the permeability. Since it is seldom possible to redrill a field, information is needed in order to choose a pattern which will result in the maximum recovery of oil within a reasonable time interval, and without excessive sacrifice in producing capability. Landrum and Crawford[31] have studied the effect of directional permeability on sweep efficiency and production capacity for the five-spot and direct line-drive (square pattern with d/a equal to one). Figure 4-47 presents data on the sweep efficiencies for a five-spot flood for a range of directional permeability values (fluid mobility equal to one). Notice how the practice of drilling an in-fill well for pilot flooding purposes can result in a shifting of the pattern by 45 degrees with a resulting drastic change in the sweep efficiencies, which may be observed in the field flooding case. Figures 4-48 and 4-49 are plots of the effect of directional permea-

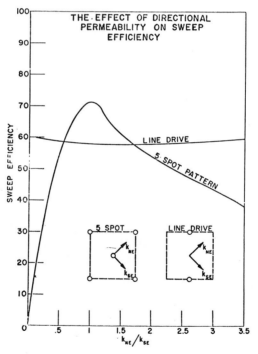

Figure 4-49. The effect of directional permeability on sweep efficiencies for varied degrees of permeability anisotropies at a fluid mobility ratio of one (after Landrum and Crawford[31]).

bility on sweep efficiency for the five-spot and direct line-drive well networks, for two relative positions of the directional permeability. Interpolation of the results should permit an estimation of the sweep efficiency to be expected when the directional permeability orientation is somewhat different from that studied, and where fluid mobility ratio is of the order of one.

Landrum and Crawford [31] have also studied the effect of directional permeability on production capacity where, due to scaling, the results are strictly applicable for a 660 foot well spacing and a 0.5-foot well bore radius. Figures 4-50 and 4-51 summarize these findings.

It is only in recent years that widespread awareness of the effect of permeability anisotropy, either beneficial or detrimental,

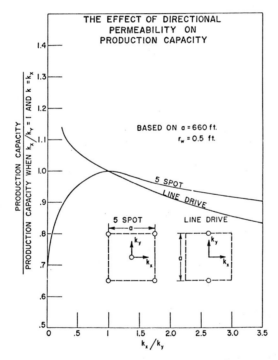

Figure 4-50. The effect of directional permeability on production capacity for varied degrees of permeability anisotropies at a fluid mobility ratio of one (after Landrum and Crawford[31]).

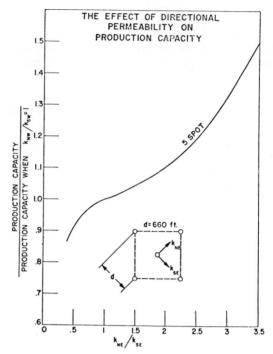

Figure 4-51. The effect of directional permeability on production capacity for varied degrees of permeability anisotropies at a fluid mobility ratio of one (after Landrum and Crawford[31]).

on the performance of secondary recovery projects has been evident. Where such directional permeabilities can be determined early in the development of a field, it will be possible to so arrange the wells that maximum recovery for the least number of drilled wells will be approached in development programs.

Off-Pattern Wells—Effect on Oil Recovery and Injection Requirements

Many times when the possibility of secondary recovery is being studied for a specific field, the question arises as to the effect of an off-pattern well(s) on oil recovery and displacing fluid injected. A well may be off-pattern for a number of reasons—surface obstacles, topography, random or not so random displacement of

holes with depth relative to the surface location, to mention just a
few possibilities. It is usually more desirable to utilize wells
presently existing than to expend large sums of money to plug the
off-pattern wells and drill new wells close by.

Prats *et al.*[32] have investigated mathematically the influence of
off-pattern wells for unit mobility ratio of the displaced and displacing fluids. The authors assume that there is an infinite array
of five-spot patterns in which there is *one* displaced producing or
injection well. The reservoir is assumed to be homogeneous as to
rock properties and to be of constant thickness. In addition, it
was necessary to assume that the initial fluid saturations were
uniform, and that the injection and producing rates are the same at
all wells in the field. Then too, the mathematics assumes that
there is a sharp boundary between the displacing and displaced
fluids. Figure 4-52 represents the mathematical location of the
producing and injection wells (in an infinite array of five-spots),
where *P* marks the position of a center-located well, while *P'*
shows the position of the displaced well. Mathematically, the

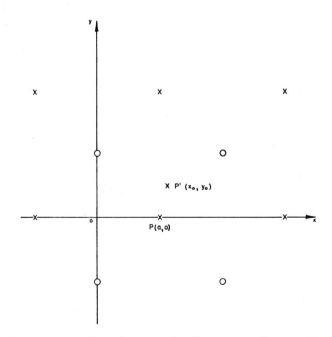

Figure 4-52. Axes for a regular five-spot well network
in an infinite array of five-spots (after Prats *et al.*[32]).

fluid pathlines may be represented by the imaginary part of the complex potential function,[33] as given by the following form:

$$\psi(x, y) = -\tan^{-1} f\left(\frac{xk}{a}\right) f\left(\frac{yk}{a}\right) - \tan^{-1}\left(\frac{y}{x-a}\right) + \tan^{-1}\left(\frac{y-y_o}{x-x_o}\right) \quad (10)$$

where

ψ = stream function expressed in radians

k = a constant, 1.854075

x_o, y_o = location of a displaced well

$$f(x) = \frac{sn\ x\ dn\ x}{cn\ x}$$

cn, dn, sn = Jacobian elliptic functions

and x, y are the coordinates of a point with respect to the axes as shown in Figure 4-52.

As one might expect, the loss in recovery increases as the well displacement, either laterally or diagonally, increases. It was found that when off-pattern wells are separated by at least one row of regular five-spots and are displaced less than one-third the pattern distance, the total overall loss for the flood would be less than one per cent. Higher losses would occur if the off-pattern wells were closer together. Even though the oil recovery losses may not be large, especially when the principles of present worth are applied to the oil produced at a future time, it was found that volumes of injected fluid required *do* increase markedly. Figure 4-54 shows the effect of displacement on the total loss in oil recovery.

Figure 4-55 depicts the oil production and displacing fluid-cut histories of a laterally-displaced production well. Notice that while the oil recovery is influenced in a minor way, the displacing fluid requirements to obtain this recovery have increased markedly. Figure 4-56 presents the same information as Figure 4-55 but for the diagonally-displaced producing well. The authors have also published information on the effect of the displaced well on the producing histories of the surrounding five-spot well configurations. In the using of Figure 4-54, the total loss in oil recovery due to a displaced well will be affected not only by the five-spot in which the displaced well is located, but also would be influenced by the surrounding five-spots. The influence of the surrounding four five-spots on oil loss may be approximated by incorporating them

Figure 4-53. Streamline patterns for a laterally-displaced and a diagonally displaced well (after Prats *et al.*[32]).

as a square of the dimensionless displacement as defined on Figure 4-54. This means that if the well is displaced by a factor of 0.25 and the surrounding patterns have zero well displacements, the calculation of average square of the displacement would be as follows:

$$\frac{(1)(0.25)^2 + 4(0)^2}{5} = 0.0125$$

from which information the total loss in oil recovery may approximate 0.4 per cent within the flood area covered by the five five-spot well networks.

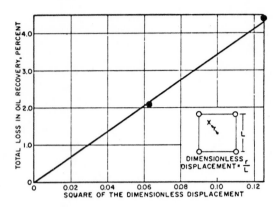

Figure 4-54. Effect of displacement of a producing well in a five-spot well network on total loss in oil recovery (after Prats *et al.*[32]).

While the aforegoing discussion would indicate that the loss of oil due to displaced wells is minor, a word of caution is necessary. The mathematical approach is for fluids of unit mobility ratio, for reservoirs with constant rock and fluid properties, and for constant injection and producing rates at each well. Without

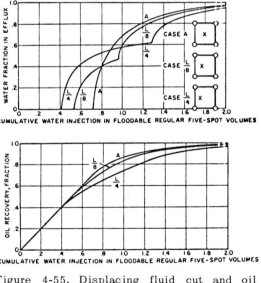

Figure 4-55. Displacing fluid cut and oil production histories for a laterally displaced production well (after Prats *et al.*[32]).

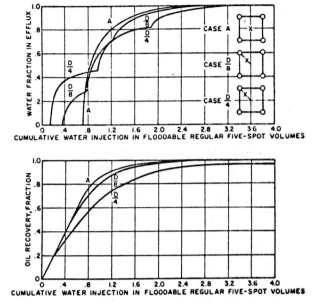

Figure 4-56. Displacing fluid cut and oil production histories for a diagonally displaced production
well (after Prats *et al.*[32]).

exception, these conditions are. seldom even approximately met in
field operations. Losses, therefore, could be considerably higher,
not to mention the increased cost occasioned by increased injection fluid requirements. This means that continued care should be
exercised in the placement of well locations, where feasible to
do so.

Dipping Beds—Effect on Areal Sweep Efficiencies

Up to the present time, most of the secondary recovery projects
using pattern well networks have been done on reservoirs having
only shallow or negligible formation dips. There has been increased interest in pattern flooding of reservoirs having steep
dips, such as are found in California and Wyoming. Peripheral injection programs are often used, but the high viscosity of some of
the crudes, with the accompanying low fluid mobility, has made
the closer spacing of pattern floods more attractive—if economical producing rates are to be attained. It should be intuitively

obvious that injected fluids would have a tendency to move preferentially down dip on dipping formations due to the effects of
gravitational forces.

Prats *et al.*[34] have made a mathematical study of a single-fluid
five-spot floods in dipping reservoirs. The approach taken was to
assume the reservoir to be initially filled with one fluid, and that
the same fluid be used for injection purposes. The result is a detailing of the results where a single fluid is used, i.e., where the
fluid mobility ratio is one. Where fluid mobilities are considerably
different than one and the displaced and displacing fluid density
difference is large, then the results will have perhaps only qualitative value. This in itself can be of more than considerable value
in the design of pattern floods in dipping reservoirs.

The authors assume that fluids are incompressible, that permeability is constant, and that the reservoir has constant thickness
and is a plane. Since for these assumptions the Laplace equation
is satisfied through the use of Darcy's Law, a mathematical
analysis based on complex potential theory can be made. The
reader is referred to the work of Prats *et al.*[34] for the mathematical
development and treatment details. The graphical results obtained
for the five-spot well network are presented as a function of three
parameters:

(a) r_w/L: where L is the length of a side of the five-spot and
r_w is the well bore radius,

(b) $R_g = \rho g h (\Delta p + \rho g h)$, a dimensionless grouping for the ratio
of the gravity head to the potential difference between the topmost
injection well and the bottommost producing well, where ρ is the
fluid density, h is the difference in reservoir elevation between
the topmost injection well and the bottommost producing well, and
Δp is the pressure difference between the same two afore-mentioned
wells in the same five-spot, and

(c) Orientation of the five-spot pattern with respect to formation
dip. When the sides of the five-spot are paralleled to the dip, the
orientation is termed parallel, when turned 45 degrees to the dip is
termed diagonal and is considered to be skew when at an angle
other than these. When the interior well, either producing or injection, is at the center of the five-spot, the pattern is considered
symmetrical. For all other positioning of the interior well, the
pattern would be considered asymmetrical.

Figure 4-57 shows the results of a mathematical analysis by
Prats *et al.*, where the streamlines are for a symmetrical five-spot
with parallel orientation and an R_g value of 0.0268. The produc-

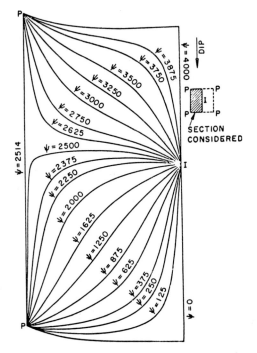

Figure 4-57. Streamlines for a symmetrical five-spot for parallel orientation, producing wells pumped off and injection sand face pressure constant at all injection wells, and $R_g = 0.0268$ (after Prats et al.[34]).

tion wells (P) were assumed to pump off and the injection pressures to be constant at the sand face for all the injection wells, (I). Figure 4-58 presents the resulting sweep efficiencies at first breakthrough of the displacing fluid as a function of the dimensionless grouping, R_g. Notice that as the R_g dimensionless grouping becomes larger, the sweep efficiency is greatly reduced from the 72 per cent areal sweep efficiency value for a five-spot in a horizontal system. This behavior is due to the increased importance of the gravity head to the potential difference between the producing well and injection wells. Notice that better sweep efficiencies are realized where the pressure difference between the producing and injection wells is maximized, i.e., where injection rates are as high as physically possible.

The areal sweep efficiency for a symmetrical five-spot well pat-

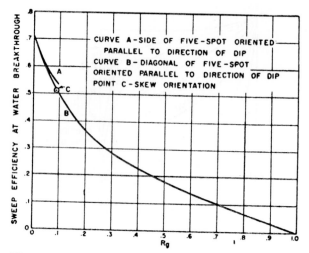

Figure 4-58. Sweep efficiency at first displacing fluid breakthrough at the producing wells for a symmetrical five-spot for values of dimensionless grouping, R_g, producing wells pumped off and injection sand face pressure constant at all injection wells (after Prats *et al.*[34]).

tern as a function of the cumulative displacing phase injected, is shown as Figure 4-59. This supplies information as to the areal sweep efficiencies that can be realized for continued fluid injection past breakthrough for three values of the dimensionless grouping, R_g, equal to 0.00633, 0.0268 and 0.0871.

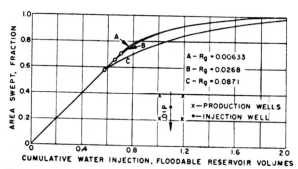

Figure 4-59. Areal sweep efficiencies for the symmetrical five-spot well network, parallel orientation, producing wells pumped off and injection sand face pressure constant at all injection wells (after Prats *et al.*[34]).

Where new wells are to be drilled for the purpose of secondary recovery and dipping formations are involved, or where infill wells are to be drilled, considerable benefit would result from knowing the proper location of each injection well in order to assure simultaneous breakthrough of the displacing fluid at each producing well. Figure 4-59 presents information helpful in selecting the location of such an infill well, where the producing wells form a square pattern diagonal to the dip of the structure. The relative distances are as defined on the figure, with the relative well position being a function of the dimensionless ratio, R_g, as might well be expected. Notice that as the R_g value approaches one, the position of the injection well becomes quite close to the topmost producing well. This result is evident from consideration of the terms in the R_g factor. When the difference in elevation between the upper producing well and the injection well is large, a steep dip is indicated. Gravity forces, represented by $\rho g h$, become predominant and the injected fluid tends to run downhill at increased rates. Displacing the injection well up structure permits the producing wells to more efficiently ''capture'' the displaced fluid—oil. Such displacement of the injection well away from the center of the five-spot well network results in asymmetry. Figure 4-60

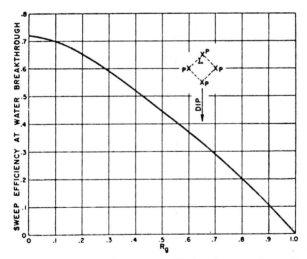

Figure 4-60. Areal sweep efficiencies at first displacing fluid breakthrough to the producing wells in an asymmetrical five-spot well network, producing wells pumped off and injection sand face pressure constant at all injection wells (after Prats *et al.*[34]).

presents the areal sweep efficiencies to be expected, within the limitations of the assumptions of the mathematical calculations, at the first displacing fluid breakthrough to the producing wells in an asymmetrical five-spot.

Since the aforegoing treatment for fluid injection and production from dipping reservoirs is for the single fluid case, a question arises as to its applicability when three immiscible phases are present, for example, oil, gas and water. For this case, the water would move down structure and gas up structure preferentially. As suggested by Prats *et al.*, three distinct periods would seem to be in evidence: (1) the first period when the displacing fluid, oil and gas are present, (2) the second period when the displacing fluid and water are present and (3) the third period when only the displacing phase is present and all of the oil has been displaced to the producing wells.

Figure 4-61 shows reservoir and fluid configurations for each of

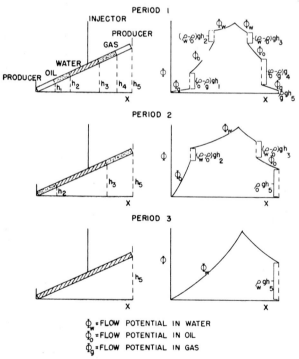

Figure 4-61. Reservoir and fluid configurations and the resulting flow potentials in a dipping reservoir for multiple immiscible fluid displacement (after Prats *et al.*[34]).

the periods and also the flow potentials which would be operative during each period. Strictly speaking, the single fluid study by the authors has application only to the third period. The authors suggest that the differences in flow potential causing a net downward flow be added together and used in the calculation of the R_g dimensionless parameter. An approximate idea of the areal sweep efficiencies, injection pore volumes necessary, and positions of injection wells may then be ascertained provided that the mobility ratio is approximately one. Figure 4-62 shows that the flow po-

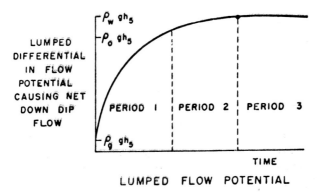

Figure 4-62. Total flow potential differential causing down dip flow as a function of injection time (after Prats *et al.*[34]).

tential difference increases as time passes, and the lighter fluid are displaced from the system. As a rough approximation, the oi density and the vertical distance between the topmost and bottom most producing wells in the five-spot may be used, i.e., $\rho_o g h_5$, a shown in Figure 4-62.

Natural and Induced Fractures—Effect of Areal Sweep Efficiency

Since the inception of induced or artificial fracturing technolog in the late forties and early fifties, an increasing concern hi been felt as to the role of this fracturing on areal sweep eff ciencies. Then too, a number of reservoirs have a natural fractu system with definite directional trends. In our efforts to recov oil from tight reservoirs, it has occasionally been necessary to i ject the displacing fluids at injection well sand face pressures exceeding the pressure to physically fracture the formation.

Crawford and Collins [35] have studied the effect of vertical fractures on secondary recovery in line-drive well patterns by means of potentiometric models. The approach taken was to construct an electrolytic reservoir, with an electrode at the center of each end piece to represent the producing and injection wells. To simulate a fractured system, a thin strip of copper was soldered to the well(s) of such orientation and length as was necessary to represent the vertical fracture system under consideration. The use of a copper strip of essentially infinite conductivity was justified by reasoning that fractures only 0.01 inches in width have a permeability approximating 5000 darcies—a large value compared to the much lower value of permeability associated with the fine pore system of the reservoir. This means that for modelling purposes, the copper strip would adequately represent the high conductivity that a fracture possesses. The authors found that if the vertical fracture extended directly toward the producing well from a fractured injection well, the areal sweep efficiency would vary from that for a line-drive well network where the fracture was short, to essential zero when the fracture extended a major portion of the distance between the wells. Figure 4-63 summarizes the findings as regards sweep efficiency as L, the dimensionless distance between the injection and producing wells, varies from zero to one. Note that the potentiometric model used is for a fluid mobility ratio of unity. All fracture orientations are not harmful to sweep

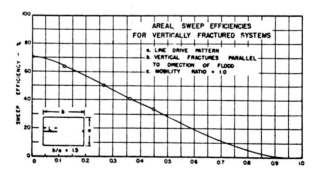

Figure 4-63. Effect of length of a fracture between producing and injection wells in a line-drive well network upon areal sweep efficiency (after Crawford and Collins [35]).

efficiency. If the vertical fracture at the injection well is oriented 90 degrees away from the direction to the producing well(s), the sweep efficiency at the displacing phase breakthrough will be increased over that realized in a non-fractured system. Figure 4-64 illustrates the results of such a fracture orientation. It is interesting to note that the sweep efficiencies would be the same regardless of whether the producing or injection wells were fractured. Fracturing both the producing and injection wells in the fracture orientation suggested by Figure 4-64 would increase the

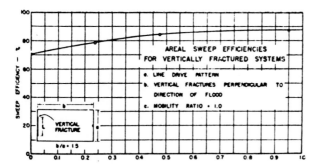

Figure 4-64. Effect of length of a vertical fracture upon sweep efficiencies in a line-drive well network when orientation is 90 degrees away from the breakthrough fluid streamline (after Crawford and Collins[35]).

sweep efficiency still further. It follows that fracturing the wells in the case represented by Figure 4-63 would only serve to further reduce the areal sweep efficiency.

The effect of fractures on sweep-out pattern in the five-spot well network has been studied by Dyes, Kemp and Caudle[36], for fluid mobility ratios ranging from 0.3 to 3.0. The shadowgraph technique was used, in which uniform pay thickness, zero gas saturation, no oil bank formed, constant absolute permeability and no formation dip are preassumed. Two fracture patterns were investigated, that where the vertical fracture is in line with the breakthrough streamline, and that where the vertical fracture is 45 degrees displaced from the breakthrough streamline, i.e., the most favorable fracture location. Figure 4-65 presents a comparison of the unfractured and fractured performance of a five-spot where the fracture orientation was favorable, the mobility ratio equal to 1.1

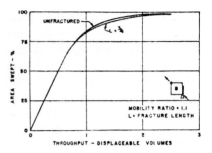

Figure 4-65. Effect upon areal sweep efficiency of a fracture of favorable orientation (after Dyes *et al.*[36]).

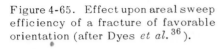

and the fracture length equal to three-fourths the distance between the producing and injection wells. Notice that the areal sweep out efficiency is affected in only a minor fashion. Figure 4-66 presents information where the vertical fracture is unfavorably oriented. In this instance, it is apparent that long fractures are detrimental due to large volumes of displacing fluid that will be required to obtain areal sweep efficiencies equivalent to those obtained without fracturing. Because of the apparent sensitivity of the five-spot pattern to unfavorably directed fractures exceeding one-half the distance between the producing and injection wells, additional studies were made of the effect of mobility ratio upon areal sweep efficiency. Figure 4-67 presents the results of this work, where the fracture length was set at one-half the distance between the producing and injection wells. The adverse effect of mobility ratios exceeding one, in the presence of an unfavorably-oriented vertical fracture, is readily apparent.

It is well to note that the effect of a fracture in the producing or injection well will not yield the same flow system of areal sweep efficiencies at the same throughout when the mobility ratio is other than unity. Figure 4-68 presents the results of work by Dyes *et al.* for areal sweep efficiency, as influenced by mobility

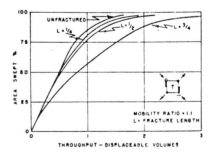

Figure 4-66. Effect upon areal sweep efficiency of a fracture of unfavorable orientation (after Dyes *et al.*[36]).

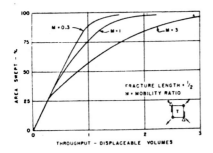

Figure 4-67. Influence of mobility ratio and an unfavorably-oriented vertical fracture upon areal sweep efficiency (after Dyes *et al.*[36]).

ratio and whether the injection well or the producing well contained the unfavorably oriented fracture. Notice that if the mobility ratio is greater than one, the sweep efficiency is improved if the vertical fracture exists in the producing well, rather than the injection well. Conversely, the sweep efficiency is somewhat improved if the fracture is in the injection well when the mobility ratio is less than one.

The aforegoing should underscore the importance of approximating the direction of fracture orientation prior to the selection of a flooding pattern if plans include the possibility of fracturing, or of injecting displacing fluids at pressures where fracturing could occur. Then too, structural considerations may well suggest the direction that natural or artificial fractures might take. Roark and Lindner[37] have published a study of a field under water flood, where extensive well fracturing, both in open hole and cased completions, was necessary to improve the injectivity and productivity of the wells. While no information was available to indicate whether vertical or horizontal fractures resulted, the 3550 foot depth of the formation would indicate that vertical fractures probably existed. Breakthrough performance of the wells indicated that

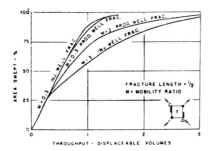

Figure 4-68. Influence of mobility ratio and location of the fracture in the wells upon areal sweep efficiency (after Dyes *et al.*[36]).

no detrimental effects upon areal sweep efficiency occurred. Since the fracture treating volumes were small, this conclusion does not imply that fracturing would always be beneficial. However, benefits from fracturing do often outweigh the possibility of reduced areal sweep efficiencies, due to improved economics and shortened project life. In addition, the results presented in Figures 4-63 and 4-64 are for sweep efficiencies at breakthrough. Ultimate oil recovery may not be reduced, since producing rates would be higher with, of course, correspondingly higher displacing phase cuts. Even though larger volumes of the displacing phase may well be handled when the fracture orientation is along the breakthrough streamline, higher producing and injection rates could offset any increase in treating expenses.

Irregularly-Bounded Flood Areas—Effect on Areal Sweep Efficiency

The previous discussion in this chapter has been concerned with confined flood patterns in, at least for the purposes of the analysis, an infinite array of wells. This is by no means always the case. Even if the drilling pattern used in the field has been regular, the small size of the reservoir may dictate a flooding pattern that will have a sweep efficiency largely different from that for a confined pattern in a large field of many wells. Then too, the areas on the edge or end of a large field occasionally require special study, in order to fix the proper well patterns and points of injection and production necessary to maximize oil recovery and producing economics.

While it may be possible, in some instances, to solve the general equations describing fluid flow in porous media for secondary recovery projects, where moving interfaces are involved [12], such an approach is very complex where irregular formation boundaries are to be treated. A computer solution will usually be necessary if such an analytical approach is undertaken. Often the size of the project and time considerations will not support such a sophisticated approach to the problem.

Frequently useful solutions to problems of this nature can be obtained by the use of electrolytic models [38,39,40], potentiometric models [41], resistance networks [17], scaled porous media models [6], or fluid mappers [42]. Even though the answers will often be semiquantitative only due to the necessary simplifying assumptions made, the relative answers as to well positions and flooding fronts, plus relative fluid recoveries can be most helpful in determining the best possible course of action.

Normally, the first step in such a study will be to determine the configuration of the reservoir or reservoir segment being studied, plus the additional required information on the pay thickness, stratification and saturation gradients that may be present. This will permit a decision to be made regarding the smallest element of the reservoir that needs to be modelled to obtain meaningful answers. Then too, the information will allow the proper selection of the modelling technique to be used. If the reservoir has limited stratification, more or less constant *net* pay thickness, no sizeable saturation gradients in evidence, negligible structural dip and a fluid mobility ratio in the vicinity of one, almost any of the modelling techniques will work satisfactorily. Fothergill[40] has described a blotter model which has good application in such a case, provided not too many injection and producing wells are involved and all wells are accepting and taking fluid at equivalent total rates. This method has the advantages that a visual record

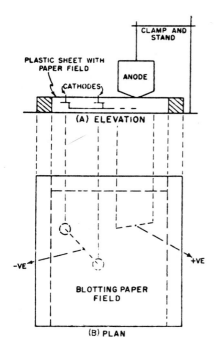

Figure 4-69. Schematic drawing of the blotter model in elevation and plan views (after Fothergill[40]).

of the front can be seen, and that very simple and easily constructed equipment will suffice. Figure 4-69 provides an elevation and a plan view of one way of laying the equipment out. One
circuit configuration that can be used is illustrated in Figure 4-70.

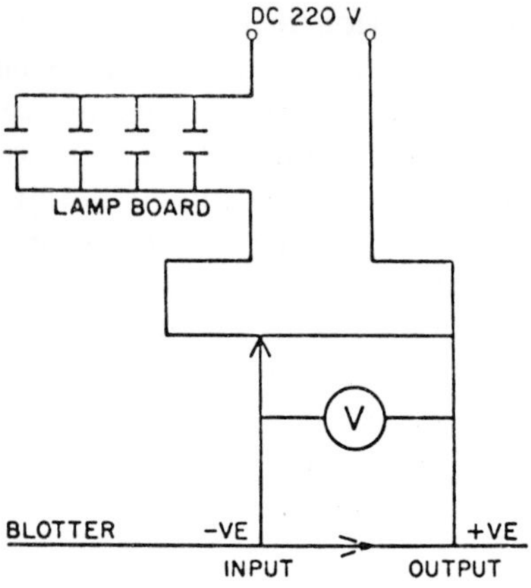

Figure 4-70. Electrical circuitry for
the blotter model (after Fothergill[40]).

A plastic base sheet is used to support the blotter, cut to the
shape of the reservoir or reservoir segment of interest. The cath-
ode(s) serve as injection wells and should be brass or other elec-
tricity conducting metal pins. The anode(s) representing the producing wells should be made from blotting paper, and have a reasonably sharp point where contact is made to the modelled reservoir. The electrolyte can be a solution of 0.2 normal potassium
sulfate, containing a small amount of phenolphthalein in a saturated alcohol solution. The indicator-bearing electrolyte can be
conveniently applied to the anode(s) and blotter model of the
reservoir with an eye dropper. A 220-volt direct current suffices
to move the hydroxyl ions or colored front from the injection to the
producing wells. A lamp board, or other resistivity device, permits control of the applied voltages. A convenient reservoir size
limit is about 4 cm. Larger model sizes require increased voltages, which result in a rapid drying out of the saturated blotter.
The purpose of the blotting paper anode is to avoid stopping of the
flood front before the anode is reached, since hydroxyl and hydrogen ions will be moving in opposite directions. When the hydrogen
ions meet the red front generated by the hydroxyl ions, the hydroxyl ions are neutralized, and the flood advance ceases.

Where the reservoir is relatively homogeneous in fluid and rock characteristics but is varied in pay thickness, the gelatin model finds good application. Usually, the vertical scale is considerably exaggerated, as compared to the horizontal scale. The gelatin used is normally agar agar and performs the same role as the blotting paper—that is, preventing excessive velocities of normal diffusion in the electrolyte. Any suitable porous medium could be used. Where thicknesses vary widely, a suitable holding base for the electrolyte containing gelatin is paraffin. The outline of the reservoir and reservoir thicknesses can be cut into the smooth surface of the paraffin. Pouring the hollowed out section full of the prepared electrolyte gelatin solution completes the reservoir. A 6 per cent by weight aqueous solution of agar agar gelatin, mixed with equal parts of the prepared 0.2 normal potassium sulfate (containing phenophthalein), works well. Again, as with the blotter model, the anodes (the injection well(s)) may be represented by pins, while the cathode(s) (the producing well(s)) may be constructed from plastic tube(s) filled with the same gelatin (containing electrolyte) as used for the reservoir. The tips of the tube(s) serve as a contact with the modelled reservoir. A circuit, similar to that represented by Figure 4-70, may be used. If the injection rates to each injection well are to be varied, it will be necessary to connect each injection well separately, through a milliammeter and rheostat and the positive side of a high voltage direct current source. A similar hook-up for each producing well can be arranged, with connections made to the negative side of the voltage supply.

The fluid mobility ratio modelled would, of course, be unity, due to the uniform electrolyte concentrations in the blotter or gelatin-supporting material. It would be possible to model mobilities other than one by varying the concentration of the electrolyte in a stepwise procedure. This is difficult, due to the concentration gradients which can result. Burton and Crawford[38] have described experiments where mobility ratios of 0.5, 0.85, 1.2 and 3.0 were modelled by using electrolytes yielding different conductivities.

General Comments on Sweep Efficiencies

Most of the treatments contained in this chapter are for rather idealized conditions. Steady-state flow is preassumed—although injection and producing operations do not often closely approximate this restrictive condition. A case in point would be the reservoir depleted by primary producing means. Since a high gas saturation would most probably exist and reservoir pressure would be low, the producing wells would represent only a very weak pressure or potential sink, whereas the injection wells would be much stronger relative sources. The result would be a much longer radial movement of the injected fluids than would be modelled by steady-state means. Only when the banked oil has reached the producing wells would the front become grossly distorted. The result would be improved sweep efficiencies over that suggested by the model studies. Then too, lateral variations in the reservoir permeability cannot be anticipated in many instances, and therefore cannot be modelled. This would influence the breakthrough times, and perhaps drastically influence the ultimate oil recovery.

Still, judicious use of the information available on areal sweep efficiency with the applicable performance prediction method will permit good, if not excellent estimates to be made of reservoir performance, under any proposed secondary recovery method.

References

1. Muskat, Morris, and Wyckoff, R. D., *Trans. AIME* **107**, 62 (1934).
2. Muskat, Morris, "Physical Principles of Oil Production," p. 648, New York, McGraw-Hill Book Co., Inc., 1949.
3. Prats, M., *J. Petrol. Technol.*, 67 (December, 1956).
4. Wyckoff, R. D., Botset, H. G., and Muskat, M., *Trans. AIME* **103**, 219 (1933).
5. Fay, C. H., and Prats, M., *Proc. World Petr. Congress, Sec. II* (1951).
6. Slobod, R. L., and Caudle, B. H., *Trans. AIME* **195**, 265 (1952).
7. Frick, Thomas C., "Petroleum Production Handbook—Volume II—Reservoir Engineering," Chap. 41, McGraw-Hill Book Co., Inc., 1962.
8. Aronofsky, J. W., and Ramey, H. J., *Trans. AIME* **207**, 205 (1956).
9. Cheek, R. E., and Menzie, D. E., *Trans. AIME* **204**, 278 (1955).
10. Dyes, A. B., Caudle, B. H., and Erickson, R. A., *Trans. AIME* **201**, 81 (1954).
11. Caudle, B. H., Erickson, R. A., and Slobod, R. L., *Trans. AIME* **204**, 79 (1955).
12. Sheldon, J. W., and Dougherty, E. L., paper no. SPE-182, AIME Dallas mtg., October 8–11, 1961.
13. Bradley, H. B., Heller, J. P., and Odeh, A. S., paper no. 1585-G, AIME Denver mtg., October 2–5, 1960.
14. Boston, Couch, Moss and White: Socony Mobil Field Research Laboratory, Dallas, private communication.

15. Craig, F. F., Geffen, T. M., and Morse, R. A., *Trans. AIME* **204**, 7 (1955).
16. Dyes, A. B., *Trans. AIME* **195**, 22 (1952).
17. Nobles, M. A., and Janzen, H. B., *Trans. AIME* **213**, 356 (1958).
18. Caudle, B. H., and Witte, M. D., *AIME Tech. Note 2047*, *J. Petrol. Technol.*, 63 (December, 1959).
19. Prats, M., Matthews, C. S., Jewett, R. L., and Baker, J. D., *J. Petrol. Technol. XI*, **3**, 98.
20. Muskat, M., "Flow of Homogeneous Fluids," p. 585, Ann Arbor, Mich., J. W. Edwards, Inc., 1946.
21. Craft, B. C., Holden, W. R., and Graves, E. D., Jr., "Well Design: Drilling and Production," p. 368, Englewood Cliffs, N. J., Prentice-Hall, Inc., 1962.
22. Paulsell, B. L., "Areal Sweep Performance of Five-Spot Pilot Floods," MS Thesis, Penn. State Univ., January, 1958.
23. Habermann, B., *Trans. AIME* **219**, 264 (1960).
24. Morel-Seytoux, Hubert J., Jr., paper no. SPE 985, presented 39th fall mtg., Houston, October 11–14, 1964, AIME (Petr.).
25. Kimbler, O. K., Caudle, B. H., and Cooper, H. E., Jr., *J. Petrol. Technol.*, 199 (February, 1964).
26. Caudle, B. H., and Loncaric, I. G., AIME paper no. 1396-G, presented October 29–30, 1959, Corpus Christi, Texas.
27. Dalton, R. L., Jr., Rapoport, L. A., and Carpenter, C. W., Jr., AIME paper 1312-G presented annual fall mtg. of SPE, Dallas, October 4–7, 1959.
28. Fischer Rosenbaum, M. J., and Matthews, C. S., *Trans. AIME* **216**, 316 (1959).
29. Muskat, Morris, "Physical Principles of Oil Production," p. 262, New York, McGraw-Hill Book Co., Inc., 1949.
30. Johnson, W. E., and Hughes, R. V., *Bull. The Penn. State College*, Min. Ind. Exp. Station 52, 180 (1948).
31. Landrum, B. L., and Crawford, P. B., AIME paper no. 1264-G, Soc. Petr. Engrs. (1954).
32. Prats, M., Hazebroek, P., and Allen, E. E., Soc. Petr. Engrs. paper no. SPE-170, presented Dallas (October 8–11, 1961).
33. Valentine, H. R., "Applied Hydrodynamics," p. 183, London, Butterworth & Co., 1959.
34. Prats, M., Strickler, W. R., and Matthews, C. S., *Trans. AIME* **204**, 160 (1955).
35. Crawford, P. B., and Collins, R. E., *Trans. AIME* **201**, 192 (1954).
36. Dyes, A. B., Kemp, A. B., and Caudle, B. H., *Trans. AIME* **213**, 245 (1958).
37. Roark, Gene E., and Lindner, J. D., *Trans. AIME* **204**, 16 (1955).
38. Burton, M. E., Jr., and Crawford, P. B., *Trans. AIME* **207**, 333 (1956).
39. Ramey, H. J., Jr., and Nabor, G. W., *Trans. AIME* **201**, 119 (1954).
40. Fothergill, C. A., *J. Petrol. Technol.*, 55 (December, 1957).
41. Botset, H. G., *Trans. AIME* **165**, 15 (1946).
42. Cheek, R. E., and Menzie, D. E., *Trans. AIME* **204**, 278 (1955).

PROBLEMS

1. A variety of modelling and analytical techniques have been used for determining the areal sweep efficiency of the five-spot. List the techniques that have been used, state the advantages of each method, and discuss which are best applied to miscible and immiscible displacements for a range of mobility ratio values.

2. The following reservoir data are available:

Oil pay thickness	18 feet
Porosity	22 per cent
Water saturation (irreducible)	23 per cent
Oil saturation	77 per cent
Residual oil saturation	28 per cent
Oil viscosity	2.6 cp
Water viscosity	0.78 cp
k_{ro} at 77 per cent oil saturation	0.82
k_{rw} at 28 per cent oil saturation	0.26
Well spacing	40 acres

If a water flood is to be instituted and water injection rates maintained at 250 BPD per injection well in an enclosed five-spot well pattern, determine:

(a) The time in days to breakthrough of the injected water to producing wells.

(b) The stock tank oil produced to breakthrough if the oil formation volume factor is 1.15.

(c) The reservoir performance curves (daily oil and water producing rates, cumulative oil and water production, and producing WOR as a function of time) to a displaceable pore volume of 2.0.

(d) The percentage recovery of oil at a producing water-oil ratio of 25, both on reservoir and stock tank bases.

(e) The injection pressures which would be encountered over the life of the project if the initial sandface injection pressure is 1200 psig.

(f) The injection pressures for "nose advance" positions of 20, 40, 60, 80, and 100 per cent by the method of Aronofsky and Ramey if wellbore radius is 0.375 feet and the initial sandface pressure is 1200 psig.

(g) Since the mobility ratio is somewhat adverse, what would be effect of the fingering phenomena which could occur on the reservoir performance as calculated in Part (c) above?

3. Use the data of the preceding problem and the method of Morel-Seytoux to calculate the performance of a direct line-drive well network. The absolute permeability of the formation is 100 md. Assume that the injection pressure at the sandface will be held constant so that the difference in pressure between the producing and injection well is 1200 psig. Plot curves of daily oil and water producing rates, cumulative oil and water production, and producing WOR as a function of time to a limiting water-oil ratio of 40.

4. An inverted nine-spot well pattern network is being considered for use in a water flooding project. If injection and producing rates are to

be balanced so that reservoir pressure is maintained at a constant level, using the data of problem 2, determine:

(a) The composite daily oil and water producing rates, cumulative oil and water production, and producing WOR as a function of time for one enclosed pattern. The side wells are to be shut in at a reservoir water cut of 95 per cent. The water injection rate is held constant at 450 BWPD.

(b) Is the theory developed strictly applicable to the reservoir problem of Part (a)? Explain.

5. Generate the performance curves which would be expected for the five-spot well pattern of Problem 2 for the case where the system is not enclosed. The rate ratio approximates 1.5.

CHAPTER 5

Immiscible Fluid Displacement Mechanisms —Buckley-Leverett Approach

The Buckley-Leverett[1] approach to the displacement of a wetting fluid by a non-wetting fluid, or vice versa, was first presented in 1941, but did not receive the attention that it merited until the late forties. The theoretical development assumes that an immiscible displacement can be modelled mathematically, based upon the relative permeability concept and the idea of the so-called "leaky" piston. This means that while the displacement can be considered as being piston-like, there is a considerable amount of bypassing of oil, due to the very irregular surface which is presented by the porous media itself. The theory permits a determination of an average pore-to-pore sweep efficiency of the displacing phase in a linear system. The Welge[2] extension of the theory permits a detailing of the sweep efficiencies that can be obtained after breakthrough of the displacing phase at the outlet end. A major limitation of the theory is the fact that it applies to a linear system which, while perhaps being the case for an edge-water (either artificial or natural) drive, peripheral water flood, gas-cap expansion, etc., is not the case for many of the injection-producing well patterns common to secondary recovery operations. The radial case encompasses a flow system which exists in many of the pattern injection fluid systems, at least over a part of the injection history. It will be shown in a following section that this objection can be largely overcome.

The Fractional Flow Formula

The development of the fractional flow formula can be attributed to Leverett[3]. Since the flow of two immiscible fluids through a porous media is to be modelled, the logical approach is to write Darcy's equation for each of the fluids:

$$\frac{q_o \mu_o}{k_o} = -\frac{\partial p_o}{\partial u} + g \rho_o \sin \alpha \tag{1}$$

$$\frac{q_D \mu_D}{k_D} = -\frac{\partial p_D}{\partial u} + g \rho_D \sin \alpha \tag{2}$$

where

Subscripts o and D refer to the displaced (oil) and displacing phases, respectively.

$\dfrac{\partial p_o}{\partial u}$ and $\dfrac{\partial p_D}{\partial u}$ = the pressure gradients in the u direction in the respective phases.

α = the angle of the fluid flow with respect to the horizontal (up-dip flow assumed positive).

q_o and q_D = displaced and displacing fluid flow rates per unit of cross-sectional area normal to u.

We could note at this point that the three causes of motion of fluids in a porous media have been included in equations (1) and (2), and are, capillarity, gravity and external applied pressure. Conceptually, it is perhaps necessary to visualize funicular saturation conditions for both fluids, so that relative permeability considerations can be pictured on a pore-to-pore basis. Recent studies of flow in porous media would indicate that while such a concept is useful to explain saturation distributions on a microscopic basis, actually, the magroscopic saturation distribution is quite different. Usually, the non-wetting fluids and wetting fluids may flow in quite separate flow channels on the microscopic level, while the overall flow may appear to justify the funicular saturation approach. The important point is that the saturations must be continuous for the relative permeability concept to have a physical basis.

Since the contact between the wetting and non-wetting fluids should be along curved interfaces, it is necessary to include the concept of capillary pressure, p_c. By convention, we shall define the capillary pressure as the difference in pressure between the displaced and displacing fluids, i.e., $p_c = p_D - p_o$. Then it follows directly that:

$$\frac{\partial p_c}{\partial u} = \frac{\partial p_D}{\partial u} - \frac{\partial p_o}{\partial u} \tag{3}$$

Equation (2) may be subtracted from equation (1) to yield:

$$\frac{q_o \mu_o}{k_o} - \frac{q_D \mu_D}{k_D} = \tau \cdot \frac{\partial p_c}{\partial u} - g(\Delta\rho) \sin \alpha \tag{4}$$

where arbitrarily we have defined $\Delta\rho = \rho_D - \rho_o$.

If the two fluids being considered are incompressible, and the system pressure will be held constant, continuity considerations require that:

$$q_t = q_o + q_D \tag{5}$$

where

q_t = total flow rate per unit cross-sectional area.

We may at this point define the fraction of the flowing stream, at reservoir conditions of pressure and temperature, which is the displacing fluid as:

$$f_D = q_D/q_t \quad \text{or} \quad q_D = f_D q_t \tag{6}$$

Similarly

$$q_o = q_t - q_D = (1 - f_D) q_t \tag{7}$$

Replacing q_D and q_o in equation (4), rearranging and solving for f_D, the fractional of the displacing phase flowing at a given point in the system, results in the final form of the fractional flow formula:

$$f_D = \frac{1 - \dfrac{k_o}{\mu_o q_t}\left[\dfrac{\partial p_c}{\partial u} + g(\Delta\rho) \sin \alpha\right]}{1 + \dfrac{k_o}{k_D} \cdot \dfrac{\mu_D}{\mu_o}} \tag{8}$$

This equation is dimensionally correct, so long as the fundamental units of darcys, centipoise, cc/sec/cm^2, atmospheres and cm are used for permeability, viscosity, total flow rate per unit cross-sectional area, pressure and for distance, respectively. If the difference in fluid density, $\Delta\rho$, has the units of gram/cc, then the gravitational constant, g, may be replaced by a constant of 1033 (cm of water per atmosphere) in the denominator of the term. This results in consistent units of atmospheres per cm in the direction u for the term $\partial p_c/\partial u$ and for the revised term $[(\Delta\rho) \sin \alpha]/1033$.

If field units common to petroleum engineering practice are desired, conversion of units readily yields the following revised form for equation (8):

$$f_D = \frac{1 - \dfrac{1.127\,k_o}{\mu_o\,q_t}\left[\dfrac{\partial p_c}{\partial u} + 0.434\,(\Delta\vartheta)\,\sin\,\alpha\right]}{1 + \dfrac{k_o}{k_D}\cdot\dfrac{\mu_D}{\mu_o}} \tag{9}$$

The units are darcys, centipoise, barrels/day/square foot, psia and feet for permeability, viscosity, total flow rate per unit cross-sectional area, pressure and for distance, respectively. The differential density term, $\Delta\rho$, of equation (8), has been replaced by $\Delta\vartheta$, the differential specific gravity.

As Pirson[4] has noted, the fractional flow equation is truly fundamental to the understanding and to the representation of the flow of two immiscible, insoluble fluids in porous media (no reaction with the porous media). Examination of the various terms of equation (9) reveals the following implicit and explicit factors, influencing the fraction of the displacing fluid flowing at a given point in the system:

(1) The displacement takes place at constant temperature and pressure, constant phase compositions and at constant total flow rate, due to assumptions made in the mathematical development. Where there is a partial miscibility of fluids involved, resulting in changing phase compositions and interfacial tensions, modification of the equation can be made, as will be illustrated in a later chapter.

(2) The explicit fluid properties included are: μ_o, μ_D, ϑ_o, ϑ_D, S_o and S_D. The implicit fluid properties which have been included are: wettability, surface and interfacial tensions and fluid saturation geometries, due to the inclusion of a capillary pressure term, p_c.

(3) The rock properties are represented through the effective permeability term, k_o, and the relative permeability relationship,

k_o/k_D. Since the rock properties of grain size, petrofabric, composition, structure and cementing materials directly influence permeability, these factors enter the fractional flow formula indirectly.

It is necessary to re-emphasize that the fractional flow equation includes in one relatively simple relationship, all the factors that affect displacement efficiency of one immiscible fluid by another immiscible fluid in porous media. All the factors in the equation are controlled by the conditions of the problem, i.e., the reservoir dip, the injection rate (either artificial or natural) of the displacing fluid, fluid viscosities, prevailing pressure and temperature and the cross-sectional area through which the displacement occurs. The only independent variable is the displacing phase saturation which, in turn, specifies the relative permeability relationship. The section on permeabilities suggests techniques by which representative values usually can be obtained for the specific reservoir being studied.

It should be noted that the sign convention, as used in the development of the fractional flow equation, would result in a suction capacity of the rock for the injected fluid, if the fluid wetted the rock. In this sense, the system would actually imbibe the injected fluid. A case in point would be that of water injection to a water-wet system, which is the most common situation in present secondary recovery operations. Here, the capillary pressure would be positive due to the sign convention adopted in the derivation, $p_c = p_w - p_o$. Intuitively, we know that if a dip exists, then the injection of the displacing fluid, if heavier, should be at the lowest elevation, as would be the case with water displacing oil or gas. Then displacement of the lighter fluid will be upward, with higher resulting displacement efficiencies. The following specialized form of equation (9) for water displacing oil shows that this would be the case:

$$f_w = \frac{1 - \dfrac{1.127\, k_o}{q_t\, \mu_o} \left[\dfrac{\partial p_c}{\partial u} + 0.434\,(\Delta \vartheta) \sin\, \alpha \right]}{1 + \dfrac{k_o}{k_w} \cdot \dfrac{\mu_w}{\mu_o}} \qquad (10)$$

Obviously, if the term $[0.434\,(\Delta \vartheta) \sin\, \alpha]$ is positive due to a positive dip angle, α, and a positive specific gravity differential, $\Delta \vartheta$, then for a particular water saturation at a given point in the reservoir, the calculated value of the water flowing as a fraction of the total flow, f_w, will be smaller than if the term were neglected, or had a negative sign. Figure 5-1 illustrates this point. It should be evident that if the fraction of water flowing is kept to a mini-

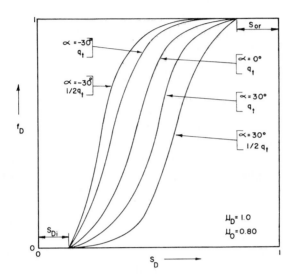

Figure 5-1. Sketch of fractional flow relationships of the injected wetting phase as a function of reservoir dip and flow rate.

mum as the average water saturation of the system increases by the proper control of the factors in equation (10), then the maximum displacement of the oil, or other fluid more valuable than the injected fluid, should result. Further examination of equation (10) indicates that forcing the term:

$$\frac{1.127 \, k_o}{q_t \, \mu_o} \left[\frac{\partial p_c}{\partial u} + 0.434 \, (\Delta \vartheta) \sin \alpha \right]$$

to be as large a positive number as possible would normally be beneficial for the case where water is displacing oil in a water-wet system. This could be done by decreasing the water injection rate, q_t, by altering the oil-water interfacial tension (perhaps by chemical means) so as to maximize the tendency for the water to be imbibed into the rock, by injecting the heaviest water available (perhaps a brine), and by being sure to inject the heavier fluid (water) low on the structure when dip angles exist.

It is evident from examination of equation (9) that the frontal displacement process is a rate-sensitive one, if either or both of the terms $\partial p_c / \partial u$ and $0.434\,(\Delta\vartheta)\,\sin\,\alpha$ have numerical values in a given problem solution. Where immiscible fluids are being used in the displacement process, then the $\partial p_c / \partial u$ term cannot be neglected. The relative influence of the terms in the numerator of equation (9), excluding the whole number one, can be controlled by means of the factor, q_t, the total flow rate per unit cross-sectional area. In secondary recovery, this control would be effected by controlling the fluid injection rates, whereas in the frontal displacement process of a natural water or gas drive, the fluid producing rates could be controlled. Figure 5-1 illustrates the relationship between the fractional flow of the displacing phase and the displacing phase saturation, as flow rate and dip are changed, where the displacing phase (wetting) is denser than the displaced phase (non-wetting).

For the case where the injected fluid is water, Figure 5-1 illustrates the relative behavior that would occur for two different injection rates, q_i and $1/2\,q_i$, and where injection is updip, downdip, or on the horizontal plane. Displacement of the curve shape to the left on the diagram would represent lower displaced fluid (oil) recoveries, at a given volume of injection, than if the curve were displaced to the right. For the water injection case, better recoveries should be effected by injection low on structure, at moderate to low rates. Just how low the injection rate would be for the maximum displacement efficiency would depend upon all the factors represented in equation (10), not just the water injection rate. For strongly water-wet systems and injection downdip, the injection rate q_t, may be quite large, and impose no restriction on the field injection rates that could be attained. In the actual field case, the problem of rate and economics are inseparable from the engineering point of view.

Where the injected fluid is gas, the equation may be modified to the following form:

$$f_g = \frac{1 - \dfrac{1.127\,k_o}{\mu_o\,q_t}\left[\dfrac{\partial p_c}{\partial u} + 0.434\,(\Delta\vartheta)\,\sin\,\alpha\right]}{1 + \dfrac{k_o}{k_g}\cdot\dfrac{\mu_g}{\mu_o}} \qquad (11)$$

where

subscripts g and o refer to the oil and gas phases, respectively:

$$p_c = p_g - p_o$$

$$\Delta \vartheta = \vartheta_g - \vartheta_o$$

The water saturation present must be at an irreducible level, or the problem would involve three-phase relative permeabilities, a condition not possible in the Buckley-Leverett development of the fractional flow equation. In this instance, the water saturation could be considered as a part of the rock matrix.

Figure 5-2 is a plot of typical gas fraction flowing diagram as gas saturation increases. Note that the abscissa represents the hydrocarbon pore volume rather than the total pore volume. Then too, the first flow of gas occurs at a zero gas saturation. It could be argued with some validity that the flow of gas should start at the critical gas saturation, a number that varies from zero to as much as 10 per cent between different reservoirs. In those reser-

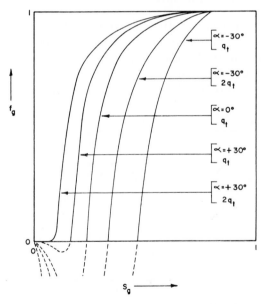

Figure 5-2. Hypothetical plot of the fractional flow of gas as a function of reservoir dip and flow rate.

voirs where the critical gas saturation exceeds several per cent, such a correction would seem important.

Figure 5-2 also presents a family of curves showing the influence of injection rate, q_t, and dip upon the fraction of gas flowing.

If the capillary pressure gradient, $\partial p_c / \partial u$, and the gravity contributing term, $0.434\,(\Delta \vartheta) \sin \alpha$, can be validly ignored, the fractional-flow equation, equation (9), takes the following form:

$$f_D = \cfrac{1}{1 + \cfrac{k_o}{k_D} \cdot \cfrac{\mu_D}{\mu_o}} \qquad (12)$$

where o and D refer to the displaced and displacing phases. This form of the fractional-flow equation indicates that the fraction is dependent only on relative permeability and viscosity ratios if the capillary and gravity force terms are neglected. A later section will show how the capillary force term is restored when the average displacing phase saturations through the system are determined.

The Rate-of-Advance Formula

Buckley and Leverett[1] first presented the rate-of-advance formula in 1941. Consider an elemental volume from a linear porous media, as shown in Figure 5-3, containing two fluids, the displaced fluid, oil, and the displacing fluid, usually gas or water. For the steady-state fluid flow case where pressure and tempera-

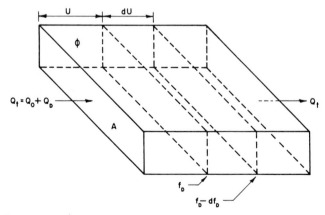

Figure 5-3. Elemental reservoir volume containing displaced and displacing fluid phases.

ture are constant, then by continuity the entering and exiting flow rates must be equal. The element has a cross-sectional area, A and a porosity, ϕ. If two phases are flowing in the element, then if an effective displacement of the oil is occurring, the displacing phase saturation in the element should be increasing with the passing of time. This means that the entering fluid should be carrying a displacing phase fraction represented by f_D, while the exiting fluid, at a distance of du from the inlet face, should be carrying a fraction of the displacing phase of $f_D - df_D$. Were this not the case, the mechanism would have little meaning, since a displacement of the oil due to continuity considerations would not be occurring. The composition of each phase must, of course, be constant. Writing the law of conservation of matter as a local or point equation results in[4]:

$$(\phi\, A du)\, dS_D = Q_t\, dt\, df_D \tag{13}$$

where the left-hand side of the equation represents the small change in the displacing phase saturation, dS_D, in the pore space, $\phi A du$, of the element of the porous media represented by Figure 5-3. The right-hand term of equation (13) expresses the decrease in the displacing phase fraction, f_D, for a movement of fluid volume, $Q_t dt$, in the time interval, dt. In view of the fact that equation (13) contains only one independent variable (the displacing phase saturation), the equation may be rewritten as follows:

$$du = \frac{Q_t}{A\phi} \left(\frac{df_D}{dS_D}\right) dt \tag{14}$$

which, due to its point or differential form, represents the advance in a linear system of a plane of constant saturation, S_D, a distance du during the time interval, dt. In view of the fact that the porosity, area and fluid rate are constant, then the partial derivative, $\partial f_D/\partial S_D$, is constant for a given saturation. This means that equation (14) affords a method by which the distance that a plane of constant saturation has advanced may be determined, i.e., since the distance will be directly proportional to time and to the value of the partial derivative, $\partial f_D/\partial S_D$. Then at time, t, the distance vector, u, could be represented by:

$$u = \frac{Q_t}{A\phi} \left(\frac{\partial f_D}{\partial S_D}\right) t \tag{15}$$

At breakthrough of the displacing phase at the system outlet, the
distance u would be equivalent to the length of the system, L,
where time would be the breakthrough value. Any consistent set
of units may be used. If barrels per day, square feet and feet are
used for Q_t, A and u, then the right-hand side of equations (14)
and (15) must be divided by 5.615.

Stabilized Zone Concept

A number of authors[5,6,7,8,9] have published treatments dealing
with the stabilized zone between the displacing and displaced
fluids in a porous media. An understanding of this concept is
necessary before a practical solution and application of the rate-
of-advance and the fractional flow formulas can be made. Ignoring
for the present the gravitational term in equation (9), the equation
may be written as:

$$f_D = \frac{1 - \dfrac{1.127\, k_o}{\mu_o\, q_t}\left(\dfrac{\partial p_c}{\partial S_D} \cdot \dfrac{\partial S_D}{\partial u}\right)}{1 + \dfrac{k_o}{k_D} \cdot \dfrac{\mu_D}{\mu_o}} \qquad (16)$$

where the $\partial p_c/\partial u$ term has been replaced by two partials, $\partial p_c/\partial S_D$
and $\partial S_D/\partial u$. This indicates that the change of capillary pressure
with distance is controlled by both the modification of capillary
pressure, with a variation in the displacing phase saturation, and
by the alteration of the displacing phase saturation, with a change
in distance. Figure 5-4 illustrates the stabilized zone concept as
it relates to equation (16). For a given porous media, a stabilized

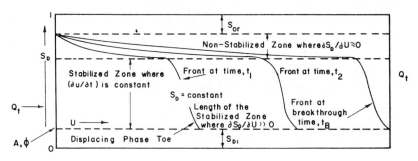

Figure 5-4. Displacing phase saturation distributions as a function of
distance and time.

zone will develop, which relates back to the idea of a leaky piston. At time, t_1, the displacing phase extends a distance into the linear system, as shown by the saturation profile. If the displacing phase already exists as an irreducible value in the porous media, S_{Di}, then permeability of the displacing phase will start at this value and extend up to, as a limit, a displacing phase saturation, equaling the total pore space less the residual oil saturation, S_{or}. Between these two displacing phase saturation limits, two zones exist, the stabilized and the non-stabilized zones. The stabilized zone is characterized by the stabilized saturation profile at time t_1 and at a later time t_2, or stated mathematically, $(\partial u/\partial t)$, is constant for all displacing phase saturations, S_D, in the stabilized zone. At breakthrough time, t_B, the profile still remains parallel to those exhibited at times corresponding to lesser amounts of displacing phase injection. The terms which are present in equations (16) or (9) all influence the shape of the saturation profile in the stabilized zone. Of particular concern, however, is the term $\partial p_c/\partial u$, since data are seldom, if ever, available to properly define it. Figure 5-4 shows that $\partial S_D/\partial u$ has a value much larger than zero in the stabilized zone, or as one might expect, that large changes in capillary pressure over relatively short distances would occur, so long as the term $\partial p_c/\partial S_D$ is non-zero. The non-stabilized zone of Figure 5-4 has been called the "drag" zone, a zone where the change of displacing phase saturation, with respect to distance, approaches zero. It follows directly that the capillary pressure-distance gradient, $\partial p_c/\partial u$, can be neglected in this region. We shall see that this part of the saturation profile contributes the subordinate or "after-breakthrough" production of the displaced oil. Figure 5-4 illustrates the case where the injection rate is sufficiently slow (always the case in an actual reservoir) so that the displacing fluid is imbibed ahead of the main displacing phase front, as shown by the "toe." This toe would not be present if the displacing phase did not wet the porous media.

The displacing phase saturation profile, as illustrated in Figure 5-4, and reproduced many times in linear systems in the laboratory, can be reproduced accurately by mathematical analysis. The position of the displacement front, both in the stabilized and non-stabilized zones, may be determined by solving equation (15) where the slope values, df/dS_D, for a range of displacing phase saturations, are known from a plot of the solution of equation (9),

where the capillary pressure-distance gradient term has been neglected due to a lack of data in this one specific regard. Then, at
time t_1, the displacement front may be as represented by Figure
5-5 where equation (15) has been solved for a range of displacing
phase saturation values and where porosity, ϕ, and cross-sectional

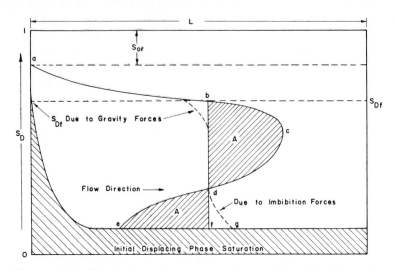

Figure 5-5. Displacing phase saturations after injection at a
constant rate over time interval t_1.

area, A, are constants of the problem. In addition, Q_t, the total
flux rate, is constant in the simplest case. The flux rate does *not*
have to remain constant over all times—but then displacement
efficiencies would perhaps change also due to rate sensitivity.

The resulting saturation profile, as shown by the line *abce*, is
obtained by plotting the derived values of u at time t_1, from the
line representing the initial displacing phase saturation, *aef*. At
this point, it is apparent that a physical difficulty occurs, since a
triple value for the displacing phase saturation at a given point
along the linear system results. This can be resolved by drawing
a vertical line *bdf*, such that the areas outlined by *bcd* and *def* are
equal. This corresponds to replacing the capillary pressure-
distance term contribution that had to be neglected due to a lack
of data. This particular aspect of the stabilized zone concept may
be more apparent by considering the graphical constructions of
Figures 5-6 and 5-7. Here the problem of multiple displacing

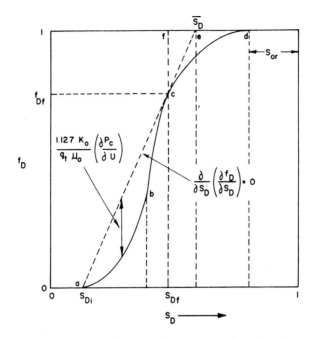

Figure 5-6. Displacing phase saturation distribution and graphical method of determining displacement efficiencies at breakthrough of the displacing phase.

phase saturations is resolved by a method suggested by Buckley and Leverett[1], which calls for balancing the areas A and areas B under the derivative, so as to obtain a single valued curve represented by the line *ghijk* in Figure 5-7. Inspection of Figure 5-6 shows that balancing areas A, in Figure 5-8, is equivalent to the construction of a tangent line, *ac*, to the S-shaped f_D curve, and letting the line *ac* represent the $f_D - S_D$ relation in this region. In addition, it is evident that the derivative of the f_D curve over the straight line section *ac* yields the constant slope value of line *ghi*, depicted by Figure 5-7. It is then apparent that the capillary pressure term contribution is represented by the distance between the straight line *ac* and curved line segment *abc*, of Figure 5-6. This means that the contribution of the capillary pressure term is sufficient change in f_D values at specific displacing phase saturations so that the stabilized zone is single-valued, as depicted by the line *bdf* on Figure 5-5. In the actual flow case, the satura-

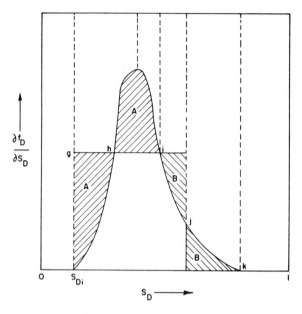

Figure 5-7. Plot of the change of slope of f_D as shown in Figure 5-6 versus S_D and the construction for the breakthrough recovery.

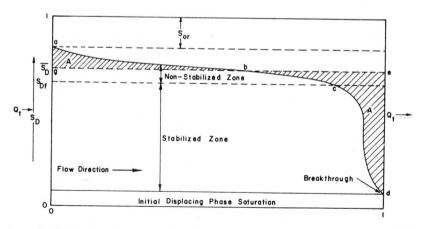

Figure 5-8. Displacing phase saturation profile at instant of breakthrough.

tion profile would not be according to the curve outlined by *abdf*, but would follow a shape shown by the line *ahdg*. The rounding of the curve at the top would be due to the influence of gravity forces where the displacing fluid is denser than the displaced fluid, i.e., when water displaces oil. The curved section *dg* occurs when the displacing fluid wets, or is imbibed into the porous media. This toe would not be present if the displacing fluid did not wet the rock. In view of the modifications to the saturation profile of Figure 5-5, good agreement with the saturation profiles of Figure 5-4 is attained.

Point *c* on Figure 5-6 is the dividing point between the line *ac* depicting the f_D relationship, over a range of displacing phase saturations from S_{Di} to S_{Df}, which permits the saturation profile segment *bf* of Figure 5-5 and the curved line *cd* (Figure 5-6) corresponding to a range of saturations from S_{Df} to S_D. Pirson[4] has shown analytically that the saturation $\bar{S}_D$ is the mean displacing phase saturation in the system, at breakthrough of the displacing phase at the outlet end. Inspection of Figures 5-6 and 5-7 shows that the extending of the tangent line from the point of initial permeability to the displacing phase, point *a*, tangent to the curve at point *c*, and on to point *e* where f_D has a value of one, results in a balancing of the areas *B* of Figure 5-7. Figure 4-8 illustrates graphically the averaging that is accomplished by extending the tangent to the f_D curve, to a value of one. Areas *A* are balanced to yield an average displacing phase saturation, $\bar{S}_D$, back through the linear system. Due to continuity conditions, knowledge of $\bar{S}_D$ results in a measure of the displaced phase recovery (oil). If the system were initially saturated to $1 - S_{Di}$ of the pore space, then, at breakthrough, fractional recovery of the original oil-in-place would amount to $(\bar{S}_D - S_{Di})/(1 - S_{Di})$. From geometric relationships evident on Figure 5-6, the following equations can be readily developed:

(1) Displacement efficiency at breakthrough

$$= \frac{\text{volume of hydrocarbon displaced}}{\text{volume of the hydrocarbon pore space}}$$

$$= \frac{S_{Df} - S_{Di}}{1 - S_{Di}} + \frac{1 - f_{Df}}{(1 - S_{Di})\left(\dfrac{\partial f_D}{\partial S_D}\right)_f} \tag{17}$$

where

S_{Df} = displacing phase saturation at the front or at the outlet end of the system at breakthrough conditions.

f_{Df} = fraction of the displacing phase flowing at the front or at the outlet end of the system at breakthrough conditions.

f = subscript referring to conditions at the front.

(2) Hydrocarbon displaced at breakthrough

$$\overline{S}_D - S_{Di} = \frac{\text{volume of hydrocarbon displaced}}{\text{volume of the pore space}}$$

$$= (S_{Df} - S_{Di}) + \frac{(1 - f_{Df})}{\left(\dfrac{\partial f_D}{\partial S_D}\right)_f} \tag{18}$$

The stabilized and non-stabilized zone concept is useful in the development of a technique by which the hydrocarbon recovery may be determined for continued injection of the displacing phase after breakthrough[2]. In this instance, it is helpful to picture that the saturation profile continues past the outflow face (or has continued past the producing well), as shown in Figure 5-9. The average displacing phase saturation through the system would correspond to $\overline{S}_D$ where areas B have been balanced. This same result can be obtained by extending a tangent line from the displacing

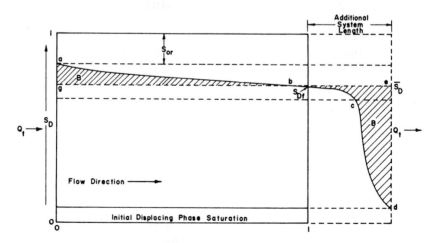

Figure 5-9. Displacing phase saturation profile in a linear system for continued injection past breakthrough.

phase saturation at the outflow face, S_{Dc}, to an f_D value of one. Pirson[4] provides an analytical justification of this construction technique.

From simple geometric considerations evident in Figure 5-10, the following equations can be developed:

(1) Displacement efficiency for continued injection past breakthrough

$$= \frac{S_{DC} - S_{Di}}{1 - S_{Di}} + \frac{1 - f_{DC}}{(1 - S_{Di})\left(\dfrac{\partial f_D}{\partial S_D}\right)_c} \tag{19}$$

(2) Hydrocarbon displaced after breakthrough

$$\bar{S}_D = (S_{DC} - S_{Di}) + \frac{1 - f_{DC}}{\left(\dfrac{\partial f_D}{\partial S_D}\right)_c} \tag{20}$$

where c is a subscript referring to conditions at the outflow face after breakthrough.

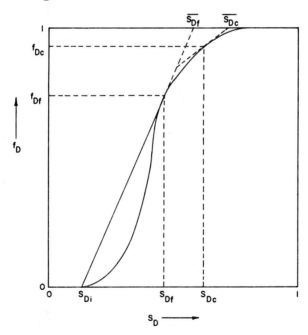

Figure 5-10. Fractional flow curve showing graphical construction for average displacing phase saturations for continued injection after breakthrough.

Practical Use of Frontal Displacement Concepts

A number of examples of the practical application of the frontal displacement concept as conceived by Buckley and Leverett have appeared in the literature[4,10,11,12,13]. Specific examples of immiscible displacement for the cases of water displacing oil and the gas displacing oil are presented in Chapters 7 and 9, respectively. Details permitting the development of production curves on a time basis are also given in those chapters. While the examples given are for only two immiscible displacement cases, any other case where one fluid is displacing another insoluble fluid in porous media may be treated. Of course, a knowledge of the factors entering equations (9) and (15), for example, must be known. The method presupposes that valid relative permeability information is at hand, and that the system being modelled mathematically is linear.

Application of Frontal Displacement Concepts to Non-Linear Systems

Inspection of the streamlines for the five-spot network, represented in Figure 7-4 (p. 191), makes evident the fact that the constant injection rate used in the previously described frontal displacement process will not yield the same flux rates at all five-spot network locations. The sensitivity of the immiscible displacement may be determined by considering the probable injection rates attainable in the actual field case under consideration. Then equation (9) may be solved for a range of q_t values, and the plots of f_D versus displacing phase saturation prepared. If considerable rate sensitivity is apparent, engineering judgment will have to be used in the choosing of a rate, which would properly model the displacement efficiency, which should occur. It will also be necessary to apply an areal coverage factor, since the injected fluid will not contact 100 per cent of the injection pattern area. Figure 5-11 illustrates the idealized representation of the five-spot quadrant, which the above-described approach uses. Similarly, approximations for a number of the other flooding patterns could be simply devised. Chapter 4 provides details as to how the effect of mobility ratio and flooding patterns will alter the results that could be expected from a linear flooding system.

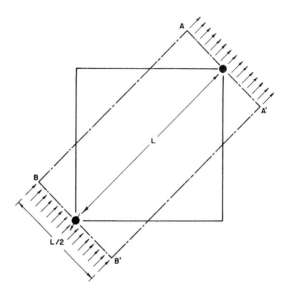

Figure 5-11. Linear system approximation to the five-spot flooding pattern (after Rapoport and Leas[7]).

References

1. Buckley, S. E., and Leverett, M. C., *Trans. AIME* **146**, 125 (1942).
2. Welge, H. J., *Trans. AIME* **195**, 91 (1952).
3. Leverett, M. C., *Trans. AIME* **132**, 149 (1939).
4. Pirson, S. J., "Oil Reservoir Engineering," p. 558, New York, McGraw-Hill Book Co., Inc., 1958.
5. Jones-Parra, J., and Calhoun, J. C., Jr., *Trans. AIME* **198**, 335 (1953).
6. Terwilliger, P. L., *et al.*, *Trans. AIME* **192**, 285 (1951).
7. Rapoport, L. A., and Leas, W. J., *Trans. AIME* **198**, 139 (1953).
8. Sheldon, J. W., Zondek, B., and Cardwell, W. T., Jr., *Trans. AIME* **216**, 290 (1959).
9. Rose, W. D., and Chanmapragada, R., Petr. Div. paper no. 1549G, AIME, presented at Denver, 1960.
10. Craft, B. C., and Hawkins, M. F., "Applied Petroleum Reservoir Engineering," p. 361, Englewood Cliffs, N. J., Prentice-Hall, Inc., 1959.
11. Higgins, R. V., and Leighton, A. J., "USBM RI 5618," 1960.
12. Higgins, R. V., "USBM RI 5568," 1960.
13. Joslin, W. J., *J. Petrol. Technol.*, 87 (January, 1964).

PROBLEMS

1. A sandstone exhibits the following relative permeability characteristics:

S_w	0.24*	0.30	0.40	0.50	0.60	0.70	0.80*	0.90	1.00
k_{ro}	0.95	0.89	0.74	0.45	0.19	0.06	0	0	0
k_{rw}	0	0.01	0.04	0.09	0.17	0.28	0.44	0.67	1.00

*Critical saturations for water and oil.

Other reservoir and fluid properties are as follows:

Oil viscosity	1.25 cp
Water viscosity	0.76 cp
Oil formation volume factor	1.12 bbl/STB
Water formation volume factor	1.03 bbl/STB

(a) If the average system pressure is held constant above the saturation pressure, develop a graph showing the fractional flow of water as a function of the water saturation.

(b) Prepare a graph of the producing water-oil ratio in surface units as a function of the system water saturation.

(c) If the system has a length of 500 feet, a cross-sectional area of 1 sq ft, and water is injected at 0.1 cu ft/hr, determine the position of the displacement front from the inlet end after 10 hours, 100 hours, 500 hours. Determine the position of the front by balancing areas to avoid the problem of triple values as illustrated in Figure 5-5.

(d) Show that the tangent line construction from the irreducible water saturation to the fractional flow curve results in a value of water saturation at the displacement front that is the same as that obtained by balancing areas in Part (c).

(e) Determine the oil recovery to breakthrough of injected water at the producing end of the system as a fraction of the recoverable oil, as a fraction of the original oil-in-place, both in reservoir and surface volumes.

(f) What would be oil recovery as a fraction on a surface and reservoir volume basis if the abandonment water-oil ratio is 25 STB/STB?

Injection Rates and Pressures in Secondary Recovery

Common to all the secondary recovery techniques presently known, or yet to be developed, is the need for a technique(s) by which accurate, or at least order of magnitude answers can be developed for injection rates or injection pressures, over the life of the projected flood. This is important for the proper sizing of injection equipment and pumps. Then too, excessive injection pressures for economic fluid injection rates may be sufficiently high to induce artificial fractures which, in turn, could be detrimental to the performance of the flood. Advance information on injection well injectivity and producing well producibility provides invaluable information to the complete engineering of a secondary recovery project. Indeed, one of the reasons for instituting a pilot flooding system in a given flooding prospect is to determine injection rates and pressures necessary for economic operations—a necessary step where a large amount of uncertainty exists.

Muskat[1] has presented equations for regular well patterns, the direct line-drive, the five-spot, the staggered line-drive, and the seven-spot, from which injection rates may be calculated for the ideal case where the fluid mobility ratio (driving to driven) is one. The mathematical arguments will not be presented here. Table 6-1 presents a summary of these equations in field units of darcys, feet, barrels per day, cp and psia. The injectivity has been termed i, and is in barrels per day at reservoir conditions of temperature and pressure. If a compressible fluid is being injected, appropri-

TABLE 6-1.[2] Injection Rates for Enclosed Well Patterns ($M = 1$)
⌀ Injection Well, ● Producing Well

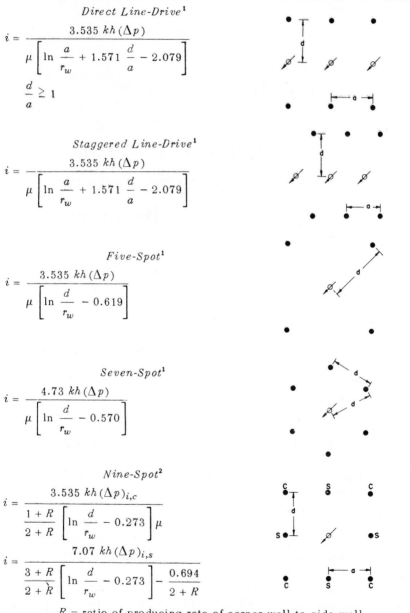

Direct Line-Drive[1]

$$i = \frac{3.535 \, kh \, (\Delta p)}{\mu \left[\ln \dfrac{a}{r_w} + 1.571 \dfrac{d}{a} - 2.079 \right]}$$

$$\frac{d}{a} \geq 1$$

Staggered Line-Drive[1]

$$i = \frac{3.535 \, kh \, (\Delta p)}{\mu \left[\ln \dfrac{a}{r_w} + 1.571 \dfrac{d}{a} - 2.079 \right]}$$

Five-Spot[1]

$$i = \frac{3.535 \, kh \, (\Delta p)}{\mu \left[\ln \dfrac{d}{r_w} - 0.619 \right]}$$

Seven-Spot[1]

$$i = \frac{4.73 \, kh \, (\Delta p)}{\mu \left[\ln \dfrac{d}{r_w} - 0.570 \right]}$$

Nine-Spot[2]

$$i = \frac{3.535 \, kh \, (\Delta p)_{i,c}}{\dfrac{1 + R}{2 + R} \left[\ln \dfrac{d}{r_w} - 0.273 \right] \mu}$$

$$i = \frac{7.07 \, kh \, (\Delta p)_{i,s}}{\dfrac{3 + R}{2 + R} \left[\ln \dfrac{d}{r_w} - 0.273 \right] - \dfrac{0.694}{2 + R}}$$

R = ratio of producing rate of corner well to side well
$(\Delta p)_{i,c}$ = pressure difference between injection well and corner well
$(\Delta p)_{i,s}$ = pressure difference between injection well and side well

ate corrections would be necessary to determine the surface volumes or injection rates at standard pressure and temperature. The equation(s) for the injection rates applicable for a nine-spot flooding network is due to Deppe.[2] Deppe also presents analytical solutions for the direct line-drive, five-spot and nine-spot well networks, where a reservoir boundary exists parallel to one side and at a distance of $d/2$, or greater away from the pattern. These solutions are specialized, and will not be reproduced here.

While exact analytical solutions can be developed for steady-state pressure distributions and the resulting injection rates where the fluid mobility ratio is one, such a method cannot be directly applied to the case where the fluid mobility ratio is other than one. It is evident from the study of laboratory models, where displacement occurs between producing and injection points, that there is a large injection interval where a region of radial flow exists, both at the producing well and the injection well. An examination of the pressure distributions shows that much of the pressure drop occurs in these regions of radial flow. Deppe[2] has concluded that the injection rates to be expected of any system can be approximated by dividing the pattern into regions where radial and linear flow predominate. The assumption that the flood front advances linearly in the linear region and radially in the radial region until breakthrough at the producing well, permits an addition of the linear and radial flow equations and a calculation of the injection rate.

This method of estimating injection rates can best be shown by considering the five-spot well pattern. Figure 6-1 shows the division of a five-spot pattern into two equal radial flow sectors having the same volume as the total pattern. The total injection rate

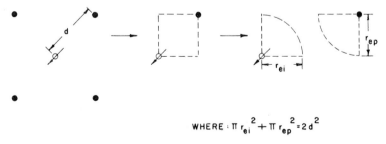

$$\text{WHERE} : \pi\, r_{ei}^{2} + \pi\, r_{ep}^{2} = 2d^{2}$$

Figure 6-1. Division of a segment of a five-spot well network into radial flow sectors (after Deppe[2]).

is then the addition of the rates at the producing and injection wells:

$$i = \frac{7.07 \; kh \, (p_{wi} - p_{wp})}{\mu \left[\ln \dfrac{r_{ei}}{r_{wi}} + \ln \dfrac{r_{ep}}{r_{wp}} \right]} \tag{1}$$

where the subscript i and p refer to conditions in the injection and producing sectors, respectively. Interestingly enough, if the producing and injection wells have the same radius, and if r_{ei} equals r_{ep}, then the equation takes the following form:

$$i = \frac{7.07 \; kh \, (p_{wi} - p_{wp})}{\mu \left[\ln \dfrac{d}{r_w} - 0.249 \right]} \tag{2}$$

Notice that equation (2) differs from the exact equation in Table 6-1 for the five-spot pattern only in the constant (0.249 as compared to 0.269) in the denominator. This rather close agreement was accomplished without including the linear flow segment, which could have been placed between the radial sections at the producing and injection sections of the well network.

If the assumption is made that the flood front moves out radially from the injection well until the boundary of the radial segment is reached, then the following equation for injection rate can be written directly:[2]

$$i = \frac{7.07 \; hk \, (p_{wi} - p_{wp})}{\left[\dfrac{1}{M} \ln \dfrac{r_f}{r_{wi}} + \dfrac{r_{ei}}{r_f} + \ln \dfrac{r_{ep}}{r_{wp}} \right] \mu} \tag{3}$$

where the areal sweep efficiency, E_s, is related to the radius of the flood front by the following relationship:

$$r_f = d \sqrt{\frac{2 E_s}{\pi}} \tag{4}$$

Again, the subscripts i and p refer to conditions in the vicinities of the injection and producing wells, respectively. The areal sweep efficiency value will be obtained from a calculation of performance—possibly the Stiles, Higgins, Craig-Stiles, or other calculation procedure. Then too, Chapter 4, on the topic of flood coverages, presents the results of laboratory studies where sweep efficiency values, as a function of fluid mobility ratio for the five-spot flood, have been presented.

Similarly, an equation may be written[2] describing the injection rate attained through the radial segment at the producing well for the pressure drop, $p_{wi} - p_{wp}$, between the injection and producing well when the mobility ratio is other than unity:

$$i = \frac{7.07 \, hk \, (p_{wi} - p_{wp})}{\left[\dfrac{1}{M} \ln \dfrac{r_{ei}}{r_{wi}} + \dfrac{1}{M} \ln \dfrac{r_{ep}}{r_f} + \ln \dfrac{r_f}{r_{wp}} \right] \mu} \tag{5}$$

and where in this instance, the radius of the flood front would be approximated as:

$$r_f = d \sqrt{\frac{2(1 - E_s)}{\pi}} \tag{6}$$

Prats *et al.*[3] have noted that the injection rate after breakthrough is controlled by the subtended angle which the injected fluid makes with the wellbore of the producing well. Deppe[2] has developed the following relationship for the angle open to flow of the injected fluid at the producing well, the fluid mobility ratio and the fraction of the producing stream from the swept region:

$$\alpha = \frac{2 \pi f}{f + M (1 - f)} \tag{7}$$

The angle α is in radians. Figure 6-2 shows the assumed shape of the flood front after breakthrough of the injected fluid to the producing well, for the five-spot well network.

The injection rate after breakthrough may be calculated from the following equation:

$$i = \frac{7.07 \, hk \, (p_{wi} - p_{wp})}{\left[\ln \dfrac{r_{ei}}{r_{wi}} + \ln \dfrac{r_{ep}}{r_f} + \dfrac{2 \pi f}{\alpha} \ln \dfrac{r_f}{r_{wp}} \right] \mu} \tag{8}$$

where

$$r_f = d \sqrt{\frac{2(1 - E_s)}{\pi \left(1 - \dfrac{\alpha}{2 \pi} \right)}} , \text{ the radius of the flood front}$$

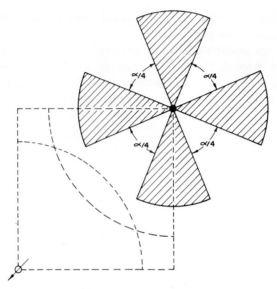

Figure 6-2. Assumed shape of the injected fluid
front after breakthrough to the producing well—
for the five-spot well network (after Deppe[2]).

$$r_{ei} = d \sqrt{\frac{2E_s}{\pi}} \text{, the radius of the radial segment at the}$$
$$\text{injection well}$$

$$r_{ep} = \sqrt{\frac{2d^2}{\pi} - r_{ei}^2} \text{, the radius of the radial segment at the}$$
$$\text{producing well}$$

$$d = \text{distance between producing and injection well}$$

All units in the equation are in the field units of barrels per day,
feet, darcys, cp and psia. Notice that the injectivity is in reservoir barrels per day, since no correction has been included for a
formation volume factor. Where the fluid being injected is compressible, a correction with the appropriate factor would be necessary to obtain injection rates at standard or surface conditions.
The calculation of reservoir performance will usually generate
areal coverage factors, E_s, from which the injection rates may be
calculated by means of equations (3), (5) and (8).

Deppe[2] has published a technique whereby the injection rates
may be determined where the well spacing is irregular, or the wells

are producing at unequal rates.   This would result in unequal
drainage areas for each producing well, not to mention the distor-
tion which would occur to the displacement front of the injected
fluid. Occasionally, it is important to be able to determine the injection rates to be expected when the five-spot pattern is close to
a reservoir boundary, or an isolating or partially isolating fault.
Deppe describes a technique developed along the lines of the preceding discussion, which will permit the calculation of injection
rates for such cases.

Prats *et al.*[3] have developed an analytical technique whereby
injection rates may be calculated for an enclosed five-spot well
pattern, where oil, gas and water saturations are present. Such a
situation will almost always exist where a field has been produced
to depletion by primary means, and is then a candidate for secondary recovery due to high remaining oil saturations. Figure 6-3
presents an idealized picture of the fluid saturations which can exist between the producing and injection well. Here the displacement has been arbitrarily divided into three regions, that from the
beginning of the flood to oil-bank interference, that from oil-bank
interference to oil breakthrough and finally the gas region. Figure 6-4 presents the dimensionless injectivity developed from ana-

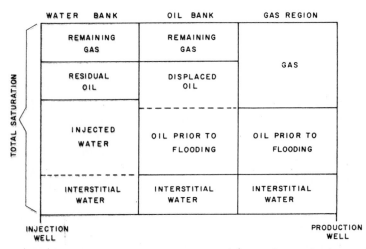

Figure 6-3. Fluid saturation regions between the producing and
injecting wells under immiscible fluid injection (after Prats
et al.[3]).

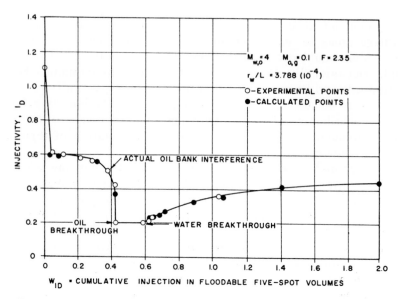

Figure 6-4. Dimensionless injectivity for a water-oil-gas system (after Prats *et al.*[3]).

lytic considerations resulting in a five-spot system. Here the dimensionless injectivity has been defined as:

$$I_D = \frac{i_w \, \mu_w}{k_w \, h \, (\Delta p)} \qquad (9)$$

for the homogeneous reservoir case. The fundamental units of the terms on the right hand side of equation (9) are those of cc/sec, cp, darcys, cm and atm. The conversion to field units could be achieved by simply multiplying the denominator by 1.127. In this instance, field units would be bbls/day, cp, darcys and feet. If the reservoir were layered, the denominator of equation (9) could be modified by replacing $k_w h$ by $\Sigma k_{wj} h_j$. Here, each separate layer is represented by the subscript j. Of course, knowing water viscosity, permeability to water, thickness intervals and pressure differential would permit a calculation of the injection rate, i_w, in consistent units. In Figure 6-4, F is a factor which takes into account the initial and residual gas and oil saturations, according to the relationship $F = 1 + (S_o - S_{or})/(S_g - S_{gr})$. The floodable pore volume in the five-spot well pattern, W_{iD}, is defined as follows:

$$W_{iD} = \frac{W_i}{L^2 h \, \phi \, (1 - S_{or} - S_{gr} - S_{wc})} \tag{10}$$

where S_{wc} is the connate water saturation, which may or may not be at the irreducible value prior to flooding. A study of Figure 6-4 shows that in the water-oil-gas system, the injectivity is initially high, due to the presence of the more permeable gas zone. A reduction of permeability is observed as the water is injected until interference between the oil banks from the offsetting five-spots is encountered. A sizeable decrease in injectivity is observed at this point until complete fill-up of the formation occurs, and oil breaks through at the producing well. In the graphical example shown, the injectivity increases after the displacing water breaks through. This would be expected, since the water-oil mobility ratio exceeds one in the example. Had the mobility ratio been less than one, injectivity would have decreased. Figure 6-5 illustrates the shapes of the various fluid banks for this three-fluid case during the various phases of the immiscible displacement. The calculation of injectivity for a homogeneous reservoir can be conveniently handled by using the three phases or periods illustrated by Figure 6-5:

(1) Injectivity in a homogeneous reservoir—from the start of the flood until oil bank interference:

The dimensionless injectivity may be closely approximated by means of the following equation by Prats *et al.*[3]:

$$I_D =$$

$$\frac{4\pi}{\ln\left(\dfrac{r_1}{r_w}\right)^2 + M_{w,o} M_{o,g}\left[2\ln\left(\dfrac{L}{r_w}\right)^2 - 3.856 - \ln\left(\dfrac{r_2}{r_w}\right)^2\right] + M_{w,o}\ln\left(\dfrac{r_2}{r_1}\right)^2} \tag{11}$$

where

$r_1 = \sqrt{W_{iD} A/\pi + r_w^2}$, the outer radius of the water bank

$r_2 = \sqrt{W_{iD} AF/\pi + r_w^2}$, the outer radius of the oil bank

W_{iD} = dimensionless water injected expressed as a fraction of the floodable pore volume, defined by equation (10)

L = length of a side of the five-spot well pattern

$M_{w,o}$ = water-to-oil mobility ratio

$M_{o,g}$ = oil-to-gas mobility ratio

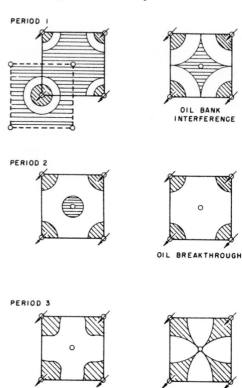

Figure 6-5. Positions of the oil, gas, and water phase during the various injection periods in a five-spot well network (after Prats *et al.*[3]).

Equation (11) is developed by assuming that flow is radial in the water and oil banks for the purposes of calculating part of the pressure drop. The pressure drop in the gas section is the pressure drop for the entire reservoir element, less the calculated pressure drop had there been gas flow in those sections where the oil and water banks exist. The term $[2 \ln (L/r_w)^2 - 3.856]$ is equivalent to $[4 \pi/\text{injectivity}]$ for the single fluid (gas) case, as shown by Prats *et al.*[4]. It should be evident that the dimensionless injectivity during the first phase can be developed by calculating values of I_D until the outer radius of the oil bank, r_2, becomes equal to $L/2$.

(2) Injectivity in a homogeneous reservoir—from interference of the oil bank to oil breakthrough:

The equation for dimensionless injectivity during this period as presented by Prats *et al.*[3] is:

$$I_D = 4\pi \div \left\{ \ln\left(\frac{r_1}{r_w}\right)^2 + M_{w,o} M_{o,g} \ln\left(\frac{r_3}{r_w}\right)^2 + M_{w,o}\left[2\ln\left(\frac{L}{r_w}\right)^2 - \right.\right.$$
$$\left.\left. 3.856 - \ln\left(\frac{r_3}{r_w}\right)^2 - \ln\left(\frac{r_1}{r_w}\right)^2\right]\right\} \quad (12)$$

where

$$r_3 = \sqrt{\frac{L^2}{\pi}(1 - W_{iD} F) + r_w^2},$$

the radius of the gas region around the producing well.

Prats *et al.* state that the lower limit of usefulness of equation (12) occurs when r_3 equals $L/6$, and when the oil-to-gas mobility ratio is equal to or less than 0.005, since in this range the gas front around the producing well should be essentially radial. Prats *et al.* also found that the lower limit of usefulness is further reduced to r_3 equal to $L/20$, when the oil-to-gas mobility ratio approximated 0.1. The upper limit of usefulness occurs either at the time of oil breakthrough, when W_{iD} equals $1/F$, or at the time of water bank interference—whichever comes first. Water bank interference would be expected to occur when r_1 equals $L/2$. Notice that the radius of the water bank, r_1, as it occurs in equation (12), remains the same as it was defined in equation (11).

(3) Injectivity in a homogeneous reservoir—after oil breakthrough to the producing well:

After the breakthrough of oil to the producing well, it usually suffices to assume that no gas region remains, and that the problem has been reduced from that where three fluids flow, to that where two fluids flow in the system. Prats *et al.*[3] have published experimental results obtained from a potentiometric analog, where injectivity has been plotted as a function of the producing water cut as a fraction, and the water-to-oil mobility ratio. Figure 6-6 presents these results. It should be noted that the results are valid only if r_w/L equals 3.788×10^{-4}. The dimensionless injectivity read from Figure 6-6 should be corrected by the following equations, when the r_w/L ratio differs from 3.788×10^{-4}:

$$I_D = 1 \div$$
$$\left\{ \frac{1}{I_D'} + \frac{1}{2\pi}[1 + M_{w,o} - f_w(M_{w,o} - 1)] \times \ln\frac{3.788 \times 10^{-4} L}{r_w} \right\} \quad (13)$$

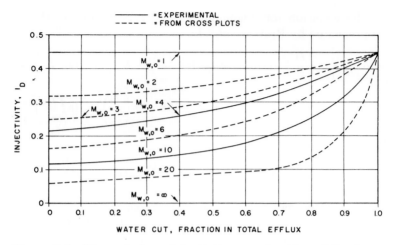

Figure 6-6. Dimensionless injectivity as a function of the fractional water cut and the water-to-oil mobility ratio (after Prats et al.[3]).

The value for I'_D is read from Figure 6-6, and the corrected value for dimensionless injectivity determined from equation (13). Injectivity is then calculated for a range of values of the fractional producing water cut, f_w. The fractional producing water cut will normally be obtained from the calculation of the flood performance. The Stiles, Buckley-Leverett, and the Craig et al. performance calculation techniques, as well as a number of other methods, yield information as regards the fractional producing water cut.

The above-outlined method will permit a determination of the injectivity behavior that could be expected during fluid displacement in the five-spot well network where water is displacing oil, and where a rather high gas saturation exists initially.

Equation (9) is useful in converting dimensionless injectivity into either sand face injection pressure (if injection rate is known), or injection rate (if injection pressure at the sand face is known). From such calculations, a good estimate of the injection rates possible at a specified injection pressure can be determined. Of course, where a pilot water flood is to be used, information will also result as regards injection pressures and injection rates.

Prats et al.[3] also describe a calculation technique, whereby the dimensionless injectivity may be determined for a stratified system, should such a treatment be necessary.

References

1. Muskat, Morris, "Physical Principles of Oil Production," p. 650, New York, McGraw-Hill Book Co., Inc., 1949.
2. Deppe, J. C., AIME paper 1472 G presented 4th biennial secondary recovery symp., May 2-3, 1960, Wichita Falls, Texas.
3. Prats, M., Matthews, C. S., Jewett, R. L., and Baker, J. D., *Trans. AIME* **216**, 98 (1959).
4. Prats, M., Strickler, W. R., and Matthews, C. S., *Trans. AIME* **204**, 160 (1955).

PROBLEMS

1. Calculate water injection rates which would be expected in a relatively homogeneous oil sand that is being subjected to a water flood. The oil pay thickness is 18 ft, the permeability to flow of water is 76 md, the applied pressure differential is 750 psig, wellbore radius is 0.375 ft, and the oil viscosity is 4.6 cp. The formation temperature is 138°F, and the injection water is fresh.

(a) For a direct line-drive well pattern where the inter-well distances are constant and spacing is one well for each 10 acres.

(b) For a staggered line-drive well pattern where the distance between the lines of producing and injection wells is double the distance between the wells in each line. The average well spacing is 20 acres.

(c) For a five-spot well pattern where wells have been drilled on 40-acre spacing.

(d) For a seven-spot well pattern where the average well spacing is 10 acres.

(e) For a nine-spot well pattern where the corner wells are produced at 1.5 times the producing rate of the side wells. The average well spacing is 20 acres.

Solve the problem by ignoring the error introduced by the fluid mobility ratio being other than unity. Would the answers obtained for the injection rates for the various well patterns be conservative or optimistic? Why?

2. Use the method of Deppe to calculate the fresh water injection rate which could be expected for a five-spot well pattern having the following conditions:

Oil viscosity	4.6 cp
Reservoir temperature	118°F
Permeability to oil flow at irreducible water saturation	150 md
Permeability to water flow at residual oil saturation	60 md
Applied pressure differential	1000 psi
Areal sweep efficiency	65 per cent
Wellbore radius	4 in
Pay thickness	27 ft
Well spacing	40 acres

(a) At the time water injection is begun.

(b) At the time of water breakthrough to the producing well(s).

(c) When water injection is continued past water breakthrough to the producing well(s), the producing water fraction is 85 per cent, and the areal sweep efficiency approximates 80 per cent.

3. Develop a graph of dimensionless water injectivity for an enclosed five-spot well pattern similar to that given by Figure 6-4, for the following available reservoir and fluid data:

Water injection rate	250 BWPD
Reservoir temperature	152°F
Permeability to water at irreducible oil saturation	100 md
Permeability to oil at irreducible water saturation	150 md
Absolute permeability	200 md
Well spacing	10 acres
Irreducible water saturation	24 per cent
Residual oil saturation	20 per cent
Residual gas saturation	5 per cent
Gas saturation at beginning of injection	15 per cent
Oil pay thickness	20 ft
Porosity	22 per cent
Wellbore radius	0.375 ft
Oil viscosity	5 cp
Gas gravity (paraffin base)	0.70

The values of the fractional producing water cut are 0.25, 0.35, 0.45, 0.55, 0.65, 0.75, 0.85, 0.95, at water saturation values of 0.300, 0.325, 0.350, 0.375, 0.400, 0.433, 0.500, 0.625, respectively. The injection water is fresh. Discuss the importance of such a graph in the prediction of water flooding performance.

Water Flooding
Performance Calculations

Water Flooding Performance Calculations—the Frontal Advance Method

The theoretical development for the frontal advance approach to the displacement of oil by water in porous media is provided in Chapter 5. The following approach will be for the case where the displacement is occurring in a linear homogeneous system of constant thickness. The method is applicable to those systems where the mobility ratio, $k_{rw}\mu_o/k_{ro}\mu_w$, is favorable, i.e., less than one, to unfavorable, perhaps ten as a limit. While the upper limit of numerical mobility ratio for satisfactory water flood performance calculations is not known, the viscous fingering approach would be preferable where the mobility ratios are extremely adverse.

Frontal Advance Method—the Linear System. A number of the artificial flooding configurations would be accurately modelled as a linear system, i.e., peripheral injection, line-drive and end-to-end sweeps. Then too, while not treated specifically herein, the frontal displacements of an advancing water front in an active water drive are essentially the same problem, in so far as pore-to-pore sweep efficiencies are concerned. The equations which apply are the fractional flow equation:

$$f_w = \frac{1 - \dfrac{1.127\, k_o}{q_t \mu_o}\left[\dfrac{\partial p_c}{\partial u} + 0.434\,(\Delta \vartheta)\sin\alpha\right]}{1 + \dfrac{k_o}{k_w}\cdot\dfrac{\mu_w}{\mu_o}} \tag{1}$$

and the frontal advance equation:

$$u = \frac{Q_t\, t}{A\,\phi} \left(\frac{\partial f_w}{\partial S_w}\right) \tag{2}$$

where the nomenclature is as described in Chapter 5. The subscripts o and w refer to the oil and water phases, respectively. It is apparent that data on fluid properties will be required, in this case, at the average pressure and temperature of the reservoir at which the displacement will take place.

The fluid properties μ_o, μ_w and $\Delta\vartheta$ may be available from reservoir sample analyses, data on a similar field, or from published correlations. Since the development of the equations was for steady-state conditions, the flux rate (q_t), in bbl/day/sq ft, will be constant, and may be determined from injection rate requirements. The injection rate requirements may be unknown at this point if there is justifiable concern as regards the rate sensitivity of the process. If this is the case, a range of q_t values may be taken and the fraction of the displacing phase flowing for varying displacing phase saturations determined. Indeed, since water flooding *is* rate sensitive, and the injection rate and the resulting producing rate are variable (for constant pressure operation), considerable care should be taken to control the rates so as to maximize oil recovery. This must, of course, be consistent with an economical operation.

The frontal advance method requires that good oil-water relative permeability information be available. This may be obtained from laboratory flow tests on *representative* reservoir rock samples from the field, from information on nearby similar fields, from production data (sometimes), or from published empirical or statistical relationships. Chapter 3 provides information as regards the selection of a good relative permeability relationship. Note that the effective permeability to oil, k_o, may be obtained in a similar fashion. Usually the value used is that existing when only oil and an irreducible water saturation are present in the system. This should be nearly the case ahead of the displacement front.

Performance before Water Breakthrough. Equation (1) may then be readily solved for a range of water saturations which determine the relative permeability ratio value, k_o/k_w, and the results plotted, i.e., as illustrated in Figure 7-1. In plotting Figure 7-1, the $\partial p_c/\partial u$ term is taken as zero and the S-shaped curve obtained.

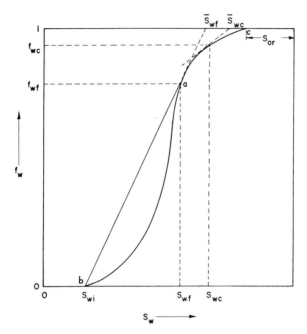

Figure 7-1. Fractional flow relationship for water displacing oil in a linear system.

Drawing the secant line from the irreducible water saturation tangent to the curve at point a restores the capillary pressure dependency of this immiscible fluid displacement. The straight line segment ba and the curved line ac define the fractional flow behavior as defined for the rock and fluid properties mathematically modelled. Extension of the straight line ba to a saturation corresponding to a fractional flow of one, f_w, results in the mean water saturation back through the system, $\overline{S}_{wf}$. If the total linear system length is L feet, and the value of the derivative, $\partial f_w / \partial S_w$, is $1/(\overline{S}_{wf} - S_{wi})$, then equation (2) may be rewritten as follows for the breakthrough time, t_B:

$$t_B = \frac{AL \, \phi \, (\overline{S}_{wf} - S_{wi})}{Q_t} \qquad (3)$$

where if the units are square feet, feet and cubic feet per day, the breakthrough time will be given in days. If the throughput rate, Q_t, has the units of bbl/month, then breakthrough time will be in months, if a conversion factor of 5.615 cu ft/bbl is placed in the numerator.

The above calculation of the breakthrough time is predicated on the throughput rate, Q_t, being of sufficient magnitude so that the length of the stabilized zone will be small compared to the total length of the linear system. In the practical field cases, this is essentially always true. In laboratory models of limited length, the stabilized zone could extend over a sizeable fraction of the total length of the system at slow throughput rates.

If the system being flooded is initially at the irreducible water saturation, there will be no water produced until water breakthrough. If the water saturation throughout the system exceeds the irreducible water saturation value, water will be produced according to the relative mobilities of the oil and water flowing to the producing well, ahead of the stabilized zone or displacement front. If the reservoir throughput rate is q_t, the total producing rate would be:

$$q_t = q_o + q_w \tag{4}$$

The corresponding water-oil ratio is:

$$(\text{WOR})_{\text{surf.cond.}} = \frac{k_w \, \mu_o \, B_o}{k_o \, \mu_w \, B_w} \tag{5}$$

where the water-oil relative permeability would be determined at the water saturation prevailing in the vicinity of the producing well. This assumes that there is no water saturation gradient between the displacing front and the producing well. This would result in a constant producing water-oil ratio until arrival of the displacing front at the outlet end of the system. The relationship between water-oil ratio at surface conditions and fraction of water flowing at reservoir conditions is as follows:

$$f_w = \frac{(\text{WOR})B_w}{(\text{WOR})B_w + B_o} \quad \text{or} \quad \text{WOR} = \frac{f_w B_o}{(1 - f_w)B_w} \tag{6}$$

Since steady-state conditions prevail, the rate of stock-tank oil production is:

$$q_o = \frac{(1 - f_w)Q_t}{B_o} = \frac{Q_t}{B_o + (\text{WOR})B_w} \tag{7}$$

and the water producing rate at surface conditions becomes:

$$q_w = q_o \, (\text{WOR}) \qquad (8)$$

An integration of the plots of oil and water rate versus time results in cumulative oil and water values, at corresponding times. Of course, if the water injection rate is constant, then multiplying the injection rate by injection time will result in the total water requirements, an important consideration in the planning for an adequate water source. The left-hand part of Figure 7-2 provides

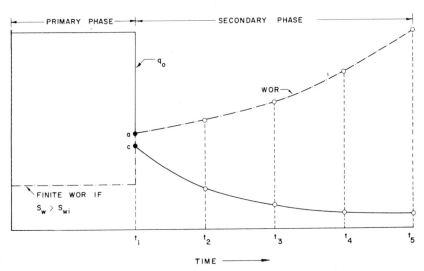

Figure 7-2. Sketch of the oil producing rate and water-oil ratio as a function of time for the frontal displacement mechanism in a linear system.

sketches of the oil producing rate and producing WOR that would be expected in the primary phase, that is, before breakthrough of the front at the producing well. The example shows the case where a mobile water saturation did exist in the system prior to water injection.

Performance after Water Breakthrough. In those reservoirs where the mobility ratio of the displacing-displaced fluid combination is favorable, the recovery at breakthrough will constitute the major fraction of the total recovery that will ultimately be recovered by flooding. This is especially true where the displacement is in an essentially linear system. Where the system is non-linear, the

areal sweep efficiency will increase, resulting in an expanded
subordinate phase following the initial breakthrough of the water.
More will be said about this in a later section. Then too, an adverse mobility ratio will result in a displacement of the f_w water
saturation relationship (see Figure 7-1, for instance) to the left or
toward lower recoveries of oil. This means that the subordinate
phase will be longer, and indeed in some cases, may contribute
more oil than was produced before water breakthrough.

Figure 7-3 presents an expanded section of the f_w water saturation relationship of interest for performance calculations relative
to the subordinate phase. If the limiting water cut that can be
handled economically is known, a simple construction will yield
the average water saturation back through the system. If, for instance, the limiting producing water cut is 0.95, then a tangent line
from the corresponding point on the f_w versus S_w curve to a saturation value when f_w equals one, yields a mean saturation, represented by $\overline{S}_{w5}$ on Figure 7-3. The difference between the mean

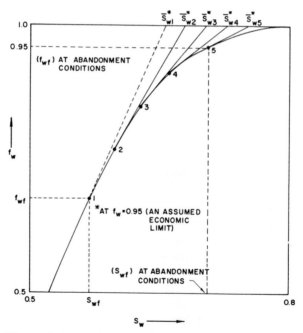

Figure 7-3. Expanded section of a fractional flow
relationship where water is displacing oil in a linear
system.

saturation at breakthrough and the saturation at abandonment conditions is a measure of the oil recovery at reservoir conditions of
temperature and pressure during the secondary or after-breakthrough
phase of the oil-water displacement mechanism in a linear system.
If a gas saturation were present, then this amount would also have
to be subtracted.

As suggested in the construction shown in Figure 7-3, the saturation increase between the mean water saturation at breakthrough
and that at abandonment may arbitrarily be subdivided as shown,
i.e., as saturations $\bar{S}_{w2}$, $\bar{S}_{w3}$, $\bar{S}_{w4}$. Then tangents drawn to the
f_w curve will provide points 2, 3, and 4 which correspond to the
f_w values at the outlet end of the system, at the time that mean
saturations back through the system were $\bar{S}_{w2}$, $\bar{S}_{w3}$, $\bar{S}_{w4}$. Equations (6), (7) and (8) provide a means by which the subordinate
phase performance may be calculated. Figure 7-2 shows a sketch
of a typical oil producing rate and WOR behavior during the secondary or subordinate phase. Graphical integration of the oil producing rate versus time relationship results in cumulative oil recovery on a time basis. Similarly, the cumulative water produced
may be determined by graphically integrating the water producing
rate-time relationship, which can be obtained through use of equation (8). The total water required for the displacement of oil down
to abandonment would be equal to the sum of the cumulative water
produced, plus the water gained by the system due to the displacement of oil. Actually the total water requirements could be little
more than the water gained by the system, if the produced water is
suitable for reinjection and is so reinjected. This assumes that
the initial water saturation is at an irreducible value. Actually,
the production of water due to a mobile water fraction would complicate the calculation of the water requirements very little. More
will be said on this in the chapter on water sources and water
treating considerations.

Conditions Necessary for the Formation of an Oil Bank. The
frontal advance method provides a means by which the conditions
necessary for the formation of an oil bank can be readily ascertained. Figure 7-1 is a plot of the fraction of water flowing *at the
front* for a given saturation, S_w. As has been previously discussed,
between points *a* and *b* the capillary pressure gradient, $\partial p_c / \partial \mu$,
has been included by the straight line construction due to the importance of the capillary pressure gradient in the stabilized or
frontal displacement zone. Between points *a* and *c*, the capillary

pressure gradient is negligible, and the non-stabilized or "non-banked" zone exists. Then, if the system water saturation exceeds the water saturation at the front, S_{wf}, as shown on Figure 7-1, no frontal displacement results. If the water saturation is between S_{wi} and S_{wf}, water will be produced before the front or oil bank arrives, but an oil bank will be formed.

Many times a water flood will be initiated in a reservoir containing a sizeable gas saturation due to primary oil recovery by the depletion drive producing mechanism. In this case, it is best to assume that the gas saturation will be filled during the "fill-up" period with the injected water. If this water saturation plus that already present due to the irreducible water saturation exceeds the S_{wf} value determined by the frontal advance calculation technique, it is unlikely that an oil bank will be formed. When no oil bank is formed during water flooding, the venture is usually uneconomical.

Application of Frontal Advance Methods to Non-linear Systems. While some systems upon which water flooding is accomplished do have linear flow characteristics, i.e., peripheral floods, line-drive floods and end-to-end floods, the majority of systems deviate significantly from such a simple representation. Figure 7-4 shows one segment of a five-spot water flooding network with the frontal position at the time of water breakthrough to the producing well, in a system where the mobility ratio is one. Streamline (1) has broken through to the producing well, while water traveling streamline path (4) has not yet covered half the distance. This situation will be even more pronounced in systems where the mobility ratio is adverse. The problem has two aspects, (1) the influence of differing flux rates on the pore-to-pore sweep efficiency, and (2) the fraction of the flood pattern that is contacted at breakthrough and at various times where injection has been continued past breakthrough. Examination of equation (1) indicates that the influence of flux rate is represented by q_t.

Recently, Higgins and Leighton[1] have published a technique to calculate two-phase flow using the Buckley-Leverett and Welge approaches for fluid displacement for any irregularly bounded porous media. The approach is best illustrated by reference to Figure 7-4, showing the isopotential and streamlines for the five-spot flooding network. Here, a potentiometric model has provided detailed information on the positions of the streamlines and isopotential lines. Notice that streamlines (1) and (2) enclose a flow

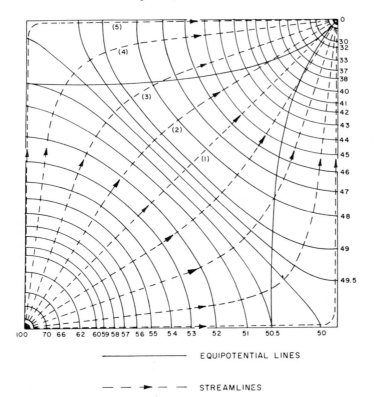

────────── EQUIPOTENTIAL LINES

─ ─ ─►─ ─ STREAMLINES

Figure 7-4. The five-spot flooding network showing streamlines and isopotential lines for unit mobility ratio (after
Muskat and Wyckoff).

channel, and that such a channel could be divided into a number of
cells of equal volume between the injection and producing wells.
Cells along a particular channel would then be of varying length
and differing effective cross-sectional area. The resistance to
flow because of the shape and size of the cells is represented by
G, the shape factor which is a measure of the geometric resistivity. Higgins, Boley and Leighton[2] have presented shape factors
and channel volumes for the direct line-drive, five-spot, seven-
spot and staggered line-drive well patterns. Tables 7-1 and 7-2
summarize these data.

In the determination of the shape factors for the five-spot element, actually one-eighth of a five-spot where the injection well
is in the middle and surrounded by four producing wells, Figure 7-4,
was used. Each channel was divided into 40 cells of equal vol-

TABLE 7-1. Shape Factors and Channel Volumes for Direct Line-drive and Five-spot well patterns (after Higgins[2])

Cell No.	Direct Line-drive Channel Number					Five-spot Channel Number			
	1	2	3	4	5	1	2	3	4
1	19.604	14.155	14.167	11.424	13.960	17.372	17.880	19.620	36.582
2	1.761	.843	1.037	1.019	.986	1.506	1.558	1.696	2.920
3	.786	.491	.666	.570	.623	.886	.900	.950	1.848
4	.510	.332	.455	.446	.438	.631	.639	.670	1.197
5	.324	.259	.348	.335	.350	.495	.495	.536	.928
6	.244	.228	.326	.288	.290	.397	.397	.440	.768
7	.174	.221	.297	.264	.266	.325	.331	.363	.643
8	.146	.175	.267	.249	.251	.286	.282	.306	.535
9	.151	.169	.247	.209	.210	.256	.245	.262	.451
10	.165	.175	.229	.196	.186	.231	.220	.229	.378
11	.193	.177	.223	.196	.173	.211	.200	.203	.310
12	.239	.182	.221	.194	.170	.196	.185	.182	.253
13	.312	.197	.215	.191	.167	.183	.172	.166	.205
14	.414	.218	.203	.185	.167	.173	.162	.152	.167
15	.530	.232	.187	.179	.164	.166	.155	.139	.139
16	.640	.245	.181	.169	.149	.161	.150	.128	.121
17	.779	.266	.181	.169	.144	.158	.147	.118	.108
18	.908	.281	.190	.190	.155	.155	.144	.110	.096
19	.931	.287	.205	.225	.171	.152	.139	.105	.088
20	.933	.288	.209	.231	.173	.150	.124	.101	.082
21	.933	.288	.209	.231	.173	.150	.124	.101	.082
22	.931	.287	.205	.225	.171	.152	.139	.105	.088
23	.908	.281	.190	.190	.155	.155	.144	.110	.096
24	.779	.266	.181	.169	.144	.158	.147	.118	.108
25	.640	.245	.181	.169	.149	.161	.150	.128	.121
26	.530	.232	.187	.179	.164	.166	.155	.139	.139
27	.414	.218	.203	.185	.167	.173	.162	.152	.167
28	.312	.197	.215	.191	.167	.183	.172	.166	.205
29	.239	.182	.221	.194	.170	.196	.185	.182	.253
30	.193	.177	.223	.196	.173	.211	.200	.203	.310
31	.165	.175	.229	.196	.186	.231	.220	.229	.378
32	.151	.169	.247	.209	.210	.256	.245	.262	.451
33	.146	.175	.267	.249	.251	.286	.282	.306	.535
34	.174	.221	.297	.264	.266	.325	.331	.363	.643
35	.244	.228	.326	.288	.290	.397	.397	.440	.768
36	.324	.259	.348	.335	.350	.495	.495	.536	.928
37	.510	.332	.455	.466	.438	.631	.639	.670	1.197
38	.786	.491	.666	.570	.623	.886	.900	.950	1.848
39	1.761	.843	1.037	1.019	.986	1.506	1.558	1.696	2.920
40	19.604	14.155	14.167	11.424	13.960	17.372	17.880	19.620	36.582
Volume	718	883	585	508	462	706	782	918	750

ume. Tests have been made to indicate that 40 cells were more than a sufficient number to insure good convergence properties during the calculation of water flooding performance. Where potentiometric data are available on an irregularly-shaped well pattern, a careful construction of the flow lines and a division into cells will result in specific values for the shape factor, G. In

TABLE 7-2. Shape Factors and Channel Volumes for Staggered Line-drive and Seven-spot Well Patterns (after Higgins[2])

Cell No.	Staggered Line-drive Channel Number						Seven-spot Channel Number		
	1	2	3	4	5	6	1	2	3
1	15.650	15.395	14.648	15.311	15.008	16.884	12.952	7.885	7.321
2	1.430	1.283	1.333	1.281	1.261	1.144	1.549	3.085	3.224
3	1.004	.779	.852	.588	.735	.603	.602	.590	.846
4	.510	.332	.455	.446	.509	.372	.387	.391	.349
5	.724	.535	.583	.433	.425	.241	.277	.335	.328
6	.648	.465	.570	.369	.399	.180	.224	.233	.246
7	.614	.430	.521	.349	.338	.126	.187	.218	.220
8	.589	.423	.498	.326	.357	.112	.157	.204	.199
9	.564	.414	.465	.313	.362	.109	.124	.185	.172
10	.538	.406	.450	.311	.363	.113	.100	.166	.169
11	.521	.404	.448	.309	.364	.140	.085	.148	.152
12	.516	.402	.439	.313	.372	.156	.082	.146	.141
13	.513	.402	.405	.327	.379	.170	.080	.139	.139
14	.512	.401	.403	.342	.383	.183	.069	.129	.129
15	.520	.232	.187	.179	.384	.215	.056	.134	.132
16	.499	.398	.392	.358	.385	.252	.051	.135	.136
17	.479	.397	.380	.360	.385	.296	.050	.137	.138
18	.465	.396	.375	.360	.386	.336	.058	.138	.139
19	.455	.394	.374	.365	.387	.376	.074	.140	.140
20	.428	.392	.373	.371	.388	.405	.113	.142	.142
21	.405	.388	.371	.373	.392	.428	.116	.148	.143
22	.376	.387	.365	.374	.394	.455	.123	.150	.146
23	.336	.386	.360	.375	.396	.465	.128	.153	.155
24	.296	.385	.360	.380	.397	.479	.132	.172	.167
25	.252	.385	.358	.392	.398	.499	.144	.188	.180
26	.215	.384	.355	.401	.398	.511	.159	.193	.185
27	.183	.383	.342	.403	.401	.512	.172	.200	.195
28	.170	.379	.327	.405	.402	.513	.206	.222	.213
29	.156	.372	.313	.439	.402	.516	.236	.239	.234
30	.140	.364	.309	.448	.404	.521	.256	.257	.252
31	.113	.363	.311	.450	.406	.538	.300	.299	.281
32	.109	.362	.313	.465	.414	.564	.323	.317	.312
33	.112	.357	.326	.498	.423	.589	.393	.349	.351
34	.126	.338	.349	.521	.430	.614	.427	.403	.407
35	.180	.399	.369	.570	.465	.648	.546	.486	.491
36	.241	.425	.433	.583	.535	.724	.639	.598	.608
37	.372	.409	.575	.720	.629	.827	.847	.815	.742
38	.603	.735	.588	.852	.779	1.004	1.234	1.092	1.022
39	1.144	1.261	1.281	1.333	1.283	1.430	2.070	3.793	4.330
40	16.884	15.008	15.311	14.648	15.395	15.650	26.358	21.040	19.708
Volume	620	494	464	464	494	620	1,346	947	863

equation form, the shape factor is:

$$G = \frac{L}{A} = \frac{L_l + L_r}{L_t + L_b} \tag{9}$$

where L = length, the subscripts l, r, t and b refer to the left side, right side, top and bottom of the cell and the channel thickness is unity. Where the cells are extremely irregular, Henley[3] has proposed a technique where the shape factor is determined by using inscribed circles. Higgins *et al.*[2] have developed a computer calculation technique whereby volumes of the streamlines and the shape factors may be determined from data taken from a potentiometric model study for any well spacing pattern. The data of Tables 7-1 and 7-2, with the exception of that for the five-spot well pattern, were developed in this fashion.

Kufus and Lynch[4] have shown that the average permeability of a cell times the shape factor determines the resistance to flow. Then the producing rate at the beginning of the flood will be:

$$q_{o_{j=0}} = k_a \, \Delta p \left/ \sum_{n=1}^{n=N} \frac{\mu_o}{k_{roIW}} \, G_n \right. \tag{10}$$

where

j = index of increments since the beginning of the flood
n = cell number index
N = total number of cells in each channel
k_{roIW} = permeability to oil at irreducible water saturation

After n number of cells have been invaded, the producing rate prior to displacing phase breakthrough (water) is:

$$q_{o_j} = (k_a \, \Delta p) \left/ \left[\sum_{n=1}^{n} \frac{1}{\left(\dfrac{k_{rw_{mean_n}}}{\mu_w} + \dfrac{k_{ro_{mean_n}}}{\mu_o} \right)} \, G_n + \sum_{i=n+1}^{i=n} \frac{\mu_o}{k_{roIW}} \, G_i \right] \right. \tag{11}$$

In order to solve equation (11), it is necessary to determine mean values for the oil and water relative permeability in each cell when j number of cells have been invaded. This is accomplished by calculating the f_w versus S_w relationship from equation (1), where the gravity term, $0.434 \, (\Delta \vartheta) \sin \alpha$, has been omitted. The present treatment does not treat a formation with dip, since the effect would be quite different between the various cells. The mean permeability of the first cell to water at the instant that water breakthrough occurs at the end of the cell is determined by dividing the slope of the f_w versus S_w curve at breakthrough, f'_{br}, by the area under a curve of the resistance to water flow, $1/k_{rw}$, versus the slope, f'. This area is that up to the value of the slope, f', corresponding to the total volume behind the flood front. Figure 7-5 illustrates these relationships. When the water front has progressed to the end of the second cell, the mean permeability to water in the first cell is one-half of the f'_{br} value, divided by the area under the resistance curve, up to the abscissa value of one-half f'_{br}. The mean permeability to water in the second cell then becomes one-half f'_{br}, divided by the remaining area under the resistance curve. The following equation presents the procedure outlined in an analytical form:

$$k_{rw_{\text{mean}_n}} = \frac{\left(\dfrac{f'_{br}}{j}\right)}{\sum \text{All areas under } 1/k_{rw} \text{ vs. } f' \text{ curve in nth cell}} \qquad (12)$$

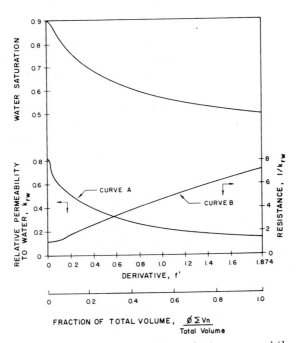

Figure 7-5. Water saturation, relative permeability to water, and resistance versus the slope term, f' (after Higgins *et al.*[1]).

A similar equation could be readily written for the mean permeability to oil. When breakthrough occurs (the subordinate phase) at the end of one of the channels, equation (11) must be modified to include a water producing rate:

$$(q_o + q_w)_j = \cfrac{k_a \Delta p}{\displaystyle\sum_{n=1}^{n=\text{no. of cells}} \left(\cfrac{1}{\cfrac{k_{rw_{\text{mean}_n}}}{w} + \cfrac{k_{ro_{\text{mean}_n}}}{o}} \right) G_n} \tag{13}$$

The mean permeabilities to water and oil in each cell will now be modified due to the decrease of the slope of the f_w versus S_w curve, as water saturations increase beyond that at water breakthrough. In equation form, the mean permeability to water in any given cell becomes:

$$k_{rw_{\text{mean}_n}} = \cfrac{\cfrac{f'_{br-m}}{N}}{\displaystyle\sum \text{All areas under } 1/k_{rw} \atop \text{vs. } f' \text{ curve in nth cell}} \tag{14}$$

where f'_{br-m} is the first derivative of the f_w versus S_w relationship from the fractional flow equation at the mth increment after water breakthrough. The equation for mean oil permeability takes a similar form.

As for the linear flow case, the water-oil ratio at reservoir conditions is determined by the water saturation at the outflow face, or at the producing well. After the mth increment of production has occurred past water breakthrough, the water-oil ratio is:

$$(\text{WOR})_{j=m+N} = \frac{k_{rw}}{k_{ro}} \cdot \frac{\mu_o}{\mu_w} \tag{15}$$

Equations (4) through (8) may be used directly to calculate the performance of the five-spot. Where the mean permeability values are used to calculate the oil and water producing rates in the subordinate phase, the instantaneous producing rates at the well may be approximated as follows:

$$(q_o)_{j-\frac{1}{2}} = \frac{(q_o)_j + (q_o)_{j+1}}{2} \tag{16}$$

$$(q_w)_{j-\frac{1}{2}} = \frac{(q_w)_j + (q_w)_{j+1}}{2} \tag{17}$$

Here, the index j is equal to m plus N, or the mth increment of production past water breakthrough, where N is the total number of cells in each channel.

Since secondary recovery by water flooding is often initiated late in the primary producing history of a field, it is common to find a rather large gas saturation in the reservoir. It has been observed in flooding such fields that a volume of water approximating that of the volume occupied by the gas saturation, must be injected before any response is noted at the producing wells. It is sometimes assumed that this volume of water necessary for "fill-up" is disseminated throughout the reservoir to the exclusion of the gas. If this approach is taken, the above-outlined technique for determining the performance of a pattern flood can be altered. The initial oil producing rate after fill-up may be written as:

$$q_o = \frac{k_a \, \Delta p}{\displaystyle\sum_{n=1}^{n=N} \frac{\mu_o}{k_{ro_c}} \, G_n} \tag{18}$$

As the encroaching water moves through the channel, both oil and water will be produced, since the water saturation will now exceed the irreducible saturation value. The equation describing the oil and water producing rate when the injected water has reached the end of a particular cell, and the increments of injection total that indicated by the index j, is the following:

$$(q_o + q_w)_j = (k_a \Delta p) \Bigg/ \Bigg[\sum_{n=1}^{n=\text{no. of invaded cells}} \frac{1}{\left(\dfrac{k_{rw_{\text{mean}\,n}}}{\mu_w} + \dfrac{k_{ro_{\text{mean}\,n}}}{\mu_o} \right)} G_n + \sum_{i=n+1}^{i=N} \frac{1}{\left(\dfrac{k_{ro_c}}{\mu_o} + \dfrac{k_{rw_c}}{\mu_w} \right)} G_i \Bigg] \tag{19}$$

The relative permeability to water is determined at the connate water saturation existing after fill-up, but ahead of the flooding front, denoted by k_{rw_c}. The performance of the water flood past breakthrough at the producing well is determined in the same manner as in the case where no initial fill-up of the reservoir is needed.

Since each channel was divided into N number of equal volume cells, the average volume of oil produced from each cell during the primary phase would be:

$$V_{o1} = \frac{V_p(S_{w_{br}} - S_{w_{IW}})}{N} \tag{20}$$

where V_p is the total pore volume of the channel.

During the subordinate phase, the Welge equation applies. The following form of the equation is convenient:

$$\overline{S}_{w_m} = S_{w_m} + \frac{1 - f_{w_m}}{f'_m} \tag{21}$$

where $\overline{S}_w$ = average water saturation in a channel, and the index m refers to the number of increments after initial breakthrough of the water at the producing well. As in the linear calculation of the frontal displacement mechanism, the oil produced during each separate step of the subordinate phase may be calculated as follows:

$$V_{o_{j=m+n}} = V_p(\overline{S}_{w_{m+1}} - \overline{S}_{w_m}) \tag{22}$$

The time for each step can be readily determined from the following equation:

$$t_j = \frac{V_{o_j}}{q_{o_{j-\frac{1}{2}}}} \tag{23}$$

Calculations are done for each channel in the pattern flood. The overall performance is determined by summing the production from each channel at the same time. It is possible to position the flood front at any given total time in a similar fashion. Notice that in this calculation technique, an areal coverage factor has not been used. Such a procedure is not required, since breakthrough at the end of particular cells in each channel defines the position of the front.

Higgins and Leighton[5] have presented the details of a computer program for a frontal displacement calculation for the five-spot water flood. The same authors[6,7] have also extended the above-described technique by presenting a method for predicting performance of five-spot water floods, where vertical stratification of the reservoir exists. The method assumes that the same relative permeability-saturation relationships apply to all layers, and that only the absolute permeability has changed. Then too, no vertical cross-flow between the channels is permitted. This would be the same as assuming that a thin impermeable membrane exists between each of the layers in the reservoir. The seriousness of such an assumption should be carefully considered in the performance calculations used to treat the stratified reservoir flow problem.

The calculation of the performance of a water flood in a pattern well network by the Higgins *et al.* method is best achieved by means of a computer, if one is available, due to the rather large number of repetitive calculations required. This becomes even more important if the stratified problem is to be treated. It should be evident that all well networks, whether regular or irregular, could be treated by means of a potentiometric study, coupled with a determination of the applicable shape factors. Such a technique has been described in detail by Higgins *et al.*[2].

Gaucher and Lindley[8] have published information on a scaled model study of water-flooding performance in a stratified five-spot well network. The models were scaled for capillary, gravitational and viscous pressure gradients, but not for the effect of viscous fingering. Such a scaling has apparent importance where the fluid mobility ratios are adverse. The purpose of the study was to determine the effect on water flood behavior where fluid rates were varied, and where the mobility ratio was changed through the use of fluids of different viscosities. The stratified models were constructed of unconsolidated sand, consisting of two laterally continuous constant thickness layers of different permeabilities. The models could be inverted to change the relative position of the more permeable layer. Figure 7-6 presents the results of one test series at two different producing rates (steady-state system), where the more permeable layer was on the bottom of the sand pack, and the oil viscosity was 2.17 cp. The effect of gravitational and imbibition forces, at the slower injection-producing rates, is clearly evident. Note too that the two sand layers were in intimate contact—i.e., with no barrier between them, as is necessary in most of the analytical calculations of performance in stratified systems. Note too that at least for this particular range of reservoir and fluid properties, no marked cross-flow between

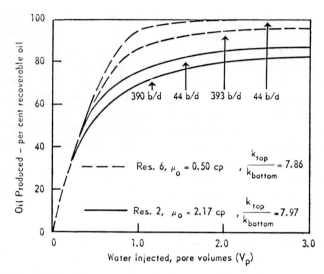

Figure 7-6. Prototype performance in a modelled five-spot water flood where two strata are in intimate contact (after Gaucher and Lindley[8]).

the two sand layers was observed. This vindicates to some extent the mathematical approaches where no accounting is made for cross-flow between the sand layers. Where there is a variation in rock wettability between layers, and where many thin layers in intimate contact exist, the performance of stratified reservoirs would still be conjectural.

Water Flooding Performance Calculations—The Viscous Fingering Method

Where the fluid mobility ratio is adverse, i.e., where the driving fluid is more mobile than the driven fluid, the interface between the fluids becomes unstable, and what has been termed as "viscous fingering" by Engelberts and Klinkenberg[9], can occur. This stability is of concern when calculations using the Buckley-Leverett and Welge methods are used, since these depend upon the relative permeability relationships applying—even though the front is unstable, and fingers of the displacing fluid have developed. While no specific work has been published, it is reasonable to assume that the Buckley-Leverett and Welge approaches to immiscible displacement calculation techniques are valid for a favorable mobility ratio, and for those adverse mobility ratios up to, and perhaps somewhat exceeding, a value of ten. Very little published information exists where the Buckley-Leverett approach has been used above a mobility ratio of ten; and then, no laboratory data under rigorously scaled conditions are available to adequately justify the resulting analytical answers.

It is in the adverse fluid mobility range of operations where the viscous fingering approach to performance calculations finds its application. van Meurs[10] has published laboratory results where transparent three-dimensional models were used to study displacement of oil by water in a linear system and a five-spot well pattern for a range of viscosity ratios. Viscous fingering was plainly evident, as illustrated by Figure 7-7, for the linear system of two layers of differing absolute permeabilities. Notice that very little evidence exists of cross-flow between the layers in this instance.

Table 7-3 presents a summary of laboratory results for a homogeneous and a stratified formation for two values of oil-to-water viscosity ratio. As has been observed in secondary recovery projects, increased volumes of injection water are necessary in stratified formations to obtain the equivalent percentage oil recovery, as compared to that obtainable from homogeneous pay sections.

TABLE 7-3.   Oil Recovery in Per cent of Pore Volume
(after van Meurs[10]).

	$\mu_o/\mu_w = 1$		$\mu_o/\mu_w = 80$	
	Homogeneous Formation	Stratified Formation	Homogeneous Formation	Stratified Formation
Breakthrough recovery	85	50	12	6.6
Ultimate recovery at 98 per cent water cut	88	64	53	47
Percent pore volume injected at 98 per cent water cut	150	250	700	700

van Meurs and van der Poel[11] have presented a theoretical description of the water-oil displacement process, where viscous fingering is involved. The approach is based upon observations made using transparent models under scaled conditions. The following analytical development closely parallels that published by the authors.

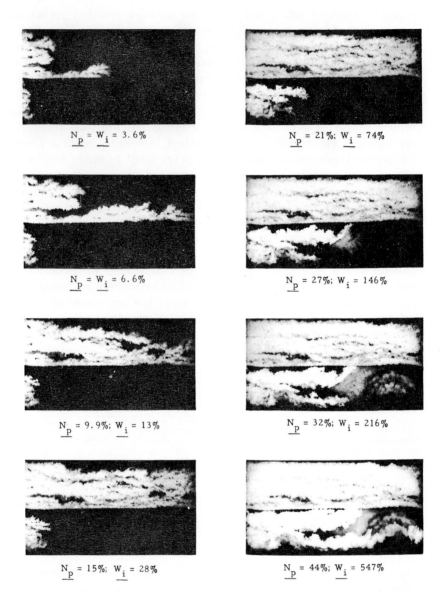

$N_p = W_i = 3.6\%$

$N_p = 21\%; \ W_i = 74\%$

$N_p = W_i = 6.6\%$

$N_p = 27\%; \ W_i = 146\%$

$N_p = 9.9\%; \ W_i = 13\%$

$N_p = 32\%; \ W_i = 216\%$

$N_p = 15\%; \ W_i = 28\%$

$N_p = 44\%; \ W_i = 547\%$

Figure 7-7. Views of linear model of stratified reservoir for an oil to water viscosity ratio of 80 (after van Meurs[10]).

In order to make an analytical solution possible where viscous fingering is involved, the following idealized system is preassumed as depicted by Figure 7-8:

(1) the flow can be divided into three parts, area 1 where oil flows unhindered, area 2 where water flows unhindered, and area 3 where both oil and water are immobile.

(2) the sizes of areas 1, 2 and 3 can be characterized by their oil and water saturations. If the immobile water of area 3 is expressed as S_{wm}, and the residual oil saturation as S_{or}, then the size of area 2 is equal to $S_w - S_{wm}$. In like fashion, the size of area 1 may be written as $S_o - S_{or}$, or $1 - S_{or} - S_w$.

(3) steady-state conditions prevail.

(4) the system is horizontal and homogeneous as to rock properties.

(5) the fractions of water and oil flowing in the system, at any given cross-section, are independent of time and place.

Since areas are being represented by saturations, the water flow rate through a unit cross-sectional area containing a number of fingers could be represented by the following equation:

$$q_w = u_w (S_w - S_{wm}) \qquad (24)$$

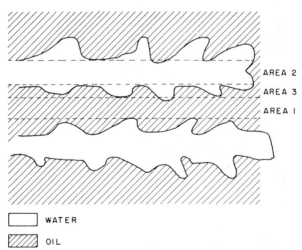

AREA 2

AREA 3

AREA 1

☐ WATER

▨ OIL

Figure 7-8. Idealized picture of viscous fingering showing the various flow areas (after van Meurs *et al.*[11]).

where u_w is the volumetric velocity or flow rate per unit area of water in area 2. Since only water is flowing in area 2, the volumetric velocity, u_w, may be defined in terms of Darcy's equation for single-phase flow:

$$u_w = \frac{-k}{\mu_w} \frac{\partial p_w}{\partial x} \tag{25}$$

in which p_w is the pressure in the water phase. Similarly, equations for the oil flow rate and velocity may be developed for area 1:

$$q_o = u_o (1 - S_{or} - S_w) \tag{26}$$

$$u_o = \frac{-k}{\mu_o} \frac{\partial p_o}{\partial x} \tag{27}$$

Notice that the absolute permeability is used in equations (25) and (27), since only oil and water flow in areas 1 and 2, respectively. Then too, since it is assumed that the viscous fingers occur on a magroscopic, rather than a microscopic scale, capillary pressure should be essentially constant, i.e., $p_o - p_w$ = constant. Then,

$$\frac{\partial p_o}{\partial x} - \frac{\partial p_w}{\partial x} = 0 \tag{28}$$

In view of equation (28), equations (25) and (27) can be combined:

$$u_w = u_o M \tag{29}$$

where M is the oil-water viscosity ratio. Assuming that the oil and water phases are incompressible, i.e., that steady-state conditions exist, an equation for the fractional flow of water can be developed using equations (24), (26), and (29):

$$f_w = \frac{q_w}{q_t} = \frac{M(S_w - S_{wm})}{(M - 1)(S_w - S_{wm}) + B} \tag{30}$$

where B is a constant equal to $1 - S_{or} - S_{wm}$, and q_t is the total fluid throughput. Since steady-state conditions exist, the equation of continuity for the water phase can be written as:

$$\frac{\partial q_w}{\partial x} + \frac{\phi \partial S_w}{\partial t} = 0 \tag{31}$$

or as:

$$\frac{\partial f_w}{\partial X} + \frac{\partial S_w}{\partial W_i} = 0 \tag{32}$$

where cumulative water injected $W_i = \frac{\int q \, dt}{\phi L}$. Here, L is the length of the linear system, and $X = \frac{x}{L}$, the dimensionless distance from the beginning point of the displacement. Since the fraction of water flowing is a function of saturation only in equation (30), equation (32) may be rewritten as:

$$\frac{df_w}{dS_w} \cdot \frac{\partial S_w}{\partial X} + \frac{\partial S_w}{\partial W_i} = 0 \tag{33}$$

Differentiating equation (30) provides an expression for the change of f_w, with respect to S_w:

$$f'_w = \frac{df_w}{dS_w} = \frac{MB}{[(M-1)(S_w - S_{wm}) + B]^2} \tag{34}$$

Equation (33) is a linear, first order differential equation, of which a general solution is:

$$X = \omega(S_w) + \frac{df_w}{dS_w} W_i \tag{35}$$

the term $\omega(S_w)$ may be omitted, since at time zero, no water has been injected, and the initial water front is assumed undisturbed (no fingers yet formed) and is perpendicular to the flow direction. An expression for the dimensionless position of the flood front, X, can now be written by replacing the differential of equation (35) by its value in equation (34):

$$X = \frac{MBW_i}{[(M-1)(S_w - S_{wm}) + B]^2} \tag{36}$$

Equation (36) can also be rearranged to provide a means of calculating the water saturation in the region where both oil and water are flowing:

$$S_w = S_{wm} - \frac{B}{M-1} + \frac{\sqrt{MB}}{M-1} \sqrt{\frac{W_i}{X}} \tag{37}$$

Of necessity, water saturation values calculated with equation (37) are limited to a maximum of $1 - S_{or}$. Substitution of this into equation (36) results in an expression for this critical distance, X_1, of:

$$X_1 = \frac{W_i}{BM} \qquad (38)$$

Up to this point, the displacement process has been described analytically as being one-dimensional, i.e., flow occurring only in the x-direction. Certainly as the fingers swell and as vertical protrusions form, a certain vertical pressure gradient is in evidence. This behavior requires that the water saturation, S_w, exceed the immobile water saturation, S_{wm}. Notice too that equation (24) breaks down when S_w is less than S_{wm}. Apparently then, there should be a critical distance, X_c, at which the water saturation, as described analytically by equation (37), is larger than S_{wm}. Figure 7-9 gives a schematic representation of water saturations through the linear system prior to breakthrough. The actual cutoff in a reservoir could be expected to be somewhat more gradual. To develop an expression for this critical distance, we can write an equation for the critical velocity, v_c, at which the critical saturation, S_c, is propagated:

$$v_c = \frac{q_t}{\phi S_c} \qquad (39)$$

The critical velocity may also be written as:

$$v_c = \left(\frac{dx}{dt} \right)_{S_w = S_c} = \frac{q_t}{\phi} \left(\frac{dX}{dW_i} \right)_{S_w = S_c} \qquad (40)$$

since $x = XL$ from which $dx = L\,dX$ and $W_i = \dfrac{\int q\,dt}{\phi L}$, which results in

$$dt = \frac{\phi L\, dW_i}{q}$$

Or from equation (35):

$$v_c = \frac{q_t}{\phi} (f_w')_{S_w = S_c} \qquad (41)$$

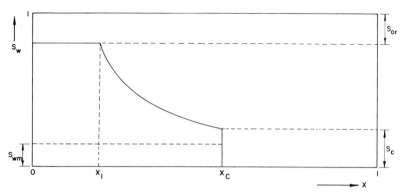

Figure 7-9. Schematic representation of water saturations through the linear system prior to breakthrough (after van Meurs *et al.*[11]).

Equations (39) and (41) may now be combined to yield:

$$\frac{q_w}{q_t} = f_w = S_c \, (f'_w) s_{w = S_c} \tag{42}$$

Substituting values for f_w and f'_w, in equations (30) and (34), respectively, and into equation (42), results in an expression for the critical water saturation at the cut-off point, X_c:

$$S_c = S_{wm} + \sqrt{\frac{B \, S_{wm}}{M - 1}} \tag{43}$$

Equation (36) may now be modified using equation (43) to yield an equation for the critical distance:

$$X_c = \frac{MBW_i}{[B + \sqrt{B(M - 1) \, S_{wm}}]^2} \tag{44}$$

Breakthrough of the displacing phase, water, occurs when the critical water saturation, S_c, arrives at the outlet end, i.e., when $X = 1$. At this time, the cumulative oil produced in reservoir units, N_{pb}, is equal to the water injected to breakthrough, $(W_i)_b$. Substituting 1 for X_c and N_{pb} for W_i in equation (44) provides an analytical expression for the oil recovered at breakthrough in terms of the residual oil saturation, the oil-water viscosity ratio and the immobile water saturation:

$$N_{pb} = \frac{[\sqrt{(M - 1) B \, S_{wm}} + B]^2}{MB} \tag{45}$$

In order to predict flooding performance, it is convenient to develop a specialized equation for the calculation of the water cut at water breakthrough at the outlet end. Equations (30) and (37) may be modified for the case where $X = 1$ at breakthrough, and the water saturation and water cut is represented as S_{wl} and f_l, respectively:

$$f_l = \frac{M(S_{wl} - S_{wm})}{(M - 1)(S_{wl} - S_{wm}) + B} \qquad (30a)$$

$$S_{wl} = S_{wm} - \frac{B}{M - 1} - \sqrt{\frac{MBW_i}{M - 1}} \qquad (37a)$$

Substituting equation (37a) into equation (30a) and simplifying results in the final equation for water cut at breakthrough:

$$f_l = \frac{M}{M - 1}\left[1 - \sqrt{\frac{B}{MW_i}}\right] \qquad (46a)$$

Reference to Figure 7-9 permits the writing of the following equation for oil recovery after breakthrough:

$$N_p = X_1(1 - S_{or}) + \int_{X_1}^{1} S_w \, dX \qquad (46)$$

where it is apparent that the second term on the right-hand side is the amount of water gained by the system at the expense of the oil saturation. It should be noted that the calculation ignores the effect of an initial connate water saturation in the porous media. This can be thought of as being a part of the rock matrix for the purpose of these calculations. Equation (46) may be modified to a more usable form by substituting in values for X_1 and S_w, from equations (37) and (38), respectively. Some manipulation of the factors yields the following equation form:

$$N_p = S_{wm} + \frac{1}{M - 1}(2\sqrt{MBW_i} - B - W_i) \qquad (47)$$

Equation (47) is valid until $X_1 = 1$, or $W_i \leq MB$. Should W_i become larger than MB, the recovery of oil reaches the limiting value of $1 - S_{or}$. This means that all the movable oil would have been swept from the linear system, as modelled analytically by the viscous fingering theory. Where the oil-water viscosity ratio is large, a sizeable volume of injected water would be required to effect this high recovery where the displacement is with an immiscible fluid.

van Meurs and van der Poel[11] determined by a comparison of laboratory and analytical data that the above theory worked very well for systems where the oil-water viscosity ratio was high, and when the immobile water saturation was taken to be 0.15. The authors did find that the experimental breakthroughs occurred sometimes slightly earlier than the above viscous theory would have predicted. This was believed to be due to the fact that it is nearly impossible to pack a truly homogeneous column in the laboratory.

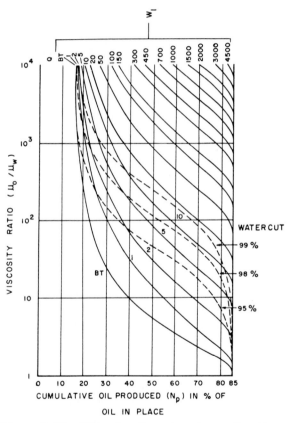

Figure 7-10. Ultimate oil produced as a function of the oil-water viscosity ratio, with the water injected and the water cut as parameters for $S_{or} = S_{wm} = 0.15$ (after van Meurs[11]).

Any inhomogeneities in the system would result in an earlier breakthrough, as was indeed observed to be the case. After breakthrough, agreement between the experimental and analytical techniques was within experimental error.

Examination of equation (47) indicates that oil recovery after breakthrough is a function only of the immobile water saturation, the residual oil saturation, the oil-water viscosity ratio and the water injected. The performance may be readily generated by varying the amount of water injected. Equation (47) may be modified for graphing purposes through the use of equations (30) and (37):

$$N_{pu} = S_{wm} + B \left\{ \frac{M - f_l^2 (M - 1)}{[M - f_l (M - 1)]^2} \right\} \tag{48}$$

where f_l is the water cut at the economic limit, and N_{pu} is the ultimate oil recovery. Figure 7-10 presents equation (48) in graphical form, where S_{or} and S_{wm} have both been fixed at 0.15. It is interesting to note the tremendous difference in the ultimate oil produced when the oil-water viscosity ratio approximates 100, as the abandonment water cut is varied from 95 to 99 per cent. Where water injection and produced water costs can be kept low, water flooding of a reservoir sand containing a viscous oil can result in good oil recoveries. Large volumes of injection water will have to be handled however. The water quantities to be injected will be further increased if the reservoir is stratified.

PROBLEM:

Given a horizontal linear system containing oil, and having a cross-sectional area of 1 square foot and a length of 1000 feet, calculate the performance under water flooding where the following data apply:

$$q_t = 10 \text{ cu ft/hr} \qquad S_{wm} = 0.15$$
$$\mu_o = 10 \text{ cp} \qquad B_o = 1.00$$
$$\mu_w = 1 \text{ cp} \qquad M = \mu_o/\mu_w = 10$$
$$S_{wi} = 0.20 \qquad S_{or} = 0.20$$
$$\phi = 0.25$$

SOLUTION:

(1) Total oil in the system $= N = 1.00 \, (1000)(0.25)(1 - 0.20) = 200$ cu ft

(2) $B = 1 - S_{or} - S_{wm} = 1 - 0.20 - 0.15 = 0.65$

(3) $S_c = S_{wm} + \sqrt{\dfrac{B \, S_{wm}}{M - 1}} = 0.254$

(4) Calculate the oil produced at breakthrough:

$$N_{pb} = \frac{[B + \sqrt{(M - 1)S_{wm}B}\,]^2}{MB} = 0.39$$

(5) At breakthrough, the position of the critical distance where all the movable oil has been swept out is:

$$X_1 = \frac{W_i}{BM} = \frac{0.39}{0.65\,(10)} = 0.06 \text{ or } 60 \text{ ft from the inlet end}$$

(6) Calculate the water cut at breakthrough:

$$f_1 = \frac{M}{M - 1} \left[1 - \sqrt{\frac{B}{M\,W_i}}\right] = 0.657$$

(7) The ultimate recovery at a water cut of 0.98:

$$N_{pu} = S_{wm} + B \frac{[M - f_1^2\,(M - 1)]}{[M - f_1\,(M - 1)]^2} = 0.740$$

(8) To include the time relationship with the water injected and the oil produced:

$$W_i = \frac{\int q\,dt}{\phi L} = \frac{10\,t}{0.25\,(1000)} = 0.04\,t$$

It would then be a direct problem to generate the performance curves as a function of time or of cumulative oil produced.

The viscous fingering approach to predicting performance in a linear system has the advantage of requiring unsophisticated data, such as the immobile water saturation, the residual oil saturation, the oil-water viscosity ratio, the throughput rates and the dimensions and porosity of the porous media. Where the system is stratified, the assumption can be made that no cross-flow occurs and a calculation made for each layer. Where pattern water flooding is being studied, the technique used by Higgins and Leighton[1] for frontal displacement can be readily modified[12] for the viscous fingering calculation technique. The shape factors can be used directly.

Rachford[13] has recently published an analytical study of the instability which can occur in the flooding of oil from water-wet porous media containing connate water. It was found that fingering might not occur at the front as postulated by the Buckley-Leverett frontal displacement theory, but could occur at a saturation behind the front. Rinehart[14] has compared the performance curves obtained from the viscous fingering and Buckley-Leverett approaches for a wide range of mobility ratios. It was found that the two techniques provide answers that agree fairly well at adverse mobility ratios in excess of 5. The viscous fingering approach apparently did not model the reservoir system at mobility ratios less than 5.

Land[15] has shown that the viscous fingering equations of van Meurs *et al.*[11] can be derived directly from the Buckley-Leverett approach. Figure 7-11 shows the oil and water relative permeabil-

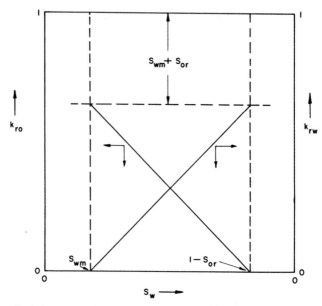

Figure 7-11. Relative permeability curves preassumed for the viscous fingering immiscible displacement theory.

ity curves which are preassumed in the development of the viscous fingering theory. It is apparent from the figure that the equation for relative oil permeability is:

$$k_{ro} = 1 - S_{or} - S_w \qquad (49)$$

while that for relative permeability to water flow is:

$$k_{rw} = S_w - S_{wm} \qquad (50)$$

If capillary pressure gradients and reservoir dip are taken to be zero, the Buckley-Leverett fractional flow equation can be written as:

$$f_w = \frac{k_{rw}/\mu_w}{k_{rw}/\mu_w + k_{ro}/\mu_o} \tag{51}$$

Substituting values from equations (49) and (50) into equation (51), and noting that in the viscous fingering development that M is equal to μ_o/μ_w, equation (30) can be developed. In a similar fashion, all the viscous fingering equations can be derived, based upon the Buckley-Leverett frontal advance equation, and the Welge equation for fluid saturations behind the displacement front after displacing fluid breakthrough.

It is apparent that the differences between the viscous fingering and the Buckley-Leverett approaches to the modelling of immiscible displacement at adverse mobility ratios is not whether fingering occurs, or if there is a frontal advance; it is simply the difference obtained by two different relative permeability curves. The viscous fingering approach does provide easily calculated reasonable answers for performance under adverse mobility conditions, and is useful for this particular reason. It is really a specialized form of the Buckley-Leverett frontal advance development, for one specific set of relative permeability curves that is a linear function of saturation.

Water Flooding Performance Calculations—Stratified Reservoirs— Stiles Method

A goodly number of water-flooding performance calculation techniques have appeared in the petroleum industry's technical literature during the past two decades. One of the methods that has been used rather widely due to its simplicity is that credited to Stiles[16]. The method indicates that the following conditions can be assumed for purposes of simplifying the calculations:

(1) The formation is composed of a number of strata of constant thickness continuous between all wells.

(2) No segregation of fluids in the layers or cross-flow between the layers occurs.

(3) The displacement occurs in a piston-like manner, with no oil produced from behind the displacement front.

(4) The system is linear, has constant porosity, constant fluid saturations and the same relative permeability to oil ahead of the front and to water behind the front in all layers, i.e., with the ex-

ception of absolute permeability, all layers have the same fluid and rock characteristics.

(5) The position of the flooding front in any given layer is directly proportional to the absolute permeability of that layer.

(6) The producing water cut depends upon the total reservoir capacity (kh), which has experienced breakthrough.

The method does, of course, assume that a frontal displacement will occur. This means that the water saturation will be sufficiently low so that continuous water permeability to the producing wells will not already exist. The primary attraction of the Stiles calculation technique is, as previously stated, its relative simplicity. Calculations may be conveniently done without the aid of a computer, and engineering answers obtained expeditiously if the above-mentioned assumptions can be tolerated. Substantial agreement with field-flooding results have been obtained where the fluid mobility ratio is favorable, or not greatly in excess of one.

In view of the assumptions of the Stiles calculation technique, it is convenient to divide the strata into layers of constant thicknesses, h_1, h_2, etc., and absolute permeabilities, k_1, k_2, etc., and arrange them in order of decreasing permeability. Such a measure simplifies the calculation procedure and the development of equations to describe the flood performance. It should be evident that the topmost bed would flood out first, the second bed second, and so on. The following analytical development will follow fairly closely that given by Craft and Hawkins[17].

When the jth bed has just experienced water breakthrough at the producing well, then the fraction of the recoverable oil that has been produced will be equivalent to the fraction of the pay section flooded out, plus the flooded parts of the strata still contributing to the production of oil. This may be written analytically as:

$$R = \frac{h_j}{h_t} + \frac{k_k}{k_j} \frac{\Delta h_k}{h_t} + \frac{k_1}{k_j} \frac{\Delta h_1}{h_t} + \ldots + \frac{k_n}{k_j} \frac{\Delta h_n}{h_t}$$

$$= \frac{h_j}{h_t} + \frac{1}{k_j h_t} [k_k \Delta h_k + k_1 \Delta h_1 + \ldots + k_n \Delta h_n] \tag{52}$$

or

$$R = \frac{h_j}{h_t} + \frac{1}{k_j h_t} [C_t - C_j] \tag{53}$$

where

h_t = total thickness of the strata being flooded.

$(k_k/k_j) \times \Delta h_k$ = oil recovered from the kth bed.

$h_j = \Delta h_1 + \Delta h_2 + \ldots + \Delta h_j$, the total feet of formation flooded out with water.

C_t = total capacity of the strata being flooded.

C_j = capacity of the formation which has been totally flooded out with water.

Notice in equation (53) that the first term on the right-hand side yields the recovery from those strata completely flooded out, while the second term gives the contribution of the strata still producing oil.

The producing water fraction at surface conditions may be determined by means of Darcy's law for flow in porous media, where the strata all have the same length and width:

$$f'_w = \frac{q_w}{q_w + q_o} = \frac{\dfrac{\Sigma(k\,k_{rw}\,\Delta h_w)}{B_w\,\mu_w}}{\dfrac{\Sigma(k\,k_{rw}\,\Delta h_w)}{B_w\,\mu_w} + \dfrac{\Sigma(k\,k_{ro}\,\Delta h_o)}{B_o\,\mu_o}} \tag{54}$$

Equation (54) may be simplified by rearranging, and letting:

$$A = \frac{k_{rw}}{k_{ro}} \times \frac{\mu_o}{\mu_w} \times \frac{B_o}{B_w} \tag{55}$$

Equation (54) then becomes:

$$f'_w = \frac{A\Sigma(k\,\Delta h_w)}{A\Sigma(k\,\Delta h_w) + \Sigma(k\,\Delta h_o)} \tag{56}$$

In equation (56), notice that the term $[\Sigma\,k\,\Delta h_w]$ represents the formation capacity producing water, C_j, while the term $[\Sigma\,k\,\Delta h_o]$ is the formation capacity still producing oil, $C_t - C_j$. The most convenient form of the equation for the producing surface water cut then is:

$$f'_w = \frac{AC_j}{AC_j + (C_t - C_j)} \tag{57}$$

Table 7-4 presents the calculations for water flooding performance for an enclosed five-spot well pattern where stratification exists. The data are that due to Stiles[16], while the tabulation is similar to that presented by Craft and Hawkins[17]. Columns (1) and (2) present the data for a stratified reservoir, where for convenience the strata have been taken as being one foot thick. Such

TABLE 7-4. Calculation of Water Flooding Performance in an Enclosed Five-spot Well Pattern by the Stiles Method.

(1) h_j ft	(2) k_j md	(3) $\sum k_j \Delta h_j$ md-ft	(4) $h_j k_j$ md-ft	(5) R	(6) $N_p = 47,600 \times$ (5), STB	(7) ΔN_p STB	(8) f_w res. cond.	(9) q_{sc} STB/day	(10) $t = 93 + \sum[(7) \div (9)]$	(11) W_i surf. cond.	(12) f_w surf. cond.
1	776	776	776	0.226	10,750	10,750	0.000	93.2	209	20,900	0.204
2	454	1230	908	0.361	17,180	6,430	0.192	75.3	295	29,500	0.312
3	349	1579	1047	0.449	21,400	3,220	0.297	65.5	344	34,400	0.391
4	308	1887	1232	0.495	23,570	2,170	0.373	58.4	381	38,100	0.457
5	295	2182	1475	0.511	24,350	780	0.439	52.3	396	39,600	0.517
6	282	2464	1692	0.526	25,040	690	0.499	46.7	411	41,100	0.573
7	273	2737	1911	0.537	25,560	520	0.555	41.5	424	42,400	0.624
8	262	2999	2096	0.549	26,110	530	0.607	36.6	436	43,600	0.672
9	228	3227	2052	0.590	28,070	1,960	0.656	32.1	497	49,700	0.713
10	187	3414	1870	0.651	31,000	2,930	0.697	28.2	626	62,600	0.745
11	178	3592	1958	0.667	31,750	750	0.731	25.1	656	65,600	0.775
12	161	3753	1932	0.697	33,180	1,430	0.762	22.2	720	72,000	0.801
13	159	3912	2067	0.700	33,300	120	0.789	19.7	726	72,600	0.827
14	148	4060	2072	0.719	34,200	900	0.816	17.1	778	77,800	0.850
15	127	4187	1905	0.758	36,100	1,900	0.841	14.8	906	90,600	0.870
16	109	4296	1744	0.798	38,000	1,900	0.862	12.9	1053	105,300	0.887
17	88	4384	1496	0.857	40,750	2,750	0.879	11.3	1296	129,600	0.900
18	87	4471	1566	0.860	40,900	150	0.893	10.0	1311	131,100	0.913
19	87	4558	1653	0.860	40,900	0	0.907	8.7	1311	131,100	0.926
20	77	4635	1540	0.887	42,250	1,350	0.921	7.4	1493	149,300	0.937
21	71	4706	1491	0.903	43,000	750	0.933	6.2	1614	161,400	0.948
22	62	4768	1364	0.929	44,200	1,200	0.944	5.2	1845	184,500	0.957
23	58	4826	1334	0.941	44,800	600	0.953	4.4	1981	198,100	0.965
24	54	4880	1296	0.952	45,300	500	0.962	3.5	2124	212,400	0.973
25	50	4930	1250	0.962	45,780	480	0.971	2.7	2302	230,200	0.980
26	47	4977	1222	0.968	46,050	270	0.978	2.1	2431	243,100	0.986
27	47	5024	1269	0.968	46,050	0	0.985	1.4	2431	243,100	0.993
28	35	5059	980	0.981	46,700	650	0.992	0.7	3359	335,900	0.998
29	16	5075	464	1.000	47,600	900	0.998	0.2	7859	785,900	1.000

data are often available from core analyses. It is interesting to note that because of the rather large range of permeability values, ten feet of the section to be flooded has 67 per cent of the formation, or producing capacity. The fraction of the recoverable oil produced as each stratum floods out is presented in column (5), where R is calculated from equation (53). The oil recovered in stock tank barrels as the strata flood out may be determined from volumetric considerations of the pore space, reservoir data and residual oil saturation information expected, after complete flushing with water. The residual oil saturation is commonly obtained from flood pot tests, from pilot water floods, from information on similar fields producing a similar crude, from frontal advance calculations and relative permeability relationships, or from published correlations. If the porosity is taken as 19 per cent, the five-spot reservoir volume as 100 acre-ft, the connate water saturation as 24 per cent, and the oil formation volume factor at initial reservoir conditions as 1.073 bbls/STB, then the original oil-in-place at surface conditions would be:

$$N = \frac{7758 \times \phi \times V \times (1 - S_w)}{B_{oi}} =$$

$$\frac{7758 \times 0.19 \times 100 \times 0.76}{1.073} = 104,000 \text{ STBO}$$

Of considerable interest to the engineer when considering a field for possible secondary recovery is the remaining oil saturation after primary depletion, or at some point during the primary producing history. Where the primary producing phase has been completed, producing history should permit the calculation of the remaining oil saturation. Where no good producing information is available, and the field producing mechanism was depletion drive, a published correlation by Wahl *et al.*[18] will usually permit a determination of the oil produced down to atmospheric conditions, from which the residual oil saturation can be readily calculated. Assuming that the residual oil saturation in this problem after primary depletion approximates 59 per cent, and that the residual oil saturation after water flooding will be 21 per cent, the recoverable or movable oil can be calculated:

$$\text{Recoverable oil} = 7758 \times \phi \times \text{A.F.} \times \left[\frac{S_{op} - B_{oi} S_{wi}}{B_{oi}} \right]$$

$$= 7758 \times 0.19 \times 100 \times \left[\frac{0.59 - 1.073 \, (0.21)}{1.073} \right]$$

$$= 50,100 \text{ STBO}$$

If the system were linear, then at breakthrough of the last and least permeable stratum, the recovery would be 50,100 STBO, and the producing water cut would be 100 per cent. In a five-spot flood, however, the coverage will be less than 100 per cent. Laboratory experiments have shown that the areal sweep efficiency, or coverage, is largely a function of the water-to-oil mobility ratio and the limiting water cut, which is in turn determined by the producing and treating costs. If the relative permeability of the water behind the displacing front is 0.20, the relative permeability of the oil ahead of the front is 0.80, and the oil and water viscosities at reservoir pressure and temperature equal to 4.34 and 0.82 cp, respectively, the mobility ratio is:

$$M = \frac{k_{rw}}{k_{ro}} \cdot \frac{\mu_o}{\mu_w} = \frac{0.20}{0.80} \times \frac{4.34}{0.82} = 1.32$$

If the economic limit is taken as occurring at a water cut of 0.90, Figure 4-7 of Chapter 4 yields an areal sweep efficiency of 95 per cent. The recoverable oil will then be 95 per cent of 50,100 STB, or 47,600 STB. Column (6) may now be calculated. Column (7) is the difference in the cumulative oil produced as given in column (6), or can be considered as the oil produced in the interval between the watering out of individual stratum. The water cut at the producing well at reservoir conditions is given in column (8), and is calculated from equation (58):

$$f_w = \frac{M \, C_j}{M \, C_j + C_t - C_j}, \quad M = \frac{k_{rw}}{k_{ro}} \cdot \frac{\mu_o}{\mu_w} \tag{58}$$

Equation (58) is developed in the same manner as equation (54), except that the flow rates for oil and water are those at reservoir conditions. The surface oil production rate can be readily determined from the following equation:

$$q_{sc} = \frac{1 - f_w}{B_o} \times q_{\text{res. cond.}} \tag{59}$$

since the fraction of the total reservoir fluid that is produced is $1 - f_w$. If the steady-state producing and injection rate is 100

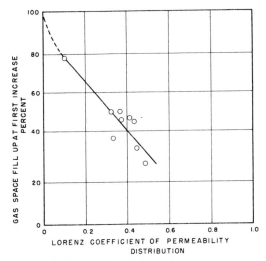

Figure 7-12. Gas space fill-up at first oil production increase versus the Lorenz coefficient of permeability distribution from field data (after Schauer[19]).

reservoir barrels per day, then the oil production in reservoir barrels is $100 \times (1 - f_w)$. The stock tank oil producing rate is obtained by dividing by the oil formation factor, B_o, corresponding to the prevailing reservoir pressure and temperature. Column (9) summarizes the calculations of oil-producing rate that is realized as each stratum goes to water production.

If the reservoir being considered for water flooding has been produced to depletion by primary means, usually a substantial gas saturation will exist throughout the oil producing section. Before oil is produced, it is usually necessary to inject a volume of water somewhat less than, or perhaps approaching the space occupied by the gas. The Stiles approach is to assume that fill-up will occur when a volume of water equaling the gas space has been injected. If the water injection rate is fixed, i.e., perhaps at 100 barrels per day, then the time for the first production of oil can be readily determined. If the reservoir is linear and homogeneous as to rock properties, such an approach would be realistic. Where large variations in the absolute permeability exist between stratum, the first oil production "kick" can be expected at an earlier time.

Schauer[19] has presented a method whereby the injection necessary for the first production increase may be calculated. The technique is illustrated by means of Figure 7-12, where the gas space fill-up at first increase is plotted, versus the Lorenz coefficient of permeability distribution. Figure 7-12 was developed from the actual performance of nine water floods under a variety of injection-producing well networks.

The Lorenz coefficient is determined from a permeability profile by a technique described by Schmalz and Rahme[20], and discussed by Pirson[21]. Figure 7-13 presents a plot of cumulative formation capacity as a fraction, versus the cumulative fractional formation thickness for the formation data of Table 7-4. The abscissa was constructed from column (1), where h_t is equal to 29 feet. Column (4) yields the data for the ordinate when reduced to a cumulative fraction. The term $h_t \bar{k}$ represents the total formation capacity or 5075 md-feet. In this instance, $\bar{k}$ is the average permeability of the section to be flooded. The Lorenz coefficient is defined as the shaded area of Figure 7-13, divided by the area ABC. If the formation were homogeneous, the Lorenz coefficient would be zero.

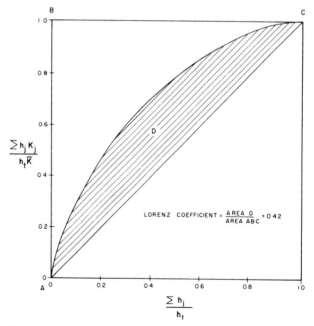

Figure 7-13. Plot of cumulative formation capacity fraction versus cumulative fractional formation thickness.

Complete heterogeneity would be represented by a Lorenz coefficient of unity. For the data of the example problem, the Lorenz coefficient is 0.42. Figure 7-12 indicates that the first production increase can be expected when a volume of water, equivalent to 37 per cent of the gas space, has been injected.

If the oil saturation after primary depletion is 59 per cent and the connate water saturation 24 per cent, the remaining 17 per cent of the pore space must be occupied by gas. The time to achieve the first production increase would then be:

$$t = \frac{7758 \times \phi \times V \times S_g \times F}{q_w} =$$

$$\frac{7758 \times 0.19 \times 100 \times 0.17 \times 0.37}{100} = 93 \text{ days}$$

where

F = fraction of gas space fill-up at first oil production increase.

This time to initial production increase is still probably pessimistic where a pattern flood is being treated. Reference to Figure 4-12 (p. 87) for the five-spot well network indicates that an areal sweep efficiency of 66 per cent could be expected at breakthrough, when the water-to-oil mobility ratio is 1.32. This would reduce the time to breakthrough to 61 days. A time to breakthrough of 93 days will be used, however, in order to partly compensate for the volumes of injection water, which will be ultimately needed. At this point, column (10) of Table 7-4 is self-explanatory.

Since steady-state conditions are assumed in the Stiles water flood performance calculation technique, the volume of water injected, W_i, as each stratum floods out, will be the injection rate, 100 barrels of water per day, times the time in days since the inception of the flood. Column (11) summarizes this calculation. The fraction of the total produced fluid, which is water at surface conditions, can be calculated by means of equation (57). Column (12) reproduces the results of such a calculation where the values obtained are those at the instant that a specific stratum contributes to the water production at the producing well. Notice that if the economic limit is reached at a surface water cut of 90 per cent, the cumulative oil production from the well pattern will be 40,750 STB of oil. This value does not check with the previous value of 47,600 STB of oil, calculated from an areal sweep efficiency of 95 per cent for the five-spot. The values should not agree, however, since in the actual flood, some 12 of the stratum have still to contribute water to the producing well. Had the economic limit been reached at 95 per cent surface water cut, the recovery would have been 43,500 STB of oil. In this instance, column (6) of Table 7-4 should have been calculated using a recoverable oil value of 48,900 STB, since the areal sweep efficiency would have been increased to 97.5 per cent. Figure 4-12 of Chapter 4 provides the areal sweep efficiencies in a five-spot well network as a function of the mobility ratio and the producing water cut.

Figure 7-14 presents the predicted performance of a five-spot well pattern for the data used where the Stiles method is employed. In actual practice, the oil producing rate curve would be smooth. The general shape of the performance curves is good and could be expected to reliably predict the performance of an enclosed pattern water flood, within the accuracy of the data and the assumptions of the calculation technique. Where the mobility ratio is not of the order of one, the Stiles calculation method probably does not adequately model the performance to be expected. This is particularly true when the mobility ratio is adverse. In this instance, Craft and Hawkins[17] have shown that the velocity accelerates in each bed as the displacement proceeds, with the most rapid acceleration occur-

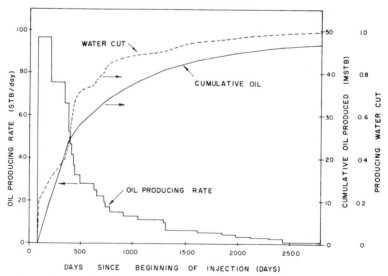

Figure 7-14. Predicted water flooding performance by the Craig-Stiles calculation technique.

ring in the most permeable beds. This results in earlier water breakthrough than the Stiles method would predict, and lower ultimate oil recoveries. When the mobility ratio is considerably less than one, the Stiles calculation method should give answers that are slightly pessimistic—at least relative to the water requirements and producing water cuts.

Comments on the Stiles Method. A number of modifications to the multilayer reservoir model for predicting reservoir performance, as originally described by Stiles[16], have been suggested in the petroleum literature. Ache[22] has presented a useful technique for including the effect of radial flow, since initially the direction of fluid flow from an injection well is radial. Essentially the modified technique divides the performance into three separate periods. The first period is radial flow, the second is a period of transition from radial to linear flow, and the third period assumes linear flow after reservoir fill-up has been achieved.

One of the disadvantages of the Stiles method is the requirement that, except for absolute permeability, all stratum have the same characteristics. Johnson[23] has presented a method for predicting water flood performance by a graphical presentation of the porosity and permeability distribution. Where sufficient data are at hand, the water saturation for each stratum may also be taken into account, so that the hydrocarbon pore volume may be directly related to the absolute permeability of that specific stratum. The technique also lends itself to the classification of a number of wells where there is considerable lateral, as well as vertical variation in reservoir characteristics. Of course, the Stiles method could be handled in a similar fashion where the permeability distribution is averaged on a statistical basis, such as suggested by Law[24], thereby taking into account the lateral variation in absolute permeability. Then too, the Stiles method could be applied to individual well patterns, and a composite performance readily determined—if sufficient data are available to justify such an expanded calculation.

Miller and Lents[25] have also presented a useful refinement of the Stiles technique where there are sufficient core analysis data on the wells to be flooded. The modification provides for the averaging of the permeability on those wells where core analyses are available for an entire field or lease at the same relative position in the formation. This method is advantageous, since such an averaging would tend to smooth out irregularities in data at specific wells. The performance calculation is the same as that presented by Stiles in all other respects.

Since the Stiles method is strictly applicable only when the water-to-oil mobility ratio is in the range of unity, another calculation technique should be used when mobility ratios are significantly different.

Water-Flooding Performance Calculations—Stratified Reservoirs—Simplified Dykstra-Parsons Method

Dykstra and Parsons[26] have presented a method for predicting the performance of a water flood in stratified reservoirs, in which a sizeable range of mobility ratio values may be tolerated. The method assumes constant values for the relative permeability to water and to oil behind and ahead of the displacement front, respectively, i.e., the displacement is piston-like, with no oil production from behind the front. The calculation technique makes use of a permeability variation plot to properly evaluate the effect of permeability upon the vertical sweep efficiency for a given mobility ratio and at a specified producing water-oil ratio. The technique, while developed for the linear system, may be readily adapted to the various well patterns through consideration of areal sweep efficiencies, as presented in Chapter 4.

Johnson[27] has presented a simplified graphical treatment of the Dykstra-Parsons water flood calculation method, which is particularly useful where quick estimates of recovery are desired. This may well be the case where certain properties with water flood potential are available, where several projects need quick comparison before detailed work is done and where properties must be evaluated for estate settlements or for the purpose of a bank loan. The technique is semi-empirical, and is based upon the correlation for four variables—the vertical permeability variation, V, the initial water saturation, S_w, the water-to-oil mobility ratio, M, and the fractional recovery of the oil-in-place at a specified water-oil ratio, R. The reader is referred to the original work of Dykstra and Parsons for the mathematical arguments and the details of the analytical development.

Figures 7-15, 7-16, 7-17 and 7-18 show plots correlating V, M, S_w and R, for water-oil ratios of 1, 5, 25, and 100, respectively. To use these graphs, values must be assigned to V, M, and S_w. The mobility ratio, M, is equal to $k_{rw}\mu_o/k_{ro}\mu_w$, where k_{rw} is the relative permeability to water behind the displacement front, where

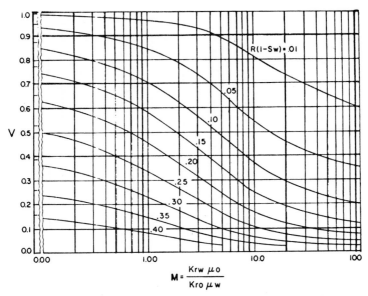

Figure 7-15. Correlation of permeability variation, mobility ratio, water saturation and fractional oil recovery for a producing water-oil ratio of 1 (after Johnson[27]).

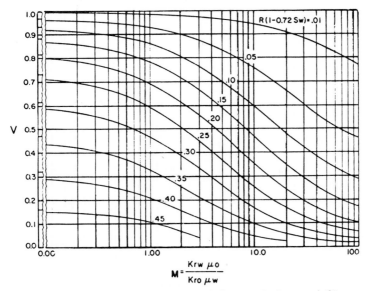

Figure 7-16. Correlation of permeability variation, mobility ratio, water saturation and fractional oil recovery for a producing water-oil ratio of 5 (after Johnson[27]).

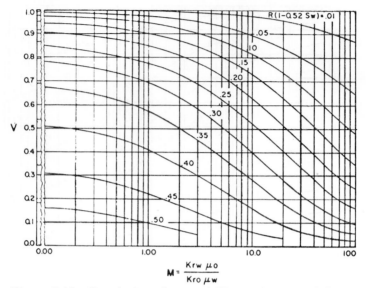

Figure 7-17. Correlation of permeability variation, mobility ratio, water saturation and fractional oil recovery for a producing water-oil ratio of 25 (after Johnson[27]).

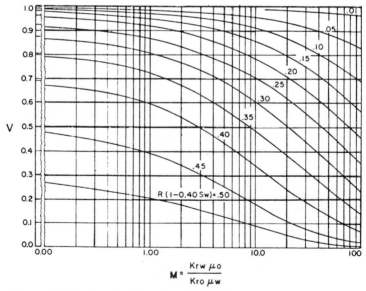

Figure 7-18. Correlation of permeability variation, mobility ratio, water saturation and fractional oil recovery for a producing water-oil ratio of 100 (after Johnson[27]).

only water flows (at the residual oil saturation), k_{ro} is the relative permeability to oil ahead of the front, where only oil is flowing (assumes zero gas saturation and an irreducible water saturation) and μ_o and μ_w are the viscosities of the oil and the flood water under reservoir conditions. The oil and water viscosities may be measured directly or taken from published correlations. The oil and water relative permeabil'ty values may be available from laboratory analyses, or can be calculated by methods outlined in Chapter 3. The water saturation, S_w, is usually available from log analysis, from information on a similar reservoir, or from published correlations.

It remains only to define the permeability variation in order that the fractional oil recovery at specified water-oil ratios may be determined. Pirson[28] has pointed out that the method of treatment of permeability variation, by Dykstra and Parsons, could just as well be applied to any other well property where a statistical measure of the non-uniformity of data has meaning. The procedure for determining the variance of a property is as follows:

(1) Arrange the values of the property in descending order of magnitude over equal intervals of perhaps 1 foot of formation.

(2) Calculate the per cent of the property magnitude exceeding each tabulated entry to obtain the "cumulated per cent greater than."

(3) Construct a plot of the data as presented in steps (1) and (2) on log probability graph paper as illustrated in Figure 7-19.

(4) Construct the best fit straight line through the plotted data. The "variance," V, or the permeability variation for the purposes of our discussion, is defined as follows:

$$V = \frac{\text{property magnitude at 84.1 cumulative per cent} - \text{median value}}{\text{median value}}$$

$$(60)$$

where the median value is an average value of the property magnitude in the series as set up in step (1), rather than a magnitude average. Information on the permeability of the formation as a function of depth would normally be taken from representative core analyses, or possibly from well log analysis.

To illustrate the utility of Figures 7-15 through 7-18, assume that the permeability variation is 0.6, the water saturation is 30 per cent and the mobility ratio is 2.6. The oil recovery by water flooding is to be determined at the assumed economic limit correspond-

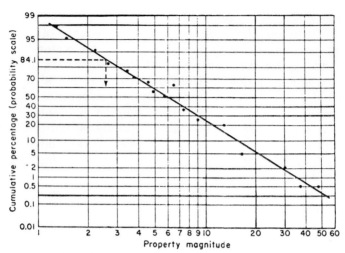

Figure 7-19. Plot for determining the variance of a property
magnitude (after Pirson[28]).

ing to a water-oil ratio of 25. Figure 7-17 yields a value for the
factor $R(1 - 0.52S_w)$ of 0.30. The fractional recovery of the oil-in-
place may then be calculated as being 0.356. If the oil-in-place
amounts to one million STB, then the recovery should approximate
356,000 STB, if the oil saturation is initially 100 per cent of the
hydrocarbon pore space, and if the oil formation volume factor is
of the order of one. If a high shrinkage crude is being recovered,
the oil recovery would be decreased according to the value of the
oil formation volume factor, at the prevailing reservoir temperature
and pressure. Then too, a further correction will be necessary if
the flood is not linear. The proper correction may be determined
from reference to Chapter 4 on flood coverages. For instance, if
the mobility ratio is 2.6, the water cut is 96 per cent, and a five-
spot flood is being studied, Figure 4-12 (p. 87) gives an areal
sweep efficiency of 96 per cent. The oil recovery for the project
would then approximate 341,000 STB.

The above-described method is a very good technique where a
rapid appraisal of a water flood project(s) is needed. The author[27]
states that the method works well where the initial oil saturation
is 45 per cent or greater. The method does assume a piston-like
displacement behind and ahead of the displacement front, with no
additional oil being produced after the passage of the front. Where
the mobility ratio exceeds roughly 10, the values would be ex-

pected to have relative significance only. At mobility ratios greater than 2, the oil produced from behind the front is no longer negligible, and should be accounted for when a detailed study of a project is undertaken. Then too, the method does not take into account the formation of an oil bank due to an initial gas saturation, such as would be formed under primary depletion. However, the method could be expected to give good relative answers. If the gas saturation throughout the system had been 15 per cent prior to the beginning of the water flood, the above-described method could be modified. Since the displaced oil would now amount to roughly 0.356 less 0.15, or 0.206 per cent of the original oil-in-place, the recovery in the five-spot would be reduced from 341,000 to 197,000 STB. Reference to the water-flooding performance calculation of Stiles should permit the reader to develop a time and oil producing rate relationship for the above-outlined performance calculation method should such be needed.

Water-Flooding Performance Calculations—Miscellaneous Calculation Methods

Space does not permit a detailing of all the calculation methods which have been proposed and published in the technical literature. Some of the methods do have specific features which make them advantageous in the solution of a given water flooding problem.

One of the earlier attempts to predict water flood performance is due to Suder and Calhoun[29], and has application if the vertical variation in permeability is limited. In the method, an average permeability for the reservoir is found by totalling the millidarcy-foot capacity for all wells on which there are core data, and dividing the total by the total sand thickness of all the wells. As in the Stiles method, the displacement of oil by water is assumed to be piston-like, which limits the method to those floods where the mobility ratio will approximate one. However, the method does differ in that the injection rate is calculated for a radial, an intermediate and a linear injection phase. The method essentially assumes a constant injection pressure and a variable injection rate.

Muskat[30] has published information on the effect of permeability stratification upon water flood performance. The development applies to systems where the stratification is either of the exponential or linear type. The theory presented assumes an idealized layering where each stratum has constant thickness, but may vary in porosity and permeability. The system is assumed linear so that no pattern sweep efficiency has been included. Then too, the development is for a piston-like displacement, with no oil production obtained from behind the front. In view of these assumptions, we could expect the method to have only limited application where the mobility ratio is markedly larger than one, since the production from behind the front could not then be validly ignored.

Craig, Geffen and Morse[31], in 1954, presented a method for calculating oil recovery performance of pattern gas or water injection in uniform, or non-stratified sands. The method incorporates the concept of an expanding swept area after water or gas breakthrough to the producing wells with continued displacing fluid injection. The method also includes the frontal advance concept, as given by Buckley and Leverett[32], and advance by Welge[33], for the case of continued oil production behind the displacement front. The basis for the Craig *et al.* method is the following equation:

$$N_p = \frac{V_p}{B_o} [(S_{wr} - S_{wi}) E_{as} - S_{gi}] \tag{61}$$

where

V_p = pore volume of the pattern
N_p = oil produced after initiation of the flood
S_{wr} = final average water saturation in the swept area
S_{wi} = beginning water saturation in the area to be flooded
E_{as} = areal sweep efficiency
S_{gi} = average gas saturation in the pattern at beginning of the flood
B_o = oil formation volume factor at reservoir pressure and temperature

The method for determining the average water saturation behind the front at and after breakthrough has been presented in a previous section of this chapter.

Abernathy[34] has improved the Craig *et al.* calculation technique by incorporating the concept of stratification, as given by Stiles[16]. In this adaptation, the effect of different thicknesses, different absolute and relative permeabilities, different oil, gas and water saturations, as well as different porosities, may be treated. The calculations are tedious, however, and are best programmed for computer solution. The specific advantage of the Abernathy improvement of the Craig *et al.* technique is in the treatment of a vertical sweep efficiency by the layering process. The method assumes that each layer is discrete, and that no cross-flow occurs. This means that the performance of each layer may be calculated independently following exactly the Craig *et al.* method, but using the data applicable to the particular stratum being treated.

The resulting composite performance for all layers combined, i.e., the position of the flood front in any given layer, the volume of injected fluid being accepted by a given layer, etc., is determined by computing the relative injection rate, as specified by Darcy's equation for the particular enclosed well pattern being treated. Table 6-1 of Chapter 6 presents the equations for injection rate for enclosed well patterns where the mobility ratio is unity. For the five-spot well pattern, it is convenient to use the five-spot conductivity ratio, as published by Caudle and Witte[35]. Figure 4-14 of Chapter 4 presents the conductance ratio as a function of mobility ratio, and the areal sweep efficiency for the five-spot well network. Equation (2) of Chapter 4 defines conductance ratio. Essentially the ratio compares the water injection rate at a given time, or amount of water injected against a base injection rate when the injection of water has just been initiated. In this instance, the oil mobility ratio still applies in the calculation of the beginning water injection rate, so that the rate may be calculated from the equation:

$$i = \frac{3.535 \, k_o \, h(\Delta p)}{\mu_o \left[\ln \dfrac{d}{r_w} - 0.619 \right]} \tag{62}$$

where the nomenclature is as defined in Chapter 6. Since the injection rate is known, the time for injection of a given volume of water to each layer may be computed independently. Summing of the performance of all layers at the same time intervals then permits the generation of an overall water flood performance for the formation. No published data are presently available for conductance ratios, except that by Caudle and Witte for the enclosed five-spot. This data may be used for other well networks not differing markedly from that of the five-spot. Hopefully, information on conductivity ratio for other well networks will appear soon in the technical literature.

Wooddy and Moore[36] have advanced a general theory for performance calculations with natural or artificial water drives, using the frontal advance equations of Buckley and Leverett[32] and Welge[33]. The method is similar in many respects to that of Craig *et al.*[31], but in the general form permits the inclusion of gravitational forces to account for reservoir dip. Hendrickson[37] has also proposed a method for performance calculations, which uses many of the same concepts employed by Craig *et al.* The concept of an increasing areal sweep efficiency for continued fluid injection past break-through, plus frontal advance theory to account for production from behind the displacement front, have been included. Hendrickson uses the areal sweep efficiencies as given by Habermann[38], although the work of other experimenters could have been used as well. The effect of reservoir stratification has also been included.

Felsenthal, Cobb, and Heuer[39] have published modifications to the Dykstra and Parsons[26] performance calculation method, to account for the resaturating of the free gas space in each layer of a stratified reservoir. This is important when an optimum formation pressure during the primary-producing phase is to be chosen at which to start a water flood, in order that oil recovery will be maximized. Other suggested modifications to the Dykstra and Parsons method allow for layer-to-layer variations of beginning and ending fluid saturations and porosity. Felsenthal *et al.* found that these latter modifications usually resulted in rather small changes to the overall reservoir performance.

References

1. Higgins, R. V., and Leighton, A. J., *J. Petrol. Technol.,* 679 (June, 1962).
2. Higgins, R. V., Boley, D. W., and Leighton, A. J., *J. Petrol. Technol.,* 1076 (September, 1964).
3. Henley, D. H., Min. Ind. Exp. Sta. Circ., No. 64, Penn. State Univ., University Park, Penn., (1962).
4. Kufus, H. B., and Lynch, E. J., *Producers Monthly* 23, 32 (December, 1959).
5. Higgins, R. V., and Leighton, A. J., "Principles and Computer Techniques for Calculating Performance of a Five-Spot Waterflood—Two-Phase Flow," USBM Report of Invest. 6305, 1963.
6. Higgins, R. V., and Leighton, A. J., paper no. 1625-G, presented October, 1960, Soc. Petrol. Engrs. of **AIME**, Pasadena, Calif.
7. Higgins, R. V., and Leighton, A. J., "Waterflood Performance in Stratified Reservoirs," USBM Report of Invest. 5618, 1960.
8. Gaucher, D. H., and Lindley, D. C., *Trans. AIME* **219**, 208 (1960).

9. Engelberts, W. F., and Klinkenberg, L. J., *Proc., Third World Petrol. Congress*, Part II, 544 (1951).
10. van Meurs, P., *Trans. AIME* **210**, 295 (1957).
11. van Meurs, P., and van der Poel, C., *Trans. AIME* **213**, 103 (1958).
12. Mulholland, E. E., "The Application of Viscous Fingering Techniques to Predict the Performance of Five-Spot Water Floods," M.S. Thesis, Univ. of Wyo., January, 1966.
13. Rachford, H. H., Jr., *Soc. Petrol. Engrs. J.*, 133 (June, 1964).
14. Rinehart, R. D., "Comparison of the Buckley-Leverett and Viscous Fingering Approaches to Water Flood Performances," Spec. Proj. Course, Univ. of Wyo., September, 1965.
15. Land, C. S., private correspondence, November, 1965.
16. Stiles, W. E., *Trans. AIME* **186**, 9 (1949).
17. Craft, B. C., and Hawkins, M. F., "Applied Petroleum Reservoir Engineering," p. 393, Englewood Cliffs, N. J., Prentice-Hall Inc., 1959.
18. Wahl, W. L., Mullins, L. D., and Elfrink, E. B., *J. Petrol. Technol.*, 132 (June, 1958).
19. Schauer, P. E., Jr., paper no. 934-G, presented October 1957, Soc. Petrol. Engrs. of **AIME**, Dallas, Texas.
20. Schmalz, J. P., and Rahme, H. S., *Producers Monthly*, 9 (July, 1950).
21. Pirson, S. J., "Oil Reservoir Engineering, p. 87, New York, McGraw-Hill Book Co., Inc., 1958.
22. Ache, Paul S., paper no. 935-G, presented October 1957, Soc. Petrol. Engrs. of **AIME**, Dallas, Texas.
23. Johnson, J. P., paper no. 918, presented October 1964, 39th Annual fall mtg. of Soc. of Petrol. Engrs. of **AIME**, Houston, Texas.
24. Law, Jan., *Trans. AIME* **155**, 202 (1944).
25. Miller, M. G. and Lents, M. R., *Drill. & Prod. Prac.*, *API*, 128 (1946).
26. Dykstra, H. and Parsons, R. L., "Secondary Recovery of Oil in the United States," Second Edition, 160, N. Y., API, 1950.
27. Johnson, C. E., Jr., *Trans. AIME* **207**, 345 (1956).
28. Reference 21, p. 88.
29. Suder, F. E., and Calhoun, J. C., Jr., *Drill. & Prod. Prac.*, *API*, 260 (1949).
30. Muskat, M., *Trans. AIME* **189**, 349 (1950).
31. Craig, F. F., Jr., Geffen, T. M., and Morse, R. A., *Trans. AIME* **204**, 7 (1955).
32. Buckley, S. E., and Leverett, M. C., *Trans. AIME* **146**, 107 (1942).
33. Welge, H. J., *Trans. AIME* **195**, 91 (1952).
34. Abernathy, B. F., *J. Petrol. Technol.*, 276 (March, 1964).
35. Caudle, B. H., and Witte, M. D., *Trans. AIME* **216**, 53 (1959).
36. Wooddy, L. D., Jr., and Moore, W. D., *J. Petrol. Technol.*, 245 (August, 1957).
37. Hendrickson, G. E., *J. Petrol. Technol.*, 745 (August, 1961).
38. Habermann, B., *Trans. AIME* **219**, 264 (1960).
39. Felsenthal, M., Cobb, T. R., and Heuer, G. J., Jr., paper no. SPE 332, presented May, 1962, Rocky Mtn. Regional mtg. of Petr. Engrs. of **AIME**, Billings, Montana.

PROBLEMS

1. The following data are available on a linear horizontal porous media system where oil is being displaced by water:

Cross-sectional area presented to flow	1 ft^2
Porosity	25 per cent
Length	1000 ft
Oil viscosity	10 cp
Water viscosity	1 cp
Irreducible water saturation	20 per cent
Oil formation volume factor	1.15 bbl/STB
Residual oil saturation	25 per cent
Water injection rate	$0.1 \text{ ft}^3/\text{hr}$

(a) Using the analytical viscous fingering approach to immiscible displacement, calculate performance for the above system. This should include curves for water cut versus cumulative oil production as a fraction, oil producing rate in ft^3/hr versus time in hours, and cumulative oil production as a fraction versus time in hours.

(b) Reconstruct the relative permeability curve which you would have had to use in order to get the same answers by the Buckley-Leverett method.

(c) Solve the problem using the Buckley-Leverett approach and compare the resulting performance curves.

2. By means of diagrams, justify the secant line construction to the fractional flow saturation curve of the Buckley-Leverett frontal displacement theory. Show the role of capillary pressure, and provide a physical justification for the approach taken.

3. Use the method of Stiles to predict the rates of oil production and water cut for the following reservoir and fluid property data:

Irreducible water saturation	22 per cent
Porosity	25 per cent
Initial oil saturation	60 per cent
Residual oil saturation	17 per cent
Oil formation volume factor	1.12 bbl/STB
Area of enclosed five-spot pattern	10 acres
Water viscosity	0.9 cp
Oil viscosity	1.6 cp
Injection rate per five-spot	200 bbl/day

Depth, ft	Permeability, md
3280	2
3282	40
3284	45
3286	120
3288	80
3290	145

Depth, ft	Permeability, md
3292	110
3294	74
3296	48
3298	5

The net oil pay is present between 3279 and 3299 feet in the wells of the pattern. The above permeability values are thought to be valid averages for all wells in the five-spot pattern area.

4. Use the simplified Dykstra-Parsons technique to obtain the fractional oil recovery which would be expected for the reservoir described in Problem 5. The relative permeability to oil ahead of the displacement front is 0.92, while the relative permeability to water at the irreducible oil saturation is 0.40. Assume that the economic limit will be reached at a producing water-oil ratio of 30.

5. The performance of the water-injection project of Problem 1 is for a linear system. How would you modify the resulting answers for the following field cases? Describe the necessary calculation technique.

(a) An enclosed five-spot well pattern?

(b) An unenclosed five-spot well pattern?

(c) Peripheral water injection to an oil-bearing anticlinal structure?

(d) A fault block of linear configuration on the site of a salt dome having steep dip but no pre-existing gas cap or connecting aquifer?

C H A P T E R **8**

Water for Water Flooding

Selection of Water Source

One of the most important items to be considered in a proposed water flooding project is the water supply itself. The quality of the water is of great importance, as is a certain source of water supply. In some parts of the country, sufficient water from a variety of sources is available so that an operator can make a decision based primarily upon cost. In other parts of the country, either a lack of water or the presence of only one source will not permit the operator to be too particular. Thornton[1] has pointed out that in many areas the only essential qualification for a water source for flooding is an adequate volume. Corrosiveness, high solids content, scaling tendencies, the presence of micro-organisms and other foreign matter in the water may only be regarded as factors which have to be considered in the design and the possible justification for water flooding a given reservoir in preference to another secondary recovery mechanism.

The basic requirements for an injection water after the consideration of the basic justification and design factors are:

(1) Availability in sufficient quantities during the life of the flood.

(2) Freedom from undissolved solids content or other suspended material.

(3) Chemically stable and relatively inactive with compounds and elements present in the injection system and the reservoir.

To achieve the above requirements of a satisfactory injection water, the treating systems will range from very simple installations to elaborate plants for processing sour, gaseous, contaminated and certain of the mineralized waters.

A number of authors have treated the problems surrounding the selection of a water source[1,2,3,4,5]. There are three main freshwater sources and three saltwater sources which have been used for water flooding purposes. The freshwater sources include surface waters, subsurface waters and waters from alluvium beds. Subsurface formations, the ocean and water produced with oil are the common sources of salt water.

Fresh Water—Surface Sources[2]. Waters from ponds, lakes, streams, and rivers have been employed extensively for water flooding. A continued increase in the use of fresh water in many parts of the country by industry and municipalities has, or will result in water shortages. Then too, one of the hazards of using surface waters is the real possibility of drought. When water shortages occur, domestic users have first priority. Usually, if fresh water is to be used, approval must be obtained from the appropriate state agency before developing the source. Where a question of continued fresh water availability exists, an operator electing to use a brine source may be choosing wisely from the technical point of view, but also would be aiding the conservation of a particularly vital natural resource. In some areas, we can expect state authorities to prohibit the injection of fresh water into oil reservoirs.

Small ponds and streams are usually unreliable sources of water for all seasons of the year. Large lakes and rivers are preferable when the flood will be sufficiently close to utilize such a source. Surface sources of injection water are usually high in oxygen content, carry much suspended matter and a variety of micro-organisms and show a wide variation in composition during the various seasons of the year. This can result in high treating equipment costs, plus a large continuing expense for chemicals and plant operation.

Fresh Water—Alluvium Beds[2]. The quality of fresh water available from river or stream beds can often be improved by using shallow wells to tap the nearby alluvium sand beds. These sources are quite reliable on a year-round basis, are usually non-corrosive, are not subject to the wide variations in turbidity during rainy seasons and provide natural filtration of the water. Then too, the organic content of the water is reasonably constant, which permits a more consistent water treating program to be followed. Earlougher and Amstutz[2] point out that sulfate-reducing bacteria, which can be expected in all surface-water supplies since they are soil bacteria, can thrive within a few feet of the surface of the alluvium bed. For this reason, such waters are often contaminated by these bacteria. Sometimes it is possible to drill deeper wells in the alluvium beds, and either totally or partially eliminate this problem.

The type of wells completed in the alluvium beds will depend upon the water rates needed, the depth and the composition of the bed. Large casings are commonly used where high rates are needed. These can approach the size of a small caisson in some instances, with horizontal shafts driven into the bed from the bottom, where extra-high productivities are required. Turbine pumps are often used in such a high water-producing rate service. Where more modest water-producing rates are needed, it will often suffice to drive small diameter sand points into the alluvium bed at close spacings, and to pump all the points with a centrally-located centrifugal pump on the surface.

The main advantages of alluvium beds as a water source are low source development costs, natural water filtration, modest pumping costs and a relatively constant water quality. Where the sulfate-reducing bacteria in particular are no problem, chemical treatment is usually minimal, and corrosion rates of piping and injection equipment low.

Fresh Water—Subsurface Formations. In some geologic provinces, fresh water may be found in adequate quantities for water flooding, from near-surface sands or limes down to formations several thousands of feet from the surface. The completion of these wells is similar to that of oil wells with the pumping equipment being either submersible centrifugals, or rod-actuated bottom hole pumps. Usually, it is advisable to complete the well in a single zone if the zone has sufficient productivity. Mixing waters from different zones will occasionally result in chemical incompatibilities and scaling, which would not occur in a single zone completion. The spacing of wells for a water supply will depend upon the permeability of the sand, the formation pressure, the pressure drawdowns required and the aquifer recharge rate. The particular subsurface sand chosen in a given area will depend upon the economics associated with the wells required and the volumes of water needed.

It is usually preferable to use closed treating and injection systems where fresh water is obtained from subsurface formations, thus eliminating, in many instances, chemical treatment and filtration before injection. Advantages of such a system are the elimination of large treating and injection facilities, plus the low corrosion rate incident to the injection of a fresh, high quality water.

Salt Water—Subsurface[2]. Due to the scarcity of good freshwater sources in many sections of the country, increased amounts of flood waters are being taken from saltwater formations, available in the immediate vicinity of the water flood. These wells can be completed in much the same fashion as shallow fresh-water wells, and have the same characteristic of being often best adapted to a closed injection and treating system. If the water contains significant quantities of hydrogen sulfide and carbon dioxide, an open system may be necessary—depending primarily on the life of the project and the type of treating plant necessary to treat the water for other contaminants. Again, the pumping equipment will be rod pumps, shaft-driven turbine pumps, or submersible centrifugal pumps, depending on well depth and water volumes to be handled.

A significant advantage of salt water for injection is to be found in those areas where the formation to be flooded is sensitive to the injection of fresh waters due to the presence of hydratable clays. The other principal advantages of salt water for injection are the high productivity of many salt-water bearing formations, modest pumping costs and adaptability to closed treating and injection systems. On the negative side is the sometimes considerable expense of drilling the supply wells in difficult drilling areas and where the best water source is at considerable depth.

Salt Water—Ocean Source[2]. Where the projected water flood is in the vicinity of an ocean, it will often be found attractive to use the water due to infinite supply, plus low development and pumping costs. Rather than withdraw the water directly from the ocean, it is usually advisable to drill shallow wells below the ocean floor when the flood is offshore. The advantages of natural filtration are thus exploited. Where the water has been naturally filtered, it is usually advantageous to use a closed injection system. Since moderate to high corrosion rates usually are associated with the use of salt water, a corrosion inhibitor should be injected at the first possible point in the pumping, treating and injection system.

Salt Water—Formation and Return Water. In some areas, sufficient produced water is available for the flooding of a particular

oil productive interval. If this water is completely compatible
with the water of the formation to be flooded, and provided that the
produced water has not become contaminated, the reinjection of
such a water is often advisable. Where the water cannot be dis-
posed of through surface pits, injection would be necessary, and
the possibility of increasing the production of the lease by such
injection is attractive. Many of the first water floods started as
"dump" floods in this fashion, until the advantages of the injection became apparent due to increased oil production at nearby producing wells.

Then too, as the flood progresses and breakthrough of the displacing water occurs, an increasing volume of this "return" water
will be available, either for disposal or for reinjection. Usually,
this return will be added to the "makeup" water from the water
source, and then reinjected. In most areas, the produced water
will be too salty for surface disposal. Even if the injected water
were fresh, the first water produced will be the connate water of
the system, which in most geologic provinces will have a salt content. Where it is planned to mix return and makeup waters, initial
planning of the flood should include water compatibility tests to
determine if the following combinations of ions will precipitate
upon mixing: iron and oxygen, iron and sulphide, calcium and carbonate, calcium and sulfate, and barium and sulfate.

The sulfate precipitates are the most difficult to handle in an
injection system, since the common solvents and acids will not remove them and necessitate expensive mechanical removal. If precipitation of iron and calcium carbonate compounds inadvertently
occurs, removal can usually be effected with one of the commercial
acids. Sequestering and chelating agents such as citric acid, one
of the polyphosphates, etc., will often permit the mixing of waters
containing barium, calcium, or iron compounds, with waters containing sulfates, carbonates, or sulphides. Where pilot water
flooding operations are projected because of uncertainties in the
displacement mechanism or the susceptibility of the reservoir to
water flooding, the test program should include in-the-field tests
of water compatibility: It may well be found that a relatively inexpensive closed injection system will be adequate, whereas preliminary planning would have suggested that a more complex and
expensive open system was needed.

As Earlougher and Amstutz[2] state, an operator can convert from
a closed system to an open system with no large loss in investment, whereas conversion from an open to a closed system will

usually result in a substantial financial loss. If a pilot flood is to be attempted before adoption of a full scale flood, it will often be possible to determine if a closed injection and treating system will be adequate. Even when the installation of a closed system is questionable, it may well be best to take a calculated risk and use the system. Then too, if the closed system is inadequate for the pilot tests, the cleaning out and acidizing of the pilot injection well(s) will constitute only a small fraction of the cost of a closed system, found to be unnecessary. Since many closed systems do not have filtering facilities, the return water will sometimes be carrying as much as 10 ppm of oil. Many floods can tolerate such suspended particles of oil, with no noticeable detrimental effects to the injection wells.

Definitions and discussions of open, closed and semi-closed water treating systems are treated in later sections of this chapter.

Estimation of Water Requirements

The quantity of water required over the life of a water flooding project will vary, according to the type of project being considered. If most of the produced or return water is to be reinjected, ultimate water requirements will approximate from 150 to 170 per cent of the pore volume of the sand to be flooded. If sizeable thief or gas sands are present which will also be accepting large quantities of water, large water volumes will be necessary. The calculation of water flood performance by any of the methods described in Chapter 7, will also provide information on water requirements, both ultimately and on a time basis. The largest volume of the makeup water will be necessary during fill-up of the reservoir, and until water breakthrough is realized at the producing wells. Where the supply of injection water is constant but limited in daily volumes, it may be necessary to progressively flood a sand in such a fashion that water requirements will be spread out over the life of the project.

Where all the produced or return water is to be reinjected, it should be possible to limit the outside or makeup water requirements to something like 50 per cent of the oil reservoir pore-volume.

Source Water Quality

Much has been said in the literature about water quality. Certainly, a desirable water is[6]:

(1) Available in sufficient quantity at the time needed.

(2) Readily available and accessible.

(3) Cheap.

(4) Chemically compatible with other possible *makeup* waters which might be used.

Of course, in meeting the above requirements, a certain tolerance is possible so that the source water need not be:

(1) of perfect quality.

(2) free of hydrogen sulfide.

(3) or even compatible with the connate water in the formation to be flooded.

Initially then, the characteristics of the available water should be determined by laboratory analysis so that the extent to which the water *might* require treatment will be evident. In general, water will be found to be poor in quality for five major reasons[4]:

(1) For containing suspended material which can quickly and effectively plug injection wells.

(2) For containing barium, strontium, iron and calcium in solution with sulfate, carbonate, sulfide and oxide radicals. The presence of these ions can result in precipitation of compounds which can plug the formation face in the injection wells. It should be noted that precipitation of compounds due to incompatibility of the injected and reservoir waters is seldom of importance, since the deposited materials are layed down at a considerable distance from the point of injection, and do not noticeably reduce the system permeability.

(3) For containing carbon dioxide, hydrogen sulfide, and oxygen which cause corrosion. Oxygen causes rapid pitting of iron pipe and plugging of injection wells with rust. Hydrogen sulfide can create pitting, scaling, blistering and hydrogen embrittlement in steels. Carbon dioxide causes black scales and pits to form on iron and steel.

(4) For containing algae, iron bacteria and capsulated bacteria. All of these organisms tend to collect as large masses of slime which can effectively plug the entire injection system and the water input wells.

(5) For containing sulfate-reducing bacteria. The resulting corrosion products from the presence of these bacteria can plug filters and the injection well sandface.

The particular combination of contaminants in a given water source will decide the type of plant which will be required to prepare water of minimum quality for the purpose of water flooding.

Water Treating Systems

The first considerations in the construction of a water-treating system are space requirements and proximity to both the primary and produced water sources and the injection point(s). A substantial saving in the cost of the plant and in the subsequent operation and maintenance of the plant, may sometimes be effected where there is sufficient natural surface relief to permit a gravitation of the water through the plant. A central location will usually permit savings in the injection water distribution system. A careful study of all the factors will usually result in the most economical system for a particular water flooding installation.

After the characteristics of the various available flooding waters and produced waters have been determined, a decision will be possible as to whether a closed, open, or semi-closed treating system should be used.

Closed Water Treating Systems. Closed water treating systems may be defined as those where the plant is designed so that the water does not come in contact with the air. This avoids oxidation-reduction reactions where precipitates may be formed, and the solution of atmospheric oxygen in the water. Figure 8-1 shows the elements of a simple closed-type water treating system. Such systems are widely used in water flood operations.

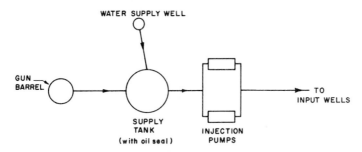

Figure 8-1. Simplified flow diagram for a closed-type water treating system (from Buck[4]).

In the indirect water treating system, the water supply may consist of a makeup water supply and a produced water supply discharged into a water storage tank, having either a natural gas blanket in the vapor space above the water, or an oil blanket to exclude air from the water. Pressure at this point is nearly atmospheric, which permits the escape of dissolved gases. In the simplest installations, the water is pumped directly to injection wells from this clear water tank. Where the water is somewhat turbid, filters using interchangeable elements constructed of silicon carbide or aluminum oxide are commonly used. These may be placed either upstream or downstream from the injection pumps. When the makeup water and produced water are compatible and stable, sequestering, sterilizing, corrosion-inhibiting and wetting agents may be added in a closed system. Chemical injection will normally take place upstream from the clear water tank.

Since the primary advantage of the closed treating system is the exclusion of air, all tanks, pumps, treaters and piping must be kept free of leaks.

Open Water Treating Systems. When the supply water is highly supersaturated or undersaturated with carbonate and requires stabilization, the open system is often used. In this instance, no effort is made to exclude air from the plant. Most plants use various aeration devices to oxidize ferrous and manganous compounds to insoluble ferric and manganic states. In addition, acidic gases are released, raising the pH value of the water, which in turn reduces the carbonate supersaturation of the water. Figure 8-2 shows a typical plant layout of the open type. In this particular

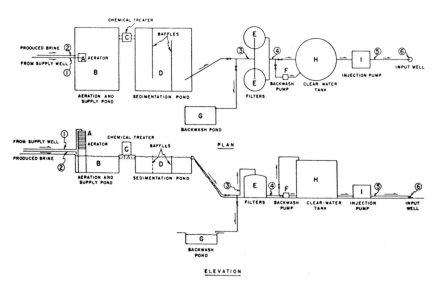

Figure 8-2. Simplified flow diagrams for an open water treating system (from Watkins *et. al.*[7]).

plant, the makeup water is aerated in a wood-slat aerator, and mixed with produced waters in a primary settling and supply pond. The water flows by gravity through a connecting dike, where necessary chemicals are added before entering the sedimentation pond containing "end-around" baffles. The water gravitates through filters into a clear water tank, from which it is pumped to the input wells. Filter backwashing facilities are also shown in the diagram.

Aeration. There are three basic types of aerators used for water conditioning: the wood-slat type, the coke tray type, and the forced-air, or degasifier type. The capacity of these aerators will depend upon the aerator type, the amount of acid gases to be removed, and the amounts of ferrous and manganous compounds to be oxidized to higher valence states. Capacities vary from 10 gpm/ square foot for coke tray aerators to 17.5 gpm/square foot for forced-draft aerators. The efficiency of this latter type comes largely from being able to vary the amount of aeration. Ideally, just sufficient air will be introduced to accomplish the purpose intended, but without dissolving excess oxygen in the water.

Sedimentation. Sedimentation ponds or tanks should be so designed that adequate detention time is provided to permit any suspended solids to coagulate and settle out. Usually such ponds or tanks are equipped with either "over-and-under" or "end-around" baffles, to assure longer detention time and more uniform movement of water through the system without channeling from the inlet to the outlet. The efficiency of the baffling system is a measure of the actual detention time, relative to the displacement time, based upon the pond capacity and the quantity of water processed over a given time interval. Inadequate detention times will result in higher filtering loads to be placed on the filters downstream.

Chemical Treatment. Many waters used for injection purposes contain sizeable quantities of suspended material and insoluble compounds, formed during the aeration phase of the water treating. If sufficient time were allowed for this material to settle out, very large detention tanks or ponds would be required. To hasten the sedimentation process, coagulating chemicals are normally used. The most common coagulant is aluminum sulfate, or alum, which reacts with alkaline salts such as calcium bicarbonate, to form a gelatinous precipitate:

$$Al_2(SO_4)_3 + 3Ca(HCO_3)_2 \longrightarrow 2Al(OH)_3 + 3CaSO_4 + 6CO_2 \quad (1)$$

This precipitate entraps the suspended matter and causes a faster rate of settling. Alum works best where the water has a pH value between 5 and 8. Where the pH of the water is 8 to 9, ferrous sulfate ($FeSO_4 \cdot 7H_2O$) works well to form an insoluble floc, ferric hydroxide. Other coagulating chemicals which have been used include ferric sulfate ($Fe_2(SO_4)_3$), ferric chloride ($FeCl_3$) and sodium aluminate ($NaAlO_2$). It is sometimes necessary to add an alkali such as lime, soda ash, or caustic soda with the coagulant, to raise the pH value to a level where more complete coagulation can take place. Lime is the most common alkali used since the chemical reaction is with free carbon dioxide, and with calcium and magnesium bicarbonates, which form calcium carbonate. The calcium carbonate is removed by flocculation, sedimentation and filtration. Since lime-treated water tends to be incrustant, it is sometimes necessary to add small quantities of the complex phosphates to sequester the calcium carbonate. A slight supersaturation of carbonates, in the amount of 10 to 15 ppm, will coat steel surfaces of the treating plant, and limit corrosion somewhat.

In those systems where algae, molds, iron bacteria, and/or sulfate-reducing bacteria are present, a variety of chemicals, biocides, or antibiotics have been used. The one most widely used is chlorine, either as a gas, or in the form of a chlorine-available compound, such as sodium, calcium hypochlorite or chlorinated lime. Chlorine will usually control bacteria, slime, molds and other organic growths in fresh water. Copper sulfate performs in a similar manner to that of chlorine. Formaldehyde is an effective bactericide and corrosion inhibitor. Some of the other structural types[8,9] used for water treating include quaternary ammonium compounds, imidazolines, phenolics, fatty amines, amine derivatives, and organic mercurials. The activity of many of these compounds in controlling the sulfate-reducing bacteria is reduced in salt waters[10].

Filtration. The types of filters used for processing flooding water include the gravity, pressure, diatomaceous earth and cartridge types. Filtration may be defined[8] as the process of passing a liquid through a medium for the purpose of removing suspended solids from that liquid. For optimum filter performance, the water from the sedimentation pond should contain a small amount of floc particles. This results in the formation of a 2 to 4 inch thick blanket of sludge, commonly termed the Schmutzdecke, in the top layer of the filter. The Schmutzdecke is composed of the floc particles and bacteria, and results in improved filter performance. Usually all particles larger than 0.5 microns will be removed where this sludge layer has developed. If larger particles are found immediately downstream from the filter, either a Schmulzdecke has not formed, or if formed, has a hole in it.

The gravity filter is illustrated in Figure 8-3. The water to be filtered enters at the top of the filter, and flows by gravity to the lower underdrain system. This type of filter is simple to construct

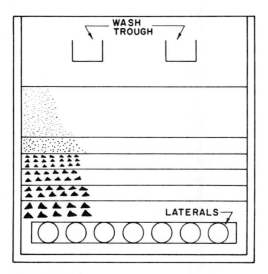

Figure 8-3. Cutaway of a gravity filter (after Kirk[9]).

and to inspect during operation, but requires additional pumps or surface relief since operation is by gravity. The filter media is usually either graded sand and gravel, or anthracite coal. Silica sand has a rough surface and is effective in filtering a poorly-coagulated water[11]. Anthracite beds do not pack down, which results in less pressure drop than would occur in a sand filter. Gravity filters are usually not equipped for backwashing. The filter is cleaned by discarding the top 2 inches of sand at regular intervals.

The pressure-type filter is similar in operation to the gravity filter, but is capable of operation under a line pressure. Pressure and gravity-type filters are designed to operate efficiently at rates from 2 to 4 gallons per minute per square foot of filter area. Figure 8-4 is a schematic drawing of a typical pressure filter. The filtering medium section is usually about 2 feet thick. The pres-

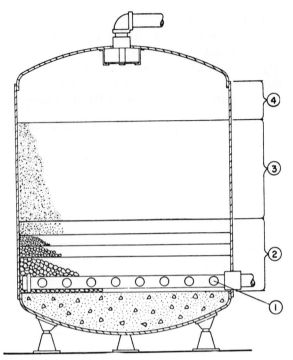

Figure 8-4.   Schematic drawing of a pressure filter showing: (1) underdrain, (2) retaining medium,
(3) filtering medium, (4) freeboard (from Cecil[11]).

sure filter is cleaned by backflowing at a rate which will not move
any of the supporting media (element 2), but will give complete
suspension of the filtering media so that the adhering floc and suspended particles will be removed.   Depending upon the type of filtering media used, rates of from 10 to 25 gallons per minute of water per square foot of filter media will be required.

Filters containing silicon dioxide or aluminum oxide elements,
and others containing diatomaceous earth[12,13] are used.   These
filters are backwashed to remove accumulated solids from the filter face.   The pressure, cartridge-type and diatomaceous earth filters may be used on either the closed, or the open water treating
systems.   Regardless of the system, the entering water should
have a turbidity less than 50 ppm, and the filtered water a turbidity less than 1 ppm, for efficient filter operation[7].

Semi-closed Water Treating Systems.   A semi-closed system for
water treating is defined as one which exhibits a variation between

the closed and open types. Usually, a semi-closed system is one in which the water is treated in an open system until the point of deaeration. From this point of the treating system to the injection well(s), the system is of the closed type. Figure 8-5 illustrates a semi-closed system. The deaeration is accomplished by applying a vacuum at the top of a packed column through which the water passes before entering the clear water tank. A blanket of oil above the water in the clear water tank, or gas in the free space above the water, will limit the amount of air which can be reabsorbed into the treated water. Where possible, it is advisable to use a gas in the place of oil at perhaps one-fourth of an ounce above atmospheric pressure. This effectively excludes air. Oil has the disadvantage of perhaps damaging injection wells if inadvertantly the water level in the tank drops below the suction level of the pump. Then too, air is readily soluble in oil, thus affording only partial protection to the injection water.

Miscellaneous Water Treating Techniques. The primary advantage of a semi-closed water treating system is the provision for the removal of dissolved oxygen, often by vacuum deaeration from the injection water. It has been observed in the field[14] that oil blankets are not effective in preventing oxygen from entering the water in the clear water tanks, in either closed or open water treating systems. On the other hand, natural gas has an absorp-

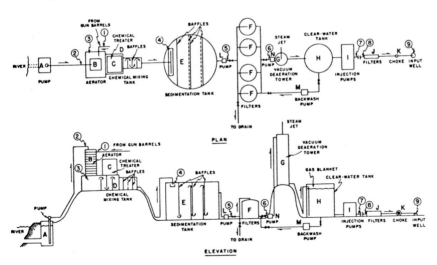

Figure 8-5. Simplified flow diagram for a semi-closed water treating system (from Watkins *et al.*[7]).

tive capacity for oxygen, and as such makes an excellent seal above stored water. Several authors[14,15] have described a technique for the desorption of oxygen from water, using natural gas for countercurrent stripping. When available to a water flood operator, the use of natural gas to remove oxygen may be attractive, since the gas can be recovered and used to fuel equipment in the treating plant. Approximately 1.75 cubic feet of natural gas is needed to reduce the oxygen content from 10 ppm to approximately 0.1 ppm per barrel of water, assuming that an efficient bubble tray column is used. Weeter provides a detailed account as to how such a tower should be designed.

Where water sources are limited, occasionally a sour makeup water must be used in secondary recovery operations. A number of hydrogen sulfide removal processes are available: forced-draft aerators or degasifiers; stripping hydrogen sulfide with exhaust gas produced by a submerged burner in a packed column[16]; and by stripping hydrogen sulfide with exhaust gas in a packed column[17]. Weeter[17] has published details relative to the design and construction of a system for stripping hydrogen sulfide by using exhaust gas in a bubble cap column. The exhaust gas generator is designed to operate at high temperature to ensure the near-total burning of all combustibles, so that a minimum of uncombined oxygen will remain in the system. The resulting exhaust gas is then also capable of removing dissolved oxygen from the water to be treated.

One of the main disadvantages of columns for the removal of oxygen and hydrogen sulfide in injection waters is the relatively high initial cost and continuing expense for operation and maintenance. Such systems can be justified where the water flooding project is large and the treating water throughput high. Small operations would perhaps not be able to justify the expense. Then too, the removal of all oxygen shifts the oxidation-reduction potential from positive to negative[18], which is conducive to the proliferation of anaerobic micro-organisms, in particular the sulfate-reducing bacteria. This can result in a corrosion problem that otherwise might have remained latent.

On the other hand, sulfate-reducing bacteria can thrive in a system which contains air. It is not unusual to find both hydrogen sulfide and oxygen present in the same system, due to the activity of sulfate-reducing bacteria underneath deposits which effectively prevent the entry of air.

While the problem of oxygen-induced corrosion has been a difficult problem in the past[19], it is now possible to inhibit such corrosion using glassy phosphates containing zinc salts, i.e., a sodium-zinc phosphate glass. This material is also effective in stabilizing or holding soluble iron in solution[20,21], and will prevent the deposition of calcium carbonate and calcium sulfate, should these components be present in the injection water. Then too, catalyzed sodium sulfite may be used as an oxygen scavenger to chemically remove the oxygen from the water. This course of action is being used successfully and economically in many water floods at the present time.

Water Compatibility Problems

Bilhartz[6] has validly questioned the need for complex water treating methods for many of the waters presently used in secondary recovery projects. In general, the best engineering design will result in a water treating facility which will provide injectior water of minimum quality at the least possible cost. This must, of course, be within the limits of safe operating practices, adequate reservoir performance and protection of the water flood facilities over the life of the project.

In the past, concern has been expressed over the need for injecting waters which are compatible with formation waters. Bernard[22] has performed laboratory experiments where the following reacting constituents would be present in cores: (1) barium and sulfate ions, (2) calcium and sulfate ions, (3) ferrous ion and hydrogen sulfide, (4) ferrous ion and oxygen, (5) ferric ion and ammonium hydroxide, and (6) magnesium ion and ammonium hydroxide. The laboratory experiments indicated that the different waters do not mix very much in the reservoir. For this reason, no detrimental effects would be expected where the flood water is incompatible with the connate water. Of course, the mixing of incompatible waters in the treating system without adequate treatment, or at or within the well, could be expected to cause plugging of the injection sandface.

Analytical Methods for Testing Injection Waters

Considerable care must be taken to obtain uncontaminated samples of the source water, treated water, or produced water when analysis for the content of unstable constituents is desired. This is particularly true when measurements of the dissolved gases such as carbon dioxide and oxygen are required. Watkins[23], of the U. S. Bureau of Mines, has prepared a detailed treatment of the analytical methods applicable to the testing of injection waters. This work will be followed closely in the following section on water testing.

The sampling jars should be clean—preferably sterile. The bottle should be rinsed three times with the water to be sampled, and then filled with a rubber tube extending to the bottom of the jar. A quantity of water equal to at least three times the capacity of the jar should be allowed to overflow and the rubber tube slowly removed. A glass stopper should be inserted in such a manner that no air bubbles are trapped. It is preferable if the sample analysis is made in the field on the freshly-caught water sample, or failing this, be transported rapidly to the laboratory and analyzed as soon as possible.

Determination of Dissolved Oxygen. Watkins[23] describes the iodine modification of the Winkler method[24], developed by Taylor and Christianson[25], for use when the water contains iodine-consuming compounds, such as hydrogen sulfide and organic matter. The modification is not necessary in uncontaminated water. The test consists of adding manganous sulfate ($MnSO_4$) and potassium hydroxide (KOH) to a glass-stoppered bottle completely filled with the test water, to obtain manganous hydroxide as a precipitate as indicated by the following chemical reaction:

$$MnSO_4 + 2KOH \longrightarrow Mn(OH)_2 + K_2SO_4 \qquad (2)$$

The manganous hydroxide combines with the oxygen dissolved in the water to form manganese hydroxide:

$$2Mn(OH)_2 + O_2 \longrightarrow 2MnO(OH)_2 \qquad (3)$$

When sulfuric acid is added in the presence of an iodide, the higher oxide of the manganese liberates a quantity of iodine, which is stoichiometrically equivalent to the dissolved oxygen present:

$$MnO(OH)_2 + 2H_2SO_4 \longrightarrow Mn(SO_4)_2 + 3H_2O \qquad (4)$$

$$Mn(SO_4)_2 + 2KI \longrightarrow MnSO_4 + K_2SO_4 + I_2 \qquad (5)$$

The amount of iodine liberated is determined by titrating a portion of the sample with a standard solution of sodium thiosulfate ($Na_2S_2O_3$) with a starch solution used as an indicator, as shown by the following reaction:

$$2Na_2S_2O_3 + I_2 \longrightarrow Na_2S_4O_6 + 2NaI \qquad (6)$$

The iodine modification of the method proposed by Taylor and

Christianson consists of converting the hydrogen sulfide to hydrogen iodide and free sulfur as follows:

$$H_2S + I_2 \longrightarrow 2HI + S \tag{7}$$

Watkins reports that the presence of iodine effectively suppresses other interfering chemicals besides hydrogen sulfide in the test water, so that an accurate determination of the oxygen content is obtained.

The stepwise procedure[23] for the determination of the water sample oxygen content is as follows:

(1) Collect the sample as previously described, and add the following chemicals with a pipette near the bottom of the test bottle, allowing the displaced water to overflow.

(2) Add an excess of 0.5N iodine solution to give the sample a yellow color, and let stand for five minutes.

(3) Add saturated hydrogen sulfide water until the sample is a light straw-yellow. The hydrogen sulfide water is prepared by saturating distilled water with hydrogen sulfide gas.

(4) Add 1 ml of starch solution as an indicator. The starch solution must be fresh.

(5) Add dilute hydrogen sulfide water until the blue color of the water disappears.

(6) Add 0.1N iodine solution until a faint blue coloration of the water reappears.

(7) Add 1 ml of manganous sulfate solution (prepared by dissolving 400 gm of manganous sulfate ($MnSO_4 \cdot 2H_2O$) in distilled water, and diluting to 1 liter).

(8) Add 1 ml of alkaline iodide solution (prepared by dissolving 135 gm of sodium iodide in distilled water, and diluting to 1 liter).

(9) Add 1 ml of concentrated sulfuric acid (sp. gr. = 1.83 or 1.84).

(10) Transfer 200 ml of the test solution from the sample bottle to a 500 ml Erlenmeyer flask.

(11) Titrate the 200 ml sample with 0.025N sodium thiosulfate solution.

The addition of starch indicator may be necessary in step (11) if the yellow color of the sample disappears before the titration drives the solution from a blue cast to colorless. When it is known that no hydrogen sulfide or other interfering constituents are present, steps (7) through (11) are sufficient for determination of the oxygen content of the water. The oxygen content of the water may be calculated from the following equation:

$$V = \frac{WX}{X - (Y - 1)} \tag{8}$$

where

V = dissolved oxygen content of the water, ppm.
W = volume of the 0.025N sodium thiosulfate solution required, ml.
X = sample bottle volume, ml.
Y = total volume of the reagents used, ml.

The number "1" in equation (8) represents the volume of sulfuric acid added. The acid added does not change the test volume of the sample, since the oxygen has already been absorbed.

The unmodified Winkler method for determination of dissolved oxygen content can be used only in waters essentially free from nitrite nitrogen, hydrogen sulfide and other interfering constituents, and containing less than 0.5 mg/liter of ferrous iron. For this reason, a large number of modifications have been proposed to the basic Winkler technique[26]. Some of these modified methods are:

(1) Rideal-Stewart (permanganate) modification[27]. To be used when samples contain ferrous iron. The method is ineffective where the sample contains oxidizable material such as sulfite, thiosulfate, polythionate, or organic material such as might be found in surface waters or sewage.

(2) Alkali-hypochlorite modification. To be used when samples contain sulfite, thiosulfate and polythionate, or large quantities of free chlorine or hypochlorite. The method is not highly accurate.

(3) Alum flocculation modification. To be used when samples contain sizeable quantities of suspended material. After treatment, one of the other oxygen quantity determination techniques is used, consistent with other contaminants present.

(4) Copper sulfate-sulfamic acid flocculation modification. To be used when the sample contains biologic flocs having high oxygen utilization rates. The technique removes the floc, and the remaining clear liquid is tested by another technique for measurement of the oxygen content, consistent with other contaminants that might be present.

(5) Short Theriault modification[28]. May be used when organic material is present which can be oxidized by a highly alkaline solution, or by free iodine in an acid solution.

(6) Pomeroy-Kirschman-Alsterberg modification[29]. May be used when the water sample has a high organic content and/or a high dissolved oxygen content.

The methods outlined above have application to most of the waters encountered in oil field operations. Some of the methods have better application to sewage waste water. However, with injection water becoming increasingly difficult to obtain in populous areas, more and more water floods will be utilizing the water from sewage treating plants.

In very recent years, a number of metering devices for the measurement of dissolved oxygen have appeared on the market. The Precision Scientific Company manufactures a galvanic cell oxygen analyzer, with a measuring range of zero to 50 ppm oxygen in liquid. The Coleman Instruments Corporation markets an analyzer which depends upon the thermal decomposition of the sample in an inert atmosphere, with a subsequent reaction of the oxygen of the pyrolysis products with hot carbon. The resulting carbon monoxide is oxidized to carbon dioxide, which can be measured by weight gain in a carbon dioxide absorption tube. The oxygen present may then be calculated and converted to a convenient set of units. Beckman Instruments, Inc., has available a polarographic line of instruments for the measurement of gaseous oxygen or dissolved oxygen, in both aqueous or non-aqueous solutions. The probe contains two electrodes connected by an electrolyte, across which a voltage is applied. In contact with the water sample (the case of interest here), the oxygen diffuses through a gas-permeable membrane, and is reduced at the cathode. The small current which is generated is proportional to the partial pressure of the oxygen present. The sensor and amplifier-meter can be calibrated to read 0 to 2.5, 0 to 10 and 0 to 50 ppm of dissolved oxygen, with an accuracy of approximately plus or minus 5 per cent. The Magna Corporation, of Santa Fe Springs, California, manufactures a similar instrument. Most of the manufacturers of laboratory instruments have a line of oxygen analyzers. The modest cost of many of these instruments makes possible their use by personnel in fields undergoing water flooding.

Determination of Carbon Dioxide. The following procedure, presented by Watkins[23] and described by Betz[30], depends upon the absorption of carbon dioxide by alkaline solutions. The stepwise procedure is as follows:

(1) Pipette 100 ml of the test water into a 250 ml Erlenmeyer flask and add 10 ml of phenolphthalein indicator. If the solution turns red, no carbon dioxide is present.

(2) If the solution does not turn red, titrate with $N/22$ sodium carbonate solution until a definite pink color develops.

The carbon dioxide content of the water in ppm is equal to the volume of the sodium carbonate solution, divided by the volume of the sample multiplied by 1000. Since other contaminants in the test water can interfere with the titration, a small error in the determination of the carbon dioxide content can result.

Determination of Hydrogen Sulfide. Watkins[23] describes a procedure for the determination of hydrogen sulfide content in water; which depends upon the reduction of iodine by the hydrogen sulfide in the water (usually brine), as illustrated by the following reaction:

$$H_2S + I_2 \longrightarrow HI + S \tag{9}$$

Due to the instability of hydrogen sulfide and the possibility of loss to the air, an excess of iodine solution is added and the test water back-titrated with a standard sodium thiosulfate solution, as shown by the following equation:

$$2Na_2S_2O_3 + I_2 \longrightarrow Na_2S_4O_6 + 2NaI \tag{10}$$

The stepwise procedure is as follows:

(1) Pipette 5 ml of 0.01N standard iodine solution into each of two Erlenmeyer flasks.

(2) Add 1 gm of potassium iodide crystals to each flask.

(3) Add 50 ml of distilled water to one flask and 50 ml of test water to the other flask.

(4) Titrate the test solutions in both flasks with standard sodium thiosulfate solution of the same normality as the iodine solution. Add 1 ml of starch indicator near the end of the titration.

The hydrogen sulfide content of the test water is calculated first by subtracting the milliliters of sodium thiosulfate solution used to titrate the test water from that used to titrate the distilled water sample. This result is in turn multiplied by 3.408, to correct for the 0.01N solutions of iodine and sodium thiosulfate that were used to obtain the hydrogen sulfide contained in the test water in ppm. If the hydrogen sulfide concentration is high, a 0.1N iodine solution may be used, in which case the multiplying factor is changed from 3.408 to 34.08 in the above-outlined determination. Heath and Lee have shown that the presence of organic matter, hydroxide and carbonates will result in an indicated hydrogen sulfide content somewhat greater than the true value. If nitrites are present in the test water, the hydrogen sulfide value calculated will be less than the true value.

The above titrimetric method for the determination of hydrogen sulfide has limitations imposed by interfering substances, and can be used only where the sulfide concentration is above 1 mg/liter. For the lower sulfide concentrations which are quite commonly found in flood waters, and for rapid testing of water samples, the colorimetric method (methylene blue method) is to be preferred. Concentrations of hydrogen sulfide as low as one-hundredth of one ppm can be detected, although the titrimetric method should be used where maximum accuracy is required[26]. The reader is referred to reference 26 for details of the colorimetric method.

Determination of pH Value. A convenient way[23] to determine the pH value of injection water is through the use of indicator solutions and color standards in a Hellige comparator. A cresol red indicator solution covers the pH range from 7.2 to 8.8, while a bromthymol blue indicator solution is valid for the pH range from 6.0 to 7.6. Where greater accuracy is required, any sensitive glass-electrode pH meter may be used.

Determination of Total and Dissolved Iron. The method described here is taken from Watkins[23], as developed by Betz[30]. The technique is based upon the acidification of the water test sample, and the oxidation of the ferrous iron to ferric iron by the addition of hydrogen peroxide. The sample is boiled to dryness, and ammonium thiocyanate and aluminum nitrate solutions added to produce the red coloring, due to ferric thiocyanate. The intensity of the red coloring is proportional to the iron content of the test water, and is measured with a colorimeter or a comparator. The analytical procedure is as follows:

(1) Filter a portion of the test water sample at the instant of sampling, if possible, and test both the filtered and unfiltered samples. (Testing at a later time may give erroneous results, since part or all of the dissolved iron precipitates when it comes in contact with air.) Testing of the unfiltered sample will yield a determination of the total and dissolved iron, while testing of the filtered sample will permit a determination of the dissolved iron present.

(2) If the iron present is in the ferrous or reduced state, treat a 10 ml sample with 1 ml of 1:1 hydrochloric acid (equal parts of concentrated HCl and distilled water) and 1 ml of hydrogen peroxide; then boil dry. This will acidify and oxidize the ferrous iron in the sample.

(3) Pipette 10 ml of distilled water and 2 ml of aluminum nitrate

solution into the flask, and dissolve the solids. (The aluminum nitrate solution is prepared by dissolving 200 gm of aluminum nitrate, $Al(NO_3)_3 \cdot 9H_2O$, in 200 ml of distilled water, and adding 15 ml of 20 per cent ammonium thiocyanate, NH_4CNS. This is extracted in a separatory funnel, with 75 ml portions of 5 to 2 isoamyl alcohol-ether, to remove any iron present. Add 200 ml of concentrated nitric acid and heat to destroy the NH_4CNS under a hood, since poisonous hydrogen cyanide gas is given off. Dilute the final solution to 500 ml with distilled water.)

(4) Add 2 ml of ammonium thiocyanate, and compare with color standards within one minute.

If a colorimeter is used, it will be necessary to devise a standardization curve, using test waters of known concentration. Laboratories making routine water analyses commonly use colorimeters, since iron content may be read rapidly once a standardization curve has been prepared.

Determination of Alkalinity and Carbonate Stability. Watkins[23] points out that the alkalinity of fresh waters is due chiefly to the presence of bicarbonate, carbonate and hydroxyl ions. The standard test for the alkalinity of water consists of titrating water samples with sulfuric acid in the presence of phenolphthalein and methyl orange indicators, with the end points corresponding to pH values of 8.3 and 4.3, respectively. The phenolphthalein indicated water alkalinity is due to the presence of hydroxides, in addition to one-half of the carbonates present. The methyl orange indicated water alkalinity is due to the presence of remaining ions contributing to the alkalinity of the water sample. Usually surface waters and oil field brines have alkalinity, as evidenced by the methyl orange indicator.

The carbonate stability test is concerned with the undersaturation or supersaturation of the water, with respect to carbonate constituents, and is based upon the addition of powdered calcium carbonate to the water. If the water is already supersaturated with carbonate constituents, these constituents will be precipitated to the extent of the supersaturation. If the water is undersaturated, carbonates will be dissolved until a saturation in carbonates is reached. These tests are of considerable interest to the water flood operator, since a supersaturation in carbonates usually results in carbonate precipitation which can plug the formation at the injection well(s). An undersaturation of carbonates in the water promotes the corrosion of metal goods. Hoover[31] has outlined a method for determining the carbonate stability of water, which is readily adapted to measurements of the property in the field.

An analytical procedure[23] for the determination of alkalinity and carbonate stability of a water to be used for injection purposes follows:

(1) Pipette 50 ml of the test sample into a white porcelain casserole.

(2) Add 4 drops of phenolphthalein indicator. Should the sample turn red, titrate to a colorless end point with 0.02N sulfuric acid. Record the ml of acid used as the "P" reading. If no initial red color occurs, proceed to step (3).

(3) Add 4 drops of methyl orange indicator. Titrate the sample with 0.02N sulfuric acid if the sample is initially orange, and record the total ml of acid used to titrate in the presence of both the phenolphthalein and methyl orange indicators as the "M" reading, which corresponds to the total alkalinity of the test water. Should the sample turn yellow-pink when the methyl orange is added, this would indicate that no hydroxyl, carbonate, or bicarbonate ions are present.

(4) Add 5 gm of powdered calcium carbonate to the test sample, and shake thoroughly.

(5) Filter off the supernatant liquid after the precipitate has settled, discarding the first 25 ml of the liquid.

(6) Repeat steps (1) through (3) using the filtered sample of step (5), and use the total acid volume used in steps (3) and (5) as the "S" reading. This corresponds to the total alkalinity of the sample when 100 per cent saturated with carbonates.

The alkalinity of the test water is calculated by multiplying the "P" reading by 20, to yield the phenolphthalein alkalinity in ppm of calcium carbonate. Similarly, multiplying the "M" reading by 20 gives the methyl orange alkalinity in ppm of calcium carbonate present. If the "S" reading is larger than the "M" reading, the difference in the values times 20 will yield the supersaturation of the test water in ppm. Should the "M" reading exceed that of the "S" reading, the difference in the values times 20 will yield the undersaturation of the water tested in ppm.

A detailed treatment of the alkalinity of water and its measurement is available in reference 26, page 44.

Stiff and Davis[32] have presented a method for predicting the scaling tendency of oil field waters due to the presence of calcium carbonate. The technique is applicable to waters containing large salt concentrations. The Langelier equation[32] is used:

$$SI = pH - pCa - pAlk - K \tag{11}$$

where

SI = stability index, positive values indicating scaling tendencies and negative values indicating corrosion

pH = pH of the water sample

pCa = negative logarithm of the calcium concentration

pAlk = negative logarithm of the total alkalinity

K = constant specified by salt concentration and temperature

Stiff and Davis made laboratory experiments to determine values of K at various ionic strengths of water samples. Figure 8-6 presents the results of these experiments. The ionic strength of a given water sample can be calculated from the following equation (12), if a standard water analysis is available:

$$\mu = 0.5 \left(C_1 V_1{}^2 + C_2 V_2{}^2 + \cdots + C_n V_n{}^2 \right) \tag{12}$$

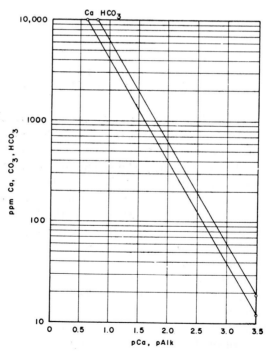

Figure 8-6. Values of K at various ionic strengths (after Stiff and Davis[32]).

where

C = ion concentration, gram ions per 1000 gm of solvent

V = valence of the ion

Table 8-1 provides a convenient list of multiplying factors which allow the conversion of water analysis values given in ppm or milliequivalents/liter to ionic strength, μ.

TABLE 8-1. Conversion Factors for Changing Water Analysis Results to Ionic Strengths (from Stiff and Davis[32]).

Ion	Factor, ppm	Factor, meq/liter
Na	2.2×10^{-5}	5×10^{-4}
Ca	5.0×10^{-5}	1×10^{-3}
Mg	8.2×10^{-5}	1×10^{-3}
Cl	1.4×10^{-5}	5×10^{-4}
HCO_3	0.8×10^{-5}	5×10^{-4}
SO_4	2.1×10^{-5}	1×10^{-3}

The sum of the individual ion strengths yields the total ionic strength of the water sample. Knowing water temperature in the treating and injection system will allow the determination of K by means of Figure 8-6. Since the pH of the water sample is a simple measurement and K has been calculated, the stability index as defined by equation (11) may be calculated if values for pCa and pAlk can be determined. The test procedures of this section will yield values for the ppm of calcium, carbonate, and bicarbonate. Figure 8-7 allows the ready conversion of these values to pCa and pAlk. The total alkalinity in ppm is the sum of the carbonate, bicarbonate and hydroxide components of the sample in ppm.

The importance of predicting scaling tendencies is not limited just to the injection water and the injection plant wells, but also is of interest in the producing wells. A knowledge of the scaling tendencies of a water will be of importance in determining what treatment prior to injection will be necessary. Stiff and Davis[34] have also published a technique for predicting the tendency of oil field waters to deposit calcium sulfate.

Determination of Hardness. Carbonate hardness is normally considered to be that up to the methyl orange alkalinity. Hardness in excess of this amount is usually considered to be due to calcium and magnesium in the test water. Since calcium and magnesium hardness are the main constituents of water which contribute to soap hardness, titration of a test sample with a standard soap solution to the point where a lather persists constitutes a valid and simple hardness test. The analytical test procedure is as follows[23]:

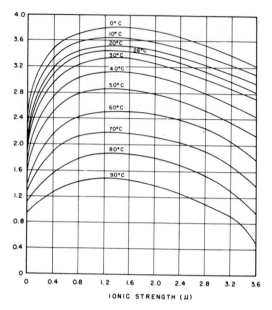

Figure 8-7. Graph for converting ppm of calcium and alkalinity to pCa and pAlk (from Stiff and Davis[32]).

(1) Pipette 50 ml of the test water into an 8-ounce stoppered bottle.

(2) Add the standard soap solution in 0.5 ml amounts and shake vigorously after every addition. As the end point is reached, reduce the additions of soap to 0.1 ml amounts. The lather in the test bottle must persist for 30 seconds after agitation before a true end point has been reached.

Should more than 7 to 8 ml of standard soap solution be required to reach an end point, the test sample should be diluted with distilled water to a 50 ml volume, and the proper multiplying factor used. Excessive precipitations of calcium and magnesium salts will otherwise mask the true end point of the titration. The hardness of the water to a soap titration can be determined from the following equation:

ppm hardness in $CaCO_3$ =

$$\frac{1000 \text{ (ml of soap solution used-lather factor in ml)}}{\text{ml of test water}} \quad (13)$$

where the lather factor is determined by adding the standard soap solution to water having zero hardness. Such water can be prepared by boiling distilled water. Usually from 0.3 to 1.5 ml of standard soap solution are necessary to develop a lather in water of zero hardness. Standard soap solution is available from all chemical supply houses.

Recently an improved, though somewhat more cumbersome water hardness determination technique[26] has been developed. The technique is widely used in commercial laboratories, but would be difficult to perform by most field personnel on a water flood site. The titration method is based upon the fact that ethylenediaminetetraacetic acid (EDTA) and its sodium salts, when added to a water containing metal cations (Ca, Mg, Sr, Fe, Al, Zn, and Mn), form a chelated soluble complex. If the pH of the water sample is increased to 10 and a dye such as chrome black T is added, a dark red coloring results. Upon titration with EDTA, complete complexing of the calcium and magnesium present will yield a blue end point. The details of reagent preparation and analytical procedure will not be developed here.

Determination of Chlorides. The standard test for the presence of a chloride ion in water is by titration with silver nitrate solution, with potassium chromate as the indicator of the end point. The analytical test procedure is as follows[23]:

(1) Pipette 50 ml of test sample into a white porcelain casserole. If the water has a high chloride ion content, dilution with distilled water to 50 ml may be necessary.

(2) Add 5 drops of potassium chromate indicator to give the sample a bright yellow color.

(3) Titrate the sample with a standard silver nitrate solution to obtain a faint red coloration. The final reading on the silver nitration solution used should be decreased by 0.2 ml, since this quantity is needed to titrate distilled water.

The calculation for the total chloride content in ppm is as follows:

ppm total chlorides =

$$\frac{(ml\ AgNO_3\ used)(AgNO_3\ strength\ in\ mg/ml)(1000)}{ml\ of\ test\ water} \quad (14)$$

Determination of Residual Chlorine. This test is necessary in those water flooding projects where chlorine is being injected to control certain micro-organisms. The test water is usually taken at the injection well head. A small residual chlorine content of the injection water would show that sufficient quantities have been added for the purpose intended. Large residual chlorine concentrations are unnecessarily destructive to unprotected steel equipment. The Bureau of Mines[23] has developed a method for determining the residual chlorine in injection water, using orthotolidine as an indicator. The yellow color produced is tested on a colorimeter which has been precalibrated by prepared standards. The orthotolidine indicator can be prepared[35], but can be conveniently obtained from most chemical supply houses. Only small quantities should be kept on hand because of its short shelf life. The analytical test procedure is as follows[23]:

(1) Pipette 100 ml of the test water into a clean, glass-stoppered bottle.

(2) Add 1 ml of the orthotolidine solution and place in dark storage for 15 minutes.

(3) Zero the colorimeter on the original untreated sample using a glass cell with a 20 mm viewing depth and a 420 millimicron blue filter.

(4) Using the treated sample, balance the instrument, and determine the chlorine content from the calibration curve prepared for this test and this colorimeter.

Figure 8-8 shows a calibration curve which was obtained using a Klett-Summerson colorimeter.

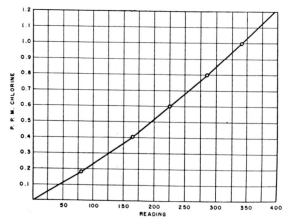

Figure 8-8. Calibration curve for a Klett-Summerson colorimeter for residual chlorine content (after Watkins[7]).

Determination of Turbidity. Watkins[23] has described the use of a Klett-Summerson colorimeter for the determination of turbidity of waters to be used in secondary recovery of oil. The method involves the use of standard solutions of Fuller's earth over the range of 0 to 400 ppm of silicon dioxide to calibrate the colorimeter. During testing of injection waters, distilled water is used to zero the instrument; the test water is then checked, and the ppm of silicon dioxide equivalent is read from the previously prepared calibration chart.

Corrosion Tests. An ever-present problem in water flooding is the corrosion of metal goods throughout the water injection and oil producing facilities of the water flooding system. Depending upon the salvage value and expected life of the project, optimum engineering of a water flood project would constitute use of only sufficient corrosion inhibiting chemicals and special alloys for the best possible economic performance of the project. Unfortunately, it is seldom possible to know exactly the life of the project, or the exact extent and severity of corrosion at every point in the system. Certainly, a treating plant should be constructed in such a manner that the nuisance-type of corrosion can be readily dealt with at a minimum of expense. Still, where the water is known to be corrosive and the life of the flood will be protracted, information on approximate corrosion rates of metal goods will be of considerable importance.

Steel coupons installed at key points in the treating system to act as steel corrosion specimens have been used for many years in the petroleum industry. The Bureau of Mines[23] has developed a variation of the technique which has some advantages in determining the corrosiveness of a particular flood water. Essentially, the method makes use of steel corrosion specimens which have been cut from 0.025-inch steel shim stock, polished with emery cloth and precorroded with dilute hydrochloric acid at 130°F for 24 hours. This procedure results in a very smooth specimen surface. The prepared samples are accurately weighed and suspended in a one-half gallon glass jar with cotton or nylon thread, thereby avoiding the setting up of an electrolytic cell with the test water. Each test jar is closed with a three-hole rubber stopper, each containing a glass tube extending nearly to the bottom of the jar. A rubber hose connected to the test water source provides a continuous supply of the water at a 5· ml per second rate, which corresponds to a velocity of 0.053 cm per second past the test specimen. The second glass tube in the jar serves as a water overflow, while the third tube, which is closed at the bottom, serves as a well for a thermometer. A convenient test period is 72 hours. The specimen is removed and thoroughly cleaned in a solution of 10 per cent ammonium citrate and 10 per cent ammonium hydroxide at 55 °C for 1 hour, and then weighed again. The loss of weight of the specimen is of course that due to corrosion by the test water. To ensure accuracy, a non-corroded specimen should be subjected to the cleaning process required of the corroded specimen, to determine the average loss caused by the cleaning process alone.

Speller's formula[36] can be used to convert the loss in specimen weight per unit area to the average penetration in inches per year:

$$P = \frac{A}{B \times C} \qquad (15)$$

where

P = average penetration per year in inches
A = loss in weight of the sample in pounds per square inch
B = weight of the metal in pounds per cubic inch
C = duration of the test period in years

It should be borne in mind that equation (15) provides a value for *average* penetration rate per year, and cannot adequately take into account the tendency for the water to pit or blister a given type of metal good in the treating and injection system. Corrosion rates are often given in mils per year, where a mil is 0.001 inches. Where information on pit, depth and frequency is of importance, the test can be extended out to 30 days, at which time the cleaned specimen can be examined for pit depth (measured in mils) and pit frequency (pits per square inch). This jar test procedure is satisfactory for the pilot testing of an injection water. For the control testing of a water flood, it is preferable to insert the metal specimen in the stream of water at a bull plug or other convenient entry point, at specific points in the treating and injection system. Information on corrosion coupons and coupon holders has been published[37].

Filter Tests. The success of most water flood projects depends upon the ability to inject water at reasonable pressures and in adequate volumes. Foreign material in the injection water will normally plug the sandface at the injection well over a period of time. Considerable expense can be incurred to restore injectivity to the well, and in severe cases can result in a loss of a well for the purposes of accepting injection water. An important tool to determine the quality of the injection water is the membrane filter[38,39]. Johnston and Castagno[40] have described apparatus, test procedures and the interpretation of filter data on injection waters. Figure 8-9 shows a diagram of the portable water quality testing equipment used by Johnston and Castagno.

The apparatus consists of a 3½ inch by 30 inch lucite cylinder, to which is attached a graduated sight glass. The filter holder is

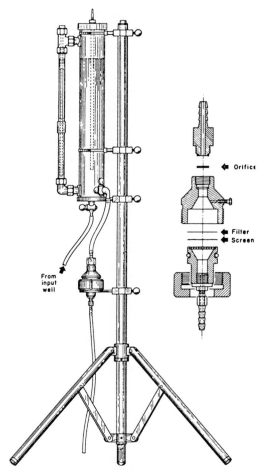

Figure 8-9. Schematic sketch of a portable water quality tester of the membrane filter type (from Johnston and Castagno[40]).

constructed of plastic and stainless steel, and is equipped with a
0.0225 inch orifice plate. A membrane filter 47 mm in diameter, having
a 0.45 micron pore opening size, a 14 md permeability to water and a
50 mesh stainless steel screen are supported by the filter holder.
The beginning point of the test is so arranged that the water level
is 87 cm above the top of the orifice plate, which yields an initial
flow rate of 50 ml per minute. Readings of the drop in fluid level
are taken at five minute intervals and the data developed so that
the water flow rate may be plotted versus test time on semi-loga-
rithmic graph paper, as shown in Figure 8-3. Notice that an arbitrary ordinate for the water quality number has been incorporated
into this figure. The authors have used a water quality numbering
system, based upon the decline in the rate of flow between the 10
and 40 minute time intervals. Since the water head is always
dropping, some drop in rate will occur even with filtered deionized
water. A calculation of the water quality number is shown on Figure 8-10. In order that the water quality number so defined will
have meaning, Johnston and Castagno have developed a classification where the water quality number is related to the absolute
gas permeability of injection well sand cores.

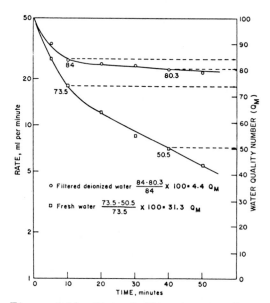

Figure 8-10. Plot of filtered water flow
rates for the determination of water quality
number, Q_M (after Johnston and Castagno[40]).

Figure 8-11 reproduces this classification. The classification shows that waters with low water quality numbers are required where the sands to be flooded are of low permeability, whereas a poorer quality water, as evidenced by a higher water quality number, may be used satisfactorily where higher permeability sands are being flooded.

Most of the cases of poor quality injection water, as indicated from the filter tests, can be traced to poor water plant filtering systems, dirty filtering media, excessive quantities of oil in the injection waters, commingling of incompatible waters and the use of excessive quantities of water treating chemicals.

Bell and Shaw[41] have classified waters using the same test filter media as Johnston and Castagno, but using a constant four foot

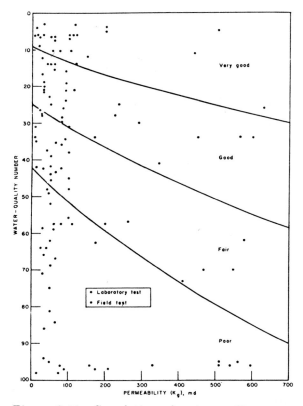

Figure 8-11. Correlation of water quality numbers with reservoir permeability (after Johnston and Castagno[40]).

water head throughout the test interval. The slope of the plot of change in flow rate versus the cumulative volume of water which has passed the membrane filter, gives a measure of the relative plugging tendency of the water tested. Figure 8-12 presents a classification of waters using this testing technique.

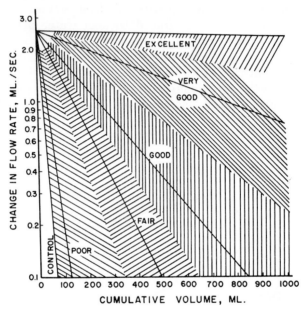

Figure 8-12. Classification of waters using a membrane filter and 4 foot water head (from Bell and Shaw[41]).

Stormont[42] has used the membrane filter test, but with 20 psi water head. This procedure probably yields a measure of the filter cake on the membrane filter, due to the screening out of particulate matter. Figure 8-13 provides a classification of waters by this technique, where the plotted values of the ordinate and abscissa provide the details on the type and form in which the test data is obtained.

Since the membrane filter is a synthetic medium constructed from cellulose esters, the trapped particulate matter can be analyzed. A variety of chemical tests, bacteriological examinations and microscopic inspections can be readily made. In addition, upon completion of these tests, the filter may be washed free of

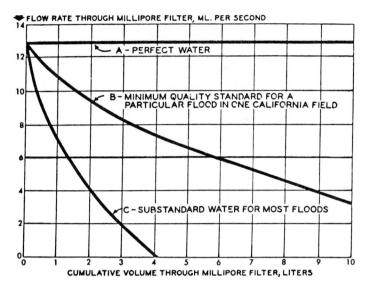

Figure 8-13. Classification of injection water quality by filter tests using 20 psi water head (from Stormont[42]).

salts, dried and then weighed to yield information on the quantity of particulate matter contained in a given quantity of water. Then too, washing of the filter in carbon tetrachloride, drying and weighing again will allow the quantity of oil in the flood water in ppm to be calculated.

It might be well at this time to point out that the term ppm has been erroneously used[43] for many years, when really milligrams per liter (mg/l) was the actual quantity measured. Where the water or oil varies from a specific gravity of one, significant errors can result from this inaccuracy.

Biological Analysis. The detection of biological growth in injection water is of considerable importance, since high bacterial counts usually go hand-in-hand with high corrosion rates of metal goods and plugging of the sandface in the injection wells. Low bacterial counts do not necessarily mean that harmful bacteria are not present[44], since active colonies may be flourishing in blind tees, slime films, or stagnant sections of the water system. Procedures for the detection of bacterial growth may be made by methods recommended by API RP 38[45]. Techniques are outlined which permit the identification of sulfate-reducing bacteria, iron bacteria, slime-forming bacteria, algae and protozoa and other sulfide-producing bacteria. Usually these tests will be conducted by a commercial laboratory, since few field laboratories would be set up to make these kinds of tests.

Sharpley[46] has outlined a method which does provide a means of determining the total bacterial count in oil field water where laboratory facilities are limited. Small quantities of the test water are filtered through a special membrane. All bacteria that are trapped are fixed and stained, the filter dried and then saturated with an immersion oil to render the filter transparent. The total bacteria present may then be counted using an oil-immersion lens under a microscope. The main disadvantage of the technique is that the total bacteria count may be many times the viable count. On the other hand, the method is rapid, and will allow the detection of bacteria that might not grow under routine culture methods.

Some types of bacteria cannot be counted in a culture system. Specific examples are iron bacteria, of which the cultural techniques determine none, and hydrocarbon-utilizing bacteria, of which the standard API media will count none.

An indirect biological analysis is possible by measuring the gain in hydrogen sulfide, from one end of the treating system to the other. A gain in hydrogen sulfide content would indicate the presence of sulfate-reducing bacteria, or other sulfide-producing bacteria. The method is not entirely satisfactory, since the hydrogen sulfide generated by bacterial growth may be removed from the water by ferrous iron, resulting in no detectable gain in the quantities of the gas present.

To insure that the biological analysis of subsurface injection waters will yield valid results, strict water sampling methods should be adhered to[45]. A clean sterile bottle must be used, and the sample must be protected from external contamination during sampling. Then too, temperature changes should be minimized between the time of sampling and time of analysis, and the sample should be cultured within 24 hours. A microscopic examination of the water should include identification of the following microorganisms[45]:

(a) Algae and protozoa—flagellates, ciliates, diatoms, and filamentous.

(b) Fungi.

(c) Bacteria—iron bacteria (sheath or stalk), slime formers, sulfur bacteria with counts of each type.

For general bacterial counts in injection waters, the following medium, which can be purchased from a biological supply house as TGE Agar, should be used[45]:

Beef extract	3.0 gm
Tryptone	5.0 gm
Dextrose	1.0 gm
Agar	15.0 gm
Distilled water	1,000 ml

The pH of the resulting media should be adjusted to 7.0 with NaOH, and sterilized for 15 minutes. The media will be adequate for those waters containing less than 20,000 ppm solids.

Testing procedures call for the preparation of dilution bottles or tubes containing 1 ml, 0.1 ml, 0.01 ml of the test water in a total dilution sample of 10 ml, by means of serial dilutions. Dilutions are accomplished using a sterile pipette, with vigorous shaking of the dilution sample bottle at each step. In preparing plates for microscopic analysis, the standard procedure[35] is to place an appropriate volume of the water sample or its dilution in a petri dish, to which a 10 ml volume of liquified agar medium is then added. Tilting and rotating the dish results in adequate sample mixing, with the gel medium and solidification over the bottom of the dish.

To match field conditions, the inoculated plates should be incubated under aerobic conditions and within 5°C of the injection water temperature. Counts should be made after 2 to 5 days of incubation.

A medium of the following composition should be used for counting sulfate-reducing bacteria, or for counting bacteria in test waters containing more than 20,000 ppm solids[45,47]:

Ferrous ammonium sulfate	$(Fe(NH_4)_2(SO_4)_2 \cdot 6H_2O)$	0.2 gm
Ascorbic acid (vitamin C)		0.1 gm
Magnesium sulfate	$(MgSO_4 \cdot 7H_2O)$	0.2 gm
Di-potassium phosphate	$(K_2HPO_4$ anhydrous)	0.01 gm
Yeast extract		1.0 gm
Sodium chloride		10.0 gm
Sodium lactate, USP		
(60 per cent syrup)		4.0 ml
Agar		15.0 gm
Distilled water		1,000 ml

The pH of the resulting medium should be adjusted to 7.3 with NaOH, 9 ml of the medium placed in each test tube and sterilized for 10 minutes in an autoclave. After sterilization, 1 ml of the test water is injected into the first test tube. The tube is stoppered under sterile conditions, and then mixed. This results in a 1 : 10 dilution. Withdrawal of 1.0 ml from the first test tube, with injection into a second test tube, results in a 1 : 100 dilution of the test water in the second bottle. This serial dilution is continued until the fifth test tube, at which point the dilution is 1 : 10,000. The culture test tubes should be kept within 5°C of the injection water of the system, for a minimum of four weeks. The tubes should be inspected on the third day and at the end of each week, for the possible appearance of intense black colonies of sulfate-reducing bacteria. Blackening of the bottled medium indicates the numbers of sulfate-reducing bacteria in the test water. For instance, if only the first bottle is blackened, there would be from one to ten bacteria present in the 1 ml water sample. Similarly, if the first and second bottles were blackened, 10 to 100 bacteria per 1 ml of test would be indicated. To avoid errors in bacterial counts, particular care should be taken in sterilizing the test tube stoppers and in accomplishing the dilutions. All work should be done in duplicate.

The presence of sulfate-reducing bacteria in the system is usually considered to be potentially serious. When general bacterial counts of less than 10,000 organisms per milliliter are found in an untreated water, it is usually considered unnecessary to institute treatment to control bacteria. If the bacterial count is in excess of 10,000, treatment may be necessary if filter plugging, loss of injectivity, or increased pumping pressures are encountered.

The biological analysis of the water used in water flooding is also useful for the evaluation of chemicals used for controlling microbial growth. The medium used for a bacteriostatic test for the chemical control of sulfate-reducing bacteria in the test water is the same as that previously described. The test culture is Mid-Continent Strain A, and may be obtained from the Department of Biology, University of Houston, Houston, Texas. The test organisms for heterotrophic bacteria have been designated as pseudomonas fluorescens (API strain) and bacillus cereus (API strain), and are also available from the above-mentioned source[45]. This type of test for evaluating the effectiveness of various chemicals in controlling biological growth in an injection water involves exposing micro-organisms, and at set time intervals, determining the numbers of the surviving organisms.

Chemical Water Analysis Patterns

Most chemical water analysis data prepared by commercial laboratories are presented in graphical form. Several different methods for presenting analytical water information have been used[48,49,50,51]. Stiff[52] has presented a graphical plotting technique, whereby the effect of dilution or concentration has been reduced to a minimum. Figure 8-14 presents the essential features of the system, where positive ions are plotted to the left and negative ions are plotted to the right of the center line, in units of milliequivalents per liter.

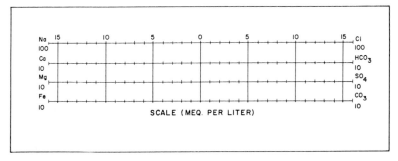

Figure 8-14. Scaling diagram for graphical presentation of chemical water analysis data (after Stiff[52]).

Should the results of a chemical analysis be in terms of parts per million, the conversion can be made by dividing by the equivalent weight in milligrams. Table 8-2 summarizes the appropriate conversion factors. The "milligram equivalent" value of a radical is

TABLE 8-2. Table of Milligram Equivalent Conversion Factors for Radicals Found in Water.

Positive Ions	Milligram Equivalent*	Negative Ions	Milligram Equivalent*
Aluminum, Al	0.1112	Bicarbonate, HCO_3	0.0164
Calcium, Ca	0.0499	Carbonate, CO_3	0.0333
Hydrogen, H	0.9922	Chloride, Cl	0.0282
Iron, Fe	0.0358	Hydroxide, OH	0.0588
Magnesium, Mg	0.0822	Nitrate, NO_3	0.0161
Potassium, K	0.0256	Sulfate, SO_4	0.0208
Sodium, Na	0.0435		
		Alkalinity, $CaCO_3$ 0.0200	
		Free CO_2, CO_2 0.0454	

*The valence of a radical divided by its atomic or molecular weight.

identical with the terms "milliequivalent" or "reacting value." The parts per million of a radical, multiplied by its milligram equivalent, results in a value for the milligram equivalents per liter. For example, if a water analysis showed the sample to contain 25 ppm of carbonate, then the milligram equivalent per liter would be 25×0.0333, or 0.8325. Figure 8-15 shows some common water patterns. The actual scales used can be changed to accentuate particular properties of water in water treatment work, in injection water studies and in corrosion control. Water analyses taken at different points through a given system, or at different times at the same sampling point, can be analyzed effectively by this graphic method.

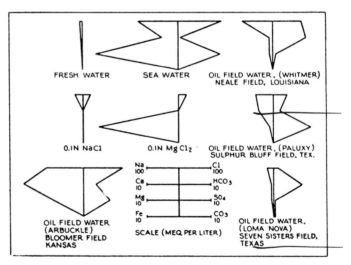

Figure 8-15. Water analysis patterns for some common waters (after Stiff[52]).

Water Quality Rating Chart

Wright[53] has presented a water flood rating chart to aid in the evaluation of a water system, from relatively inexpensive tests on the injection water. Table 8-3 is a reproduction of this chart. The rating values used are: 1 - excellent, negligible or none; 2 - very good, very low, or minor; 3 - good, low or minor; 5 - acceptable or moderate; 10 - fair, large, high, or deep; and 20 - excessive.

It can be noted that the proposed rating is based upon a number of tests: the membrane filter test, tests to show total sulfide increases, total iron count increases and total bacterial count in-

TABLE 8-3. Water Flood Rating Chart (from Wright[55]).

Rating	1	2	3	5	10	20
Membrane filter test (0.45 μ filter slope)	0-0.09 Excellent	0.10-0.29 Very good	0.30-0.49 Good	0.50-0.99 Acceptable	1.00-1.79 Fair	1.80+ Excessive
Filtered solids mg/l	0-0.4 Negligible	0.5-0.9 Very low	1.0-2.4 Low	2.5-4.9 Moderate	5.0-9.9 Large	10.0+ Excessive
Total sulfide increases lb/day/1,000 sq ft	0 None	0.001 Very low	0.002-4 Low	0.005-9 Moderate	0.01-0.019 Large	0.02+ Excessive
Iron-count increases lb/day/1,000 sq ft	0 None	0.001-0.011 Very low	0.012-0.11 Low	0.12-0.59 Moderate	0.60-1.1 Large	1.2+ Excessive
Sulfate-reducing bacteria colonies/ml	0 None	1-5 Very low	6-9 Low	10-20 Moderate	30-90 Large	100+ Excessive
Total bacteria count colonies/ml	0 None	1-99 Very low	100-999 Low	1000-9999 Moderate	10,000-99,999 Large	100,000+ Excessive
Corrosion rate (30 days) (insulated coupon) mils/year	0 None	0.01-0.09 Very low	0.10-0.99 Low	1.00-4.9 Moderate	5.0-9.9 High	10.0+ Excessive
Pit depth (30 days) (insulated coupon) mils	0 None	1 Shallow	2-3 Minor	4-5 Moderate	6-10 Deep	10+ Excessive
Pit frequency (30 days) (insulated coupon) pits/sq in	0 None	1 Very low	2 Low	3 Moderate	4 High	5+ Excessive

creases, and corrosion coupon tests. In order to properly interpret the results of such tests on a specific injection water, the type of chemical treatment being used at the time of the tests must be kept in mind[53]. Water-soluble nonfilming agents and water-dispersible chemicals may give increased membrane slopes (as defined by Figure 8-11) due to the presence of the chemical in the second instance, or due to the erosion or temporary increase of bacteria, iron counts, total sulfides, or filtered solids. Then too, the use of filming agents in the water system may cause ferrous sulphide to be deposited on the system walls, thus preventing a true measure of sulfide and iron count increases. Oxidizing agents can react with the sulfide ion to convert it to sulfur, or even the sulfate ion. The current type of chemical treatment can affect the values of the variables in Table 8-3. This should be kept in mind when interpretating the results of the various tests applicable to injection waters.

The system tolerance to hydrogen sulfide increases, reported in Table 8-3 as the sulfide increases in lb/day/1000 square feet of exposed system, is placed lower than that of iron count increases, since sulfide generation usually results in pitting-type corrosion. The method for determining hydrogen sulfide content of injection water has been explained in a previous section. The iron count increases at a given point in the system can be determined by mounting the corrosion specimen in the system. The specimen is kept from contacting the system metal goods by means of an insulator. The iron count increases are determined over extended time periods by this method, or by the United States Bureau of Mines technique, which has already been described. The presence of substantial and deep pitting on the corrosion coupons usually can be interpreted to mean that a potentially serious condition exists, or is developing in the system. Depending upon the life of the project and the cost of metal goods, treatment to decrease the corrosion rate would be indicated. The iron count increase over the entire system can only be determined by the analytical methods for the total iron content, both dissolved and suspended in the water, at the entrance to the system and at the well head or exit to the system.

References

1. Thornton, W. M., *Petrol. Engr.*, 50 (September, 1963).
2. Earlougher, R. C., and Amstutz, R. W., "Water Flooding," Petr. Trans., Reprint Series No. 2, p. 30, Dallas, **AIME**, 1959.

3. Torrey, P. D., "Water Flooding," Petr. Trans. Reprint Series No. 2, p. 22, Dallas, **AIME**, 1959.

4. Buck, J. R., *Petrol. Engr.*, B-95 (February, 1956).

5. Water Flooding, Vocational Training Series, Univ. of Okla., 7 (1956).

6. Bilhartz, "Water Flooding," Petr. Trans., Reprint Series No. 2, p. 33, Dallas, **AIME**, 1959.

7. Watkins, J. W., Willett, F. R., Jr., and Arthur, C. E., "Conditioning Water for Secondary Recovery in Midcontinent Oil Fields," U. S. Bureau of Mines, Rep. of Invest. 4930, 1952.

8. Voss, N. A., and Nordell, E., *Oil Gas J.*, 174 (November 9, 1950).

9. Kirk, J. W., *J. Petrol. Technol.*, 1220 (November, 1964).

10. Lada, A., "Handbook of Modern Secondary Recovery Methods," p. 32, Tulsa, *The Petroleum Publishing Co.*, 1959.

11. Cecil, L. K., *World Oil*, 176 (August, 1950).

12. Alciatore, A. F., Harris, M. B., and Wallin, W. E., *Petrol. Engr.*, B-57 (May, 1955).

13. Bell, G. R., and Jackson, T. M., Jr., *World Oil* (September, 1956).

14. Weeter, R. F., Soc. Petr. Engrs. paper no. SPE 933, **AIME**, presented Houston (October 11-14, 1964).

15. Brewster, P. M., Jr., Dibble, K. M., Jordan, G. S., and Neenan, A., *Producers Monthly*, 18 (July, 1955).

16. Hart, W. J., and Wingate, R. G., *Soc. Petrol. Engrs.* paper no. 515-G, **AIME**.

17. Weeter, R. F., *Petrol. Engr.*, 51 (May, 1963).

18. Baumgartner, A. W., *Soc. Petrol. Engrs.* paper no. SPE 805, presented Fort Worth (March 23-24, 1964).

19. Hatch, G. B., and Ralston, P. H., preprint, N.A.C.E. South Central Regional Mtg. (October 15-19, 1963).

20. Hatch, G. B., and Rice, O., *Ind. Eng. Chem.* 37, 710 (August, 1945).

21. Rice, O., and Hatch, G. B., *Ind. Eng. Chem.* 32, 1572 (1940).

22. Bernard, G. G., Symp. on Waterflooding at Urbana, Illinois, Bull. 80, Ill. State Geol. Sur., 98 (1957).

23. Watkins, J. W., "Analytical Methods of Testing Waters to be Injected Into Subsurface Oil-Productive Strata," U. S. Bureau of Mines, Rep. of Invest. 5031, February, 1954.

24. Winkler, L. W., *Die Bestimmung des im Wasser Golesten Sauerstoffes*, Ber. 21, 2843 (1888).

25. Taylor, S. S., and Christianson, L. F., "Application of Sand Filters to Oil-Field Brine Disposal Systems," U. S. Bureau of Mines, Rep. of Invest. 3334, 1937.

26. American Public Health Association, "Standard Methods for the Examination of Water and Sewerage," 11th Ed., Albany, N.Y., Boyd Printing Co., 1960.

27. Ruchhoft, C. C., Moore, W. A., and Placak, O. R., *Ind. Eng. Chem.*, Anal. Ed. 10, 711 (1940).

28. Theriault, E. J., and McNamee, P. D., *Ind. Eng. Chem.*, Anal. Ed. 4, 59 (1932).

29. Pomeroy, R., and Kirschman, H. D., *Anal. Chem.* 17, 715 (1945).

30. Betz, W. H., and Betz, L. D., "Betz Handbook of Industrial Water Conditioning," Phila., W. H. and L. D. Betz Co., 1945.

31. Hoover, C. P., *Water Works and Sewerage* **77**, 287 (1930).
32. Stiff, Henry A., Jr., and Davis, L. E., *Petr. Trans. AIME* **195**, 213 (1952).
33. Lewis and Randall, "Thermodynamics and the Free Energy of Chemical Substances," p. 373, New York, McGraw-Hill Book Co., Inc., 1923.
34. Stiff, Henry A., Jr., and Davis, L. E., *Petr. Trans. AIME* **195**, 25 (1952).
35. American Public Health Association, "Standard Methods for the Examination of Water and Sewerage," p. 286, 9th Ed., N.Y., 1946.
36. Speller, F. N., "Corrosion Causes and Prevention," p. 621, N.Y. McGraw-Hill Book Co., Inc., 1926.
37. "Corrosion of Oil and Gas Well Equipment," p. 17, API and Nat. Assoc. Of Corrosion Engrs., 1958.
38. Doscher, T. M., and Weber, Leon, *API Drlg. & Prod. Prac.*, 169 (1957).-
39. Felsenthal, M., and Carlberg, B. L., *Petrol. Engr.*, B-53 (November, 1958).
40. Johnston, K. H., and Castagno, J. L., "Evaluation by Filter Methods of the Quality of Waters Injected in Waterfloods," U. S. Bureau of Mines, Rep. of Invest. 6426, 1964.
41. Bell, W. E., and Shaw, J. K., *Oil Gas J.*, 84 (January 12, 1954).
42. Stormont, D. H., *Oil Gas J.*, 84 (November 3, 1958).
43. Case, L. C., and Riggin, D. M., *Oil Gas J.*, 72 (January 6, 1949).
44. Wright, C. C., *API Drlg. & Prod. Prac.*, 134 (1960).
45. "API RP 38, Recommended Practice for Biological Analysis of Water-Flood Injection Waters," N.Y., API, 2nd Ed., June, 1965.
46. Sharpley, J. M., "Devel. in Ind. Microbiology," V. 1, p. 253, N.Y., Plenum Press, 1960.
47. Moore, B. H., *Soc. Petrol. Engrs.* paper no. 887-G, **AIME**, presented Dallas (October 6-9, 1957).
48. Tickell, E. G., *Report of the California State Oil and Gas Supervisor* **6**, No. 9, 5 (1921).
49. Reistle, C. E., "U. S. Bureau of Mines Technical Paper 404," 1927.
50. Parker, J. S., and Southwell, C. A. P., *Jour. of Inst. of Petr. Technologists* **15**, 138 (1929).
51. Corps, E. V., *Proc. World Petr. Congress* **1**, 338 (1938).
52. Stiff, H. A., Jr., *J. Petrol. Technol.*, 15 (October, 1951).
53. Wright, C. C., *Oil Gas J.*, 154 (May 20, 1963).

PROBLEMS

1. A water flood is projected for a section of the country where water is in short supply. It has been suggested that the effluent from a sewage treating plant in the vicinity be considered for use. What reservoir, effluent quality, treating, and equipment considerations would be required in order to make a decision as to the suitability of such a water source?

2. Water is to be filtered through 18 in. of uniform sand of 42 per cent porosity and having an average grain diameter of 2.29×10^{-3} ft. If the

water temperature is 55°F and the throughput is 2.5 gpm/sq ft, what would be the approximate head loss through the filter in psig? Assume the sand grains are spherical.

3. A water which is being considered as a possible injection water has the following analysis:

Sodium and potassium	18,423 ppm
Calcium	1,009 ppm
Magnesium	162 ppm
Sulfate	2,010 ppm
Chloride	28,940 ppm
Carbonate	0 ppm
Bicarbonate	435 ppm
Iron	0.33 ppm
pH	7.1

Determine:

(a) Scaling or corrosion tendency of the water.

(b) Total alkalinity in milliequivalents per liter, ppm, and ionic strength.

(c) The most feasible water treating system where 500 BWPD is to be injected.

(d) In view of the treating system chosen in Part (c), what tests should be made to maintain water quality, and at what approximate time intervals?

4. Which of the more common bacteria that have been found in oil field waters could exist in the water of Problem 3? What routine steps should be taken to insure that bacteria will not multiply in the system consistent with an economical operation of the water injection facilities?

5. Discuss the differences which could exist between a satisfactory oil field injection water and a potable water.

CHAPTER **9**

Gas Injection—
Immiscible Displacement

The idea of using a gas for the purpose of restoring the productivity of oil wells was suggested in the patent literature as early as 1864—just 5-years after the drilling of the Drake well. The first actual use of gas for increasing oil production can be credited to James D. Dinsmoor. Dinsmoor was working as a roustabout on the William Hill property of Venango County, Pennsylvania, in 1888, when a company operating an offsetting property left a gas sand in contact with the oil producing sand of the area, until casing and a packer could be run. The oil producing rate of nearby oil wells was noticeably improved by the unintentional repressuring with gas. In 1890 or 1891, Dinsmoor had the opportunity to intentionally turn gas from a gas sand into an oil sand in a common well bore. The oil production from the property was more than doubled. In 1895, Dinsmoor combined gas repressuring and the use of vacuum pumps in the oil producing wells. The pump used provided a vacuum on its suction side, while the compression side was used to increase gas pressure for the injection operation. This could be considered as the first use of a compressor for the purpose of gas injection.

Many of the early gas injection projects utilized air as the injection medium. The first recorded instance of air injection was apparently by I. L. Dunn, who used the process on a Cow Run Sand field near Chesterhill, Ohio, in 1911. Since that time, air injection has been practiced continuously. Only in the past 20 years has a consistent engineering understanding of the gas drive mechanism (including air) been formulated.   Research and field trials continue to enlarge the understanding of the injection process.

Unlike the water flooding process where only an immiscible displacement of the oil can occur, the gas injection process may result in either an immiscible, or miscible displacement process; under certain operations, both mechanisms could occur. Chapter 10 will treat the miscible displacement problem. This chapter will be devoted entirely to the discussion of the gas-oil immiscible displacement process.

Within this limitation, a number of displacement types are possible: under dispersed gas injection operation, gas will usually be injected to a number of wells in the reservoir, so that the gas will be uniformly distributed to the reservoir. The approach assumes that no banking of oil will occur, and that the gas will contact the entire reservoir. This assumption can be modified through the use of a conformance factor which will represent the fraction of the total oil reservoir volume contacted by the injected gas. A further modification may also be included where a saturation gradient does occur at the point(s) where the gas is injected.

The conventional gas injection project would occur where gas is injected on a well pattern basis where, hopefully, a banking and frontal displacement of an oil bank will occur. Since gas is considerably lighter than oil, the procedure will be applicable to relatively thin sands having low or zero structural relief. In thick sands, or in sands having sizeable formation dip, the injected gas would be expected to override or to move to the higher structural positions, if sufficient vertical permeability exists. Above a directional permeability of approximately 200 millidarcys, vertical segregation of gas and oil can occur.

External gas injection may be employed to good advantage in those reservoirs having dip. In this application, gas is injected to the top of the structure, with a resulting driving force applied to the oil column. Injection may or may not be into an already existing primary or secondary gas cap. This method has advantage over the dispersed or pattern-type of gas injection, since the beneficial effects of gravity can be utilized. A variation of this process is downdip gas injection, or "attic oil recovery." If a reservoir has sufficient dip and permeability, gas injected into a lower structural position should migrate upward and create a secondary gas cap. The oil displaced downward can then be recovered with wells already drilled. A further variation of external gas injection involves the use of inert gases in those instances where a source of gas does not exist, or where gas has a high present market value.

It is interesting to note that even though gas injection often yields less ultimate oil recovery than another competitive secondary recovery technique, such as water flooding, sufficient economic advantage may be present to make the technique preferable. The return of gas to the formation early in the primary producing history of a field will permit oil production at higher rates with obvious economic benefits resulting. Then too, many producing areas prorate production based upon maximum gas-oil ratios. The reinjection of the excess gas produced, above the maximum gas allowable, will usually result in no penalty to the producing rate of the well. This fact is of particular value where a reservoir may not be considered for other more efficient secondary recovery techniques because of severe reservoir stratification problems with swelling clays, or other similar problems.

The injection of gas to oil-bearing formations that have been depleted by primary means can also serve as both a secondary recovery and a gas storage operation. This is especially true where storage reservoirs are needed for peaking periods, and where the subject oil reservoir is properly located. The use of a field as a storage vessel will usually defer the time when the oil is recovered. The ends of conservation are well served by the procedure, since oil recovery, which would not be economically possible by any other presently known means, may be accomplished.

Dispersed Gas Injection

Dispersed gas injection is normally practiced in fields where solution gas drive is the predominant producing mechanism. If the injection of gas is to occur at a number of points through the field, and if no saturation gradients result, i.e., no banking of oil occurs, then the calculation of the benefits of dispersed gas injection may be determined with sufficient accuracy. In this application, dispersed gas injection can occur early in the life of the primary producing history of the field, and as such could be considered a pressure maintenance project, rather than a secondary recovery process.

A number of authors[1,2,3,4,5] have treated the problem of solution gas drive in detail. The work of Pirson[5] has the advantage of relative simplicity, and of being solvable within a reasonable period of time, either by desk calculations or through the use of high speed computers. In addition, the equations may be readily adapted for the dispersed gas injection case, for a gas saturation gradient and for partial contact rather than entire contact of the reservoir by the injected gas.

The material balance equation for the case where there is no water influx, no water production and no gas cap, may be written as follows[5]:

$$N = \frac{N_p [B_o + B_g (R_c - R_s)]}{B_{gb} (R_{sb} - R_s) - (B_{ob} - B_o)} \tag{1}$$

where the nomenclature and units are the same as presented in Chapter 2. Two additional equations are needed if the reservoir performance of a solution gas drive type of reservoir is to be predicted; the instantaneous gas-oil ratio equation:

$$R_p = R_s + \frac{B_o}{B_g} \frac{k_g}{k_o} \frac{\mu_o}{\mu_g} \tag{2}$$

and the saturation equation:

$$S_l = S_w + S_o = S_w + (1 - S_w) \frac{(N - N_p)}{N} \frac{B_o}{B_{ob}} \tag{3}$$

where

k_g = effective permeability to gas flow, darcys
k_o = effective permeability to oil flow, darcys
μ_o = oil viscosity, cp
μ_g = gas viscosity, cp
S_l, S_w, S_o = liquid, water, and oil saturation, respectively, fraction.

It is convenient to rewrite equation (1) at the beginning and at the end of a particular finite interval of stock tank oil production, ΔN_p. The material balance equation then can be written in the finite difference form[5]:

$$\Delta N_p = \frac{(1 - N_{pi}) \Delta \left(\dfrac{B_o}{B_g} - R_s \right) - B_{ob} \left(\Delta \dfrac{1}{B_g} \right)}{\left(\dfrac{B_o}{B_g} - R_s \right)_{i+1} + R_{av}} \tag{4}$$

where

ΔN_p = fraction of the original oil-in-place, N, produced as the reservoir pressure declines from p_i to p_{i+1}, fraction

$$\Delta \left(\frac{B_o}{B_g} - R_s\right) = \left(\frac{B_o}{B_g} - R_s\right)_{i+1} - \left(\frac{B_o}{B_g} - R_s\right)_i$$

$$\Delta \frac{1}{B_g} = \left(\frac{1}{B_g}\right)_{i+1} - \left(\frac{1}{B_g}\right)_i$$

N_{pi} = cumulative oil production to the beginning of the interval at stock conditions, fraction.

$R_{av} = (R_i + R_{i+1})/2$

$i, i+1$ = subscripts denoting the term value at the beginning and end of the finite interval, respectively.

b = subscript denoting a value at the bubble-point

Equation (4) can be readily modified to account for the reinjection of a constant fraction of the produced gas as a dispersed gas phase:

$$\Delta N_p = \frac{(1 - N_{pi}) \Delta \left(\frac{B_o}{B_g} - R_s\right) - B_{ob} \left(\Delta \frac{1}{B_g}\right)}{\left(\frac{B_o}{B_g} - R_s\right)_{i+1} + R_{av}(1 - I)} \qquad (5)$$

where I = constant fraction of the produced gas which is reinjected to the oil reservoir.

A solution of the depletion drive problem where a constant fraction of the produced gas is reinjected would require simultaneous solution of equations (2), (3), and (5).

Where the injected gas is not dispersed to 100 per cent of the oil reservoir volume, a conformance factor, e, may be used to represent the fraction of the reservoir which is contacted by the injected gas. If past performance on the reservoir or on a similar reservoir is available, a fitting of the calculated performance for several values of e to the actual production performance history should yield the apparent value of e. Structural considerations, or the position of the gas injection well(s), may also suggest the fraction of the oil zone volume which would be contacted by the gas. The performance of an oil reservoir where a constant fraction of the gas is reinjected, and where the reinjected gas contacts only a fraction of the total reservoir, may be predicted from the following equations:

$$\Delta N_p = \frac{(1 - N_{pi}) \Delta\left(\dfrac{B_o}{B_g} - R_s\right) - B_{ob}\left(\Delta \dfrac{1}{B_g}\right)}{\left(\dfrac{B_o}{B_g} - R_s\right)_{i+1} + (R_e)_{av}(1 - I)} \tag{6}$$

$$\Delta N_p = (1 - e)\Delta N_{pD} + \Delta N_{pe}$$

$$R_e = R_s + \frac{B_o\,\mu_o}{B_g\,\mu_g}\left(\frac{k_g}{k_o}\right)_e \tag{7}$$

$$S_l = S_w + S_o = S_w + \frac{(1 - S_w)(e - N_{pe})B_o}{eN_{pe}B_{ob}} \tag{8}$$

where

$$(R_e)_{av} = \frac{(R_e)_i + (R_e)_{i+1}}{2}$$

ΔN_{pe} = fraction of the original oil-in-place, produced from the section of the reservoir contacted by injected gas as the reservoir pressure declines from p_i to p_{i+1}, fraction.

ΔN_{pD} = fraction of the original oil produced from the section of the reservoir not contacted by the injected gas, fraction.

The terms $(k_g/k_o)_e$ represent the simultaneous flow properties of the oil and gas in the parts of the reservoir contacted by the injected gas. The direct determination of the proper relationship for this gas-oil relative permeability relationship by laboratory measurement, from petrophysical considerations, or from performance data, is not entirely satisfactory. The usual approach is to assume that the gas-oil relative permeability performance will be the same as that for the simple solution gas drive case where no gas is injected, which is a simplification introducing some error. Chapter 3 outlines the several techniques from which valid relative permeability information may be derived for this simpler case.

Figure 9-1 presents a computer flow diagram for the solution gas

288

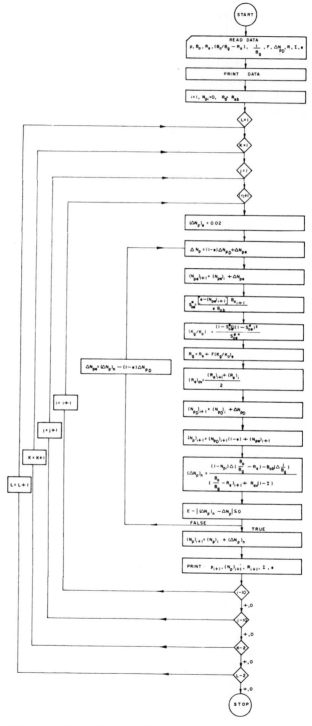

Figure 9-1. Computer flow diagram for solution gas drive with gas injection and correction for fraction of reservoir contacted (after Pirson[6]).

drive case, where gas injection and incomplete dispersion of the injected gas to oil reservoir volume has been included. A study of the diagram will outline the calculation procedure which is necessary, and will indicate the procedures required for the preparation of a computer program to solve the problem.

In the first box, the required data are listed. It is apparent that if a card input is to be used, one card will be required for the data which apply at each reservoir pressure, for which a calculation is to be accomplished. It will be found convenient to divide the difference between the bubble-point pressure and atmospheric pressure into 10 or more equal pressure intervals. Data for the card input would be prepared, corresponding to each of the resulting reservoir pressures. It will be necessary to have at hand a solution of the solution gas drive performance for the case where no gas is injected. This information can be generated by solving equations (1), (2), and (3). It will be seen that a program prepared according to the flow diagram of Figure 9-1 would also yield, with minor alteration, a determination of the simple solution gas drive performance for the special case when e and I are one and zero, respectively. A particular petrophysical relationship for the gas-oil relative permeability has been included in the flow diagram. Any other applicable relationship could be substituted. Pirson[6] has published a Fortran program for the solution gas drive performance calculation.

Figures 9-2 and 9-3 show typical results obtained by dispersed gas injection for the case where the bubble-point pressure of the reservoir was 2000 psia, where a constant fraction of the gas was reinjected and where the pressure at which the reinjection was initiated was also varied. For the conditions of the example it is evident that:

(1) Maximum recovery occurs when pressure maintenance is initiated early in the life of a field.

(2) Increases in oil recovery will be obtained even though the project is begun late in the primary producing history of the field.

(3) The highest producing gas-oil ratios occur when the gas is reinjected early.

(4) At a given percentage oil recovery, the reservoir pressure will be higher and well productivity maximized when the maximum amount of gas is reinjected early in the primary producing life of the field.

For a given field where dispersed gas injection is being con-

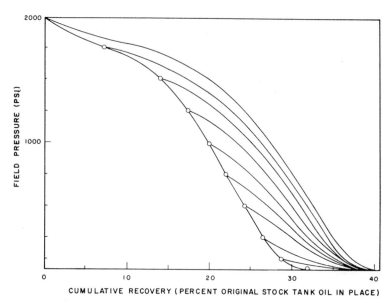

Figure 9-2. Pressure behavior of a typical solution gas drive reservoir subjected to dispersed gas injection beginning at different initial pressures (after Pirson[6]).

sidered, the preparation of performance curves of the type illustrated in Figures 9-2 and 9-3 will allow well-based conclusions to be made regarding the amount of gas to reinject, and the reservoir pressure at which the project should be initiated. It is evident that benefits derived from dispersed gas injection will be reduced if only a limited portion of the field is contacted by the injected gas. The project still may be justified if no market exists for the produced gas, or if the field is to be used for the purpose of gas storage. In addition, if a ceiling on the producing gas-oil ratio exists due to available compression and producing facilities, or due to producing regulations, it is evident that initiating the project at a lower reservoir pressure would result in lower peak producing gas-oil ratios.

Pirson[5] has outlined a procedure to account for a gas saturation gradient in gas injection projects. Such gradients will serve to increase the oil recovery beyond that predicted for the dispersed gas injection case for the same reservoir, fluid and injection conditions. Where dispersed gas injection is initiated early in the

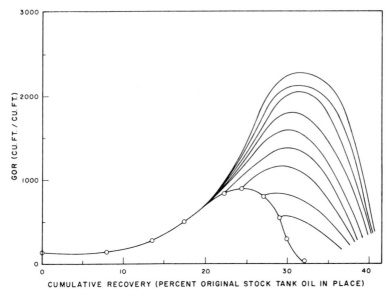

Figure 9-3. Gas-Oil ratio behavior of a typical solution gas drive reservoir subjected to dispersed gas injection beginning at different initial pressures (after Pirson[6]).

life of the field, a gas saturation gradient will necessarily exist, since continuous permeability to gas will not have developed between the producing and injection wells. Projects initiated late in field life, when pressure has dropped markedly below the bubble-point, would be expected to more closely approximate the dispersed gas injection case, since a large and continuous gas saturation will already exist in the reservoir.

Nolan and Locker[7] have reported results of a dispersed gas injection program in a massive dolomitic reservoir, the Fullerton Clearfork reservoir, in west Texas. The porosity is of a fractured and vulgar type. Approximately 4 per cent of the original oil-in-place was recovered by gas injection, with additional benefits of lower operating expenses and reduction of allowable penalties due to high gas-oil ratios. Pressure behavior of the field indicates that little of the reinjected gas went back into solution in the oil, as evidenced by the substantial maintenance of the reservoir pressure which resulted. This observation lends support to the theory presented earlier in this section, since no allowance for a resaturation of the oil with gas was included in the treatment. The authors assumed that approximately 50 per cent of the reservoir was contacted by the dispersed gas injection. No evidence presented demonstrated that a gas saturation gradient was developed in the system. No frontal displacement of oil would have been expected since the initial gas saturation at the start of gas injection was approximately 18.5 per cent, with continuous permeability to gas existing between the wells.

External or Gas Cap Gas Injection

Where sufficient reservoir dip and adequate permeability exist in a reservoir, it will often be preferable to inject or reinject gas at the highest structural positions in a reservoir. Such injection may be to an already present primary gas cap, or to a secondary gas cap, if sufficient primary producing history has occurred for such a cap to develop. In this instance, a frontal displacement of the gas-oil contact occurs with fair to good recovery of the oil-in-place. The technique has the possible advantage of good areal coverages if sufficient dip exists for gravity drainage benefits to occur, and perhaps nearly 100 per cent contact of the oil reservoir volume by the injected gas. A smaller number of gas injection wells will normally be necessary, thus leaving a larger number of oil producing wells which would be an advantage in unprorated states, or where the producing allowables of the injection wells can not be transferred to the producing wells. Due to the excellent mobility of the injection gas, most injection wells will be able to accept large volumes of gas at reasonable injection pressures.

The external gas injection technique will have best application to those reservoirs which have a very low, or non-existent gas saturation at the beginning of the project. A large gas saturation would result in no formation of a saturation gradient or front, since a continuous permeability to gas would be present between the injection and producing wells. If sufficient reservoir dip exists, and if the vertical permeability exceeds approximately 200 millidarcys, gravity segregation of the gas and oil will usually limit the gas saturation build-up during primary producing operations, and allow the formation of a front.

External Gas Injection Performance Calculations—
The Frontal Advance Method

The theoretical development for the frontal advance method approach to the problem of displacement of oil by gas in porous media has been presented in Chapter 5. The approach which follows will apply to systems where the displacement is occurring in a linear homogeneous system of constant oil-pay thickness. As previously stated, the prediction method works best when the mobility ratio, $k_{rg}\mu_o/k_{ro}\mu_g$, is favorable, i.e., less than one, or one. This will never be the case when gas is injected to an oil reservoir, since the mobility ratio will considerably exceed one. However, the frontal displacement method for predicting external gas injection performance is rigorous in mathematical development, and yields the best answers of any of the calculation methods which could be used. The influence of the adverse mobility ratio in causing gas channelling or fingering is accentuated in horizontal systems, but is reduced considerably in systems where sufficient dip exists for the influence of gravity to be felt.

Frontal Advance Method—The Linear System. The linear system approach to the frontal advance displacement method provides an accurate modelling of the gas cap gas injection case. In addition, the method models horizontal systems where peripheral injection, line-drive, or end-to-end sweeps are to be used. While not specifically treated here, the approach is directly applicable to frontal displacements occurring due to a naturally expanding gas cap drive. Chapter 5 presents the analytical development of the theory necessary for the performance calculations. The specialized form of the fractional flow equation for the gas-oil displacement case is:

$$f_g = \frac{1 - \dfrac{1.127\,k_o}{q_t\mu_o}\left[\dfrac{\partial p_c}{\partial u} + 0.434\,(\Delta\vartheta)\,\sin\,\alpha\right]}{1 + \dfrac{k_o}{k_g}\dfrac{\mu_g}{\mu_o}} \tag{10}$$

The frontal advance equation takes the form:

$$u = \frac{Q_t t}{A\phi}\frac{\partial f_g}{\partial S_g} \tag{11}$$

where the nomenclature is the same as that used in Chapter 5. The subscript g is used to refer to a property of the injected gas at the pressure and temperature of the system. Although the pressure in the system will vary between the points of injection and production, the fluid properties data should be those prevailing at the average pressure and temperature of the reservoir. Even though the injected gas may have a temperature which differs from the formation temperature, heat transfer in the wellbore will result in the temperature at the sand face differing very little from the formation temperature, except in rare cases. Even then, the temperature would not differ greatly from the reservoir value. Heat transfer in the wellbore is discussed in Chapter 12.

The reader is referred to Chapter 7 for the companion treatment of water-flooding performance calculations by the frontal advance method in linear systems, and for details and comments which are common to both water and gas injection, for the purposes of secondary recovery of oil.

The frontal advance method requires that the gas-oil relative permeability relationship for the system be known, or at least reasonably approximated. It is important that the relationship be carefully chosen, since the pore configuration, system wettability and capillary forces of the reservoir rock and fluid are being modelled. These data may be obtained from production information, from laboratory tests on representative rock samples, from information on a similar field, or from petrophysical and statistical relationships. Chapter 3 provides a detailed treatment of the subject of permeability concepts and determinations. The effective permeability to oil, as used in the fractional flow equation, is that value applying when only oil and an irreducible water saturation are present in the system. This should be a correct modelling of the system, at least immediately ahead of the displacement front. In some cases, it may be advisable to use the effective permeability to oil at the critical gas saturation, if such information is available.

Performance Before Gas Breakthrough. As in the water-flooding case, equation (10) may be solved for a range of gas saturation values which specify the applicable oil-gas relative permeability ratio values, k_o/k_g. Since the capillary pressure gradient, $\partial p_c/\partial u$, will not be known, the contribution of the term is temporarily neglected, and the S-shaped curve of Figure 9-4 is generated. Drawing the secant line from the origin of the curve tangent to the S-

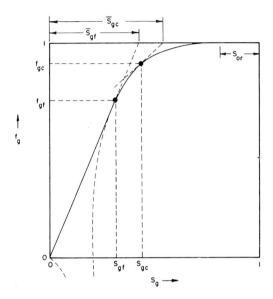

Figure 9-4. Fractional flow relationship for
gas displacing oil in a linear system.

shaped curve replaces the capillary pressure gradient term for
arguments that are presented in Chapter 5. The point of tangency
yields both the fraction of the flow at the front which is gas at
reservoir conditions, and the gas saturation at the front. Extension of the secant line to the point where f_g is one results in a
value for the mean gas saturation, $\bar{S}_{gf}$, behind the displacement
front. The time of breakthrough of gas in a system of length L can
be determined by noting that in equation (11), the $(\partial f_g/\partial S_g)$ term
can be replaced by its value, $1/\bar{S}_{gf}$. Equation (11) then takes the
specialized form:

$$t_B = \frac{AL\phi\bar{S}_{gf}}{Q_t} \tag{12}$$

where the breakthrough time will be in days if the throughput rate
(Q_t), cross-sectional area (A), and system length (L), have the
units of cubic feet per day, square feet, and feet, respectively.
Any consistent system of units could be used.

Figure 9-5 presents a diagrammatic sketch of the gas-drive performance in a linear system where the initial gas saturation is
less than the value where a free flow of the gas can occur ahead

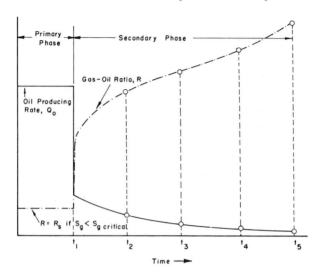

Figure 9-5. Sketch of the oil producing rate and gas-oil ratio as a function of time for frontal displacement in a linear system.

of the displacement front, i.e., S_g is less than the critical gas saturation. In the idealized case, the primary phase of the project will result in a constant oil producing rate, which would be equal to the throughput rate, Q_t, divided by the prevailing oil formation volume factor, B_o. If a free water saturation were to exist in the system, and a water production result, the present theory would not be strictly applicable, since a three-phase permeability to flow would result. Minor water production could be handled in the practical case without causing large errors from the predicted behavior in the application to the field situation. If sizeable water production is to occur because of a large mobile water saturation, an approximation can be made by considering the water and oil as one phase, thereby reducing the problem again to one of two-phase flow.

For the case where no water is produced, if the reservoir throughput rate is q_t and steady-state conditions prevail, then:

$$q_t = q_o + q_g \tag{13}$$

where all the producing rates are measured at reservoir conditions.

The gas producing rate at surface conditions would be:

$(q_g)_{sc}$ = Free Flow Gas + Gas in Solution in the Produced Oil

$$= q_g \times \frac{p}{p_{sc}} \frac{T_{sc}}{T} \frac{1}{z_{sc}} + \frac{q_o R_s}{B_o} \tag{14}$$

where any consistent set of units could be used. The producing gas-oil ratio at any time can be determined from the instantaneous gas-oil ratio equation:

$$R = R_s + \frac{B_o}{B_g} \frac{k_g}{k_o} \frac{\mu_o}{\mu_g} = R_s + \frac{B_o}{B_g} \left(\frac{f_g}{1 - f_g} \right) \tag{15}$$

where the gas-oil relative permeability would be determined at the gas saturation prevailing in the vicinity of the producing well. Before breakthrough of the front, it would be necessary to assume that no saturation gradient existed between the displacement front and the producing well.

If the capillary and effects of the system are neglected, the time of arrival at the producing end of the system of any given gas saturation larger than that at the displacement front, i.e., after initial gas breakthrough, can be determined from the following form of equation (11):

$$t = \frac{AL\phi}{Q_t \left(\dfrac{\partial f_g}{\partial S_g} \right)} \tag{16}$$

The value for the slope or gradient, $\partial f_g / \partial S_g$, is taken from tangent constructions to the fractional flow figure applicable to the system, such as illustrated in Figure 9-4, at the gas saturation of interest. At each gas saturation, the stock tank oil producing rate can be determined from the following expression:

$$(Q_o)_{sc} = \frac{Q_t}{B_o} (1 - f_g) \tag{17}$$

since if only gas and oil are flowing in the system, the fraction of oil flow will be represented by $(1 - f_g)$. In view of the relationship presented as equation (15), equation (17) can be developed to yield the following function:

$$(Q_o)_{sc} = \frac{Q_t}{\left[\left(\dfrac{B_o}{B_g} - R_s\right) + R\right]B_g} \tag{18}$$

Notice in equation (18) that Q_t will determine the units of Q_o, which means that the units of the terms in the denominator must have dimensionless units. Again, any convenient consistent set of units may be used.

If a constant fraction of the produced gas is reinjected (I), equation (18) may be readily modified to incorporate such a term. If Q_o units of oil are produced in a given time interval, then the total gas produced at standard conditions is Q_oR. The amount reinjected would be IQ_oR, which, at reservoir conditions, would be represented by B_gIQ_oR. Then from equation (18):

$$Q_o + B_gIQ_oR = \frac{Q_t}{\left[\dfrac{B_o}{B_g} - R_s + R\right]B_g} \tag{19}$$

or:

$$(Q_o)_{sc} = \frac{Q_t}{\left[\left(\dfrac{B_o}{B_g} - R_s\right) + R(1 - I)\right]B_g} \tag{20}$$

Equations (12) through (20) can be used to generate the complete performance of a gas injection project for the cases where outside gas is injected, and where the produced gas is reinjected. The performance curves of Figure 9-5 were generated in this manner. Values for the cumulative oil and gas produced can be obtained by integration of the oil rate and gas-oil ratio curves, or by measurement of the area under the curves of Figure 9-5. A measure of the injection gas requirements of the life of a project can be obtained by multiplying the injection rate by the life of the project, so that plans for a source of gas can be formulated. If the produced gas is to be recycled, as is normally the case, the outside gas requirements will be correspondingly decreased.

Figure 9-5 illustrates the gas-oil ratio which will result when the reservoir, ahead of the displacement front, contains a gas saturation which is equal to, or less than, the critical gas saturation, i.e., no permeability to gas flow exists. If the primary-producing history of the reservoir is sufficiently advanced so that the gas saturation exceeds the critical gas saturation at the initiation of the project, then the producing gas-oil ratio may be calculated directly from the instantaneous gas-oil ratio equation, if the gas saturation in the vicinity of the producing well(s) is known. Equation (15), the first form, can be used for this purpose. In practice, as the oil bank approaches the producing well(s), the gas-oil ratio will drop back to the solution gas-oil ratio value (R_s), and then increase rapidly, as the injected gas breaks through.

Performance after Gas Breakthrough. External gas injection with the creation of a frontal displacement is a rate sensitive process. This is illustrated in Figure 5-2 of Chapter 5. If the gas injection is at a restricted rate and if the injection is high on a steeply-dipping structure, then the oil recovery and pore-to-pore displacement efficiency, during the primary phase of the project (before gas breakthrough to the producing wells), may be a major fraction of the total recovery that will be obtained. If the system is horizontal, or the gas injection rate high, then the breakthrough of gas to the producing wells will occur rapidly, and the oil recovery will be low. This behavior is due to the adverse mobility ratio, which occurs when the mobile gas phase is displacing oil. When oil recovery is low at the time of gas breakthrough, the subordinate phase (see Figure 9-4) will be longer, and in many cases will contribute more oil than was obtained before gas breakthrough. It should be noted that the more adverse the displacement becomes, the more the fractional flow curve of Figure 9-4 will be shifted to the left, and the lower the oil recovery will become.

Figure 9-6 shows an expanded section of a fractional flow curve, with the constructions which are necessary to generate the information needed to calculate the subordinate phase of the gas-oil displacement. If the limiting gas-oil ratio is known or can be calculated from economic considerations, then the instantaneous gas-oil ratio equation can be used to determine the gas-oil relative permeability relationship, which prevails at the producing end of the system, i.e., from equation (2). Since the pressure and temperature of the system will be known, the gas formation volume factor, B_g, and the gas in solution, R_s, will be known; the oil and gas viscosities, μ_o and μ_g, and the oil formation volume factor, B_o, plus a knowledge of the limiting producing gas-oil ratio, R, will allow the calculation of the limiting gas-oil relative permeability, k_g/k_o. The relative permeability relationship, applicable to the particular problem at hand, will provide a relationship be-

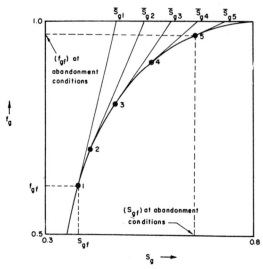

Figure 9-6. Expanded section of a fractional
flow diagram where gas is displacing oil in a
linear system.

tween gas saturation and gas-oil relative permeability, from which
the gas saturation at abandonment conditions can be determined.

Figure 9-6 shows the construction which will yield the average
gas saturation back through the system. The gas saturation at the
front, S_{gf}, determines the point of the f_g versus S_g curve, at which
a tangent line should be drawn to the top of the figure where f_g is
equivalent to one. This yields an average gas saturation throughout the linear system, represented in the illustration by $\overline{S}_{g5}$. The
difference between $\overline{S}_{g5}$ and $\overline{S}_{g1}$ is the fractional recovery of the
oil-in-place at reservoir conditions of pressure and temperature
after breakthrough of the injected gas. Total recovery of oil as a
fraction of the total pore space would be equal to $\overline{S}_{g5}$, less whatever gas saturation may have existed before initiation of the project. Volumetric considerations will allow easy determination of
the oil recovery, in either reservoir or stock tank barrel units.

As shown by Figure 9-6, the arbitrary subdivision of the satura-
tion interval, between $\overline{S}_{g1}$ and $\overline{S}_{g5}$, yields values of $\overline{S}_{g2}$, $\overline{S}_{g3}$ and
$\overline{S}_{g4}$. Tangents drawn to the f_g curve from each of these values,
provide points 2, 3 and 4, corresponding to the fraction of gas
flowing at the outlet end of the system when the average gas satu-
rations back through the system were $\overline{S}_{g2}$, $\overline{S}_{g3}$, and $\overline{S}_{g4}$, respec-

tively. Equations (15), (16), and (17) can be conveniently used to generate the producing performance of the system along the lines of the relationships presented in Figure 9-5. Graphical integration of the oil-producing rate curve will result in values for a plot of the cumulative oil recovered on a time basis. Similarly, the graph of gas-oil ratio can be used to generate a plot of cumulative gas produced versus producing time.

The preceding discussion will allow the estimation of the total pore space that will be occupied by gas at the economic limit. The net pore space that was occupied by oil, prior to the injection of gas, will have to be replaced by the injected gas. If the produced gas is reinjected along with make-up gas, the make-up gas will constitute the total gas which will be required during the life of the project. Usually, it is sufficient to determine the reservoir volume of oil produced on a time basis, then convert this volume at reservoir temperature and pressure to the standard volumes of gas required. Any volumes of gas consumed in the producing operation must also be supplied. Such calculations require only a simple knowledge of the natural gas law. At the end of the operation, most of the injected gas may be recovered by pressure depletion, provided a market is available for the produced gas. Where no market exists, the gas serves as a ready reserve of known volume. In the United States, where natural gas has become increasingly a major source of energy, a market for the gas would not normally be deferred for too long.

Conditions Necessary for the Formation of an Oil Bank. The comments made on the same topic in Chapter 7 for the water flooding case also apply to the gas injection case. It will usually be found that the maximum gas saturation which can exist will be considerably less than that for the water flooding case. This can be traced to the fact that in water-wet oil reservoirs, the gas is non-wetting, and occupies the center portions of the large pores and flow channels. The water is found in the small pores and at the grain contacts (in the case of a sandstone), and has less tendency to flow due to capillary forces. Field experience has shown that the frontal displacement mechanism breaks down when the gas saturation exceeds the critical gas saturation possible in the porous media. This may be visualized as attributable to the high mobility of the gas phase. Since the critical gas saturation marks the lower limit of gas saturation at which gas has an established flow path between injection and production wells, the extremely mobile gas phase is transmitted quickly down the channels, without the formation of an oil bank.

Figure 9-4 may be used to illustrate the requirements necessary for the formation of an oil bank. In this instance, the critical gas saturation would correspond to S_{gf}, the gas saturation value at the front, at which point the capillary pressure gradient necessary for a front becomes equal to zero. At higher gas saturations, the capillary pressure gradient is negligible and the non-banked, or subordinate phase of the displacement, would occur. In actuality, the fractional flow curve of Figure 9-4 would correspond to a system where either gas is injected high on a dipping reservoir, and the gravity term results in a shifting of the curve to the right (see equation (10)), or where the gas-oil viscosity ratio is only moderately adverse. Where conditions are less favorable, the existence of 5 per cent gas saturation could result in immediate gas breakthrough, continuously increasing producing gas-oil ratios, and the formation of no oil bank during the displacement process.

The Foam-Drive Process for Oil Recovery

The injection of an immiscible gas phase for the secondary recovery of oil has a number of drawbacks, as of course does each of the presently-known secondary recovery techniques. Since the driving-to-driven fluid mobility ratio is adverse, the pore-to-pore sweep efficiency, the areal sweep efficiency and the vertical sweep efficiency in a stratified reservoir are usually all relatively low. As the total displacement efficiency is the product of these three sweep efficiencies, a great deal of research has been expended in an effort to develop new techniques or to improve old techniques used in the recovery of oil, in an attempt to increase these efficiencies.

One of the processes which has the potential to reduce some of the drawbacks of immiscible gas injection and has received attention in very recent years is the foam-drive process. The technique has some advantage over the injection of an aqueous solution of surfactant which changes the wettability of the system and increases oil recovery (but unfortunately is retained on the sand grain surface), since the foaming agent remains almost entirely in the foam, and can be moved through the system. Where the formation is oil-wet or partially oil-wet, the surfactants used to generate the foams should also be beneficial in displacing the oil from the sand grain surfaces.

Bond and Holbrook[9] proposed in a 1958 patent that the recovery of oil be effected by a mixture of surfactant solution and gas. In 1961, Fried[10] published a detailed study of the foam-drive process for the recovery of oil. Much of the research accomplished since that time has been directed to answering the questions proposed or generated by Fried. The process is still not completely understood, although the understanding of the foam-drive mechanism apparently is to the point where trials in the field would be justified.

Foams can be defined as agglomerations of gas bubbles separated from each other by thin liquid films. The quality of foam is defined as the ratio of the gas volume to the total volume of the foam. Raza and Marsden[11] have found that the practical upper limit of foam quality is 0.96, since about 4 per cent liquid is needed to produce the fine films required. The theoretical minimum liquid requirement is 3 per cent, if it is assumed that foam is made up of poly-disperse 18-facet bodies[12]. Foams of 0.80 quality or higher are considered to be of the dry type, whereas foams of quality less than 0.70 can be considered to be wet. The wet foams are characterized by the presence of large cylindrical gas bubbles, separated by liquid slugs. The "dry" foams have a better dispersion of the phases, and are thus more stable. Most investigators have used the high quality foams in the laboratory investigations of the use of the foam-drive process for oil recovery.

Fried[10] has shown that the foams exhibit much larger viscosities than either the liquid or gas from which they are composed. This property improves the driving-to-driven fluid mobility ratio over that which occurs when an immiscible gas displaces oil, since the mobility ratio in this latter case is in the range from 10 to 100. The injecting of a foam into a porous media results in the introduction of a large number of resilient interfaces, which exert a piston-like force on the oil to be displaced. This phenomenon is well known to the reservoir engineer as the Jamin[13] effect. The mechanism is particularly beneficial, since the injected foam moves first into the largest pore channels. The high viscosity of the foam, along with the blocking of connecting pores along the channel, results in an increase in resistance to flow to the point where foam is injected into the next smaller size of pore channel. This sequence is followed until the entire permeable section is accepting foam. An improvement in the vertical sweep efficiency results. It has been known for a number of years that the presence of the gas phase has the effect of reducing the residual oil saturation. This is especially true when the system is water-wet, and an intergranular-type porosity exists that can trap the gas phase. The foam is also beneficial in this regard, since the effect of a trapped gas saturation also results.

Fried[10] points out that no fundamental or empirical method is presently available from which the size of an optimum bank length may be determined for the linear system, or the more complicated well networks. This statement is still valid after seven years. It remains for field trials to yield an insight into the minimum bank size needed.

Recent studies indicate that the proper choice of surfactant for a given fluid system can result in the generation of a foam which will remain stable under static conditions for at least one month[14]. Laboratory experiments have indicated that where the system's absolute permeability is less than 1 darcy and dynamic conditions prevail, a foam bank can probably be maintained for considerably longer periods. The foam stability was observed to increase as the absolute permeability of the porous media decreased. Since foam is dynamically unstable, most systems will require continuous foam generation. This could be achieved by injecting small quantities of the surfactant either as slugs, or continuously into the injection gas stream.

Most investigators have found that the generation and maintenance of a foam is not difficult when the fluid to be displaced is water, or even when the water contains a variety of salts. Oil, however, serves as a partial foam depressant for many of the foams prepared, using any of a number of the surface-active agents. The exact behavior of a foam can not normally be predicted with any certainty short of actual laboratory runs through cores to be flooded which contain the water and oil saturations as they will exist under flooding conditions in the oil reservoir. Even then, the understanding of the scaling of the laboratory results to the field case is not well defined. The laboratory tests will indicate a foam system that will result in a recovery of oil, but the tests on short cores will not allow definite conclusions to be made on the slug size required, the vertical sweep efficiency to be expected, or the areal sweep efficiency that will result. No published information is available which will allow the direct estimation of the areal sweep efficiency. Chapter 4 presents information for areal sweep efficiencies as a function of pore volumes injected, and the displacing-displaced phase mobility ratio for a number of well networks. However, the treatment assumes that the mobility of the displacing phase behind the front is constant. If the foam front exists as a band across which a large pressure drop occurs due to high foam viscosity, and this front is in turn driven by a more mobile gas, such information is not strictly applicable. As a first approximation, it will usually suffice to use the areal sweep efficiency corresponding to the mobility ratio, evidenced by the foam-free gas and the reservoir oil. Such an approach to a problem should result in conservative answers, and could be expected to be roughly accurate if the foam bank is not long. The existence of a large foam bank would result in much improved areal sweep efficiencies, with resulting high oil recoveries.

The improvement of the driving-to-driven fluid mobility ratio can be observed from a consideration of Figure 9-7, which represents a plot where foaming agent solutions of different concentrations were injected at a number of injection rates. It can be observed that at a 1 per cent foaming agent concentration, the permeability to gas

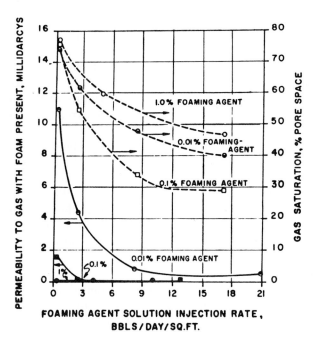

Figure 9-7. Plot of the permeability to gas and the gas saturation as a function of the injection rate and the foaming agent concentration (from Bernard and Holm[14]).

was reduced essentially to zero in a porous media having an absolute permeability of 3,890 millidarcys. Even at very low injection
rates and a foaming agent concentration of 0.01 per cent, the permeability to gas was only a very small fraction of the absolute
permeability of the system, even when the gas saturation of the
system ranged from 40 to over 70 per cent of the pore space.

The almost complete blockage of flow can in itself become a
problem. It is desirable to reduce the permeability of the porous
media to gas flow, but only to the point where the mobility ratio of
the driving-to-driven fluids becomes favorable, i.e., perhaps in the
range of one. Further decrease in the permeability to gas flow
could result in complete blockage. This phenomenon of blockage
in porous media was first reported by Fried. Raza and Marsden[11]
found that the flow of foam is similar to that observed in pseudo-
plastic materials and, as such, can be approximated by an empirical power function. It was further found that the large pressure
differential required to initiate fluid flow through certain of the
foams when injected into a porous media was a function of the
quality of the foam, of the pore diameters of the porous media, and
of the electrochemical nature of the foam-producing surface-active
agent. The electrochemical properties of the system result in
streaming potentials due to interfacial phenomena, and to the presence of electrical changes on the solid surfaces of the porous media. Raza and Marsden have shown that the streaming potential is
directly proportional to the pressure differential applied to the
system. The authors also found that foams produced from ionic
surfactants had lower streaming potentials than those produced
from the non-ionic surfactants. The lower the quality of the foam,
the lower the resulting streaming potential. It was determined
that by the addition of an electrolyte to the surfactant solution, or
by the right choice of surfactant solution, the streaming potential
could be decreased to nearly zero. Table 9-1 provides a listing of
some of the foam-producing surfactants which have been used in
laboratory studies of the foam-drive process. No recommendations
can be made as to the best types of surfactant that should be used,
the concentrations required, or the stability of the foam resulting.
At the present, the suitability of a particular foam should be determined in the laboratory. Careful extrapolation to the field case
should result in improved recovery. Of course, the foam-drive
process must be evaluated in comparison with other available secondary oil-recovery methods, so that the most efficient and economical method is used in application to a given reservoir.

TABLE 9-1. Chemical Identification of Surface-Active Agents Used in Foam Experiments (from Fried[10]).

Trade Name	Manufacturer	Chemical Name	Type
Aerosol C-61	American Cyanimid Co.	Ethanolated Alkyl guanidine-amine complex	Cationic
Aerosol OS	American Cyanimid Co.	Isopropyl naphthalene sodium sulfonate	Anionic
Aerosol OT	American Cyanimid Co.	Dioctyl sodium sulfosuccinate	Anionic
Arquad 2C	Armour and Co.	Dicoco dimethyl ammonium chloride	Cationic
Arquad T	Armour and Co.	Tallow trimethyl ammonium chloride	Cationic
Drench EP-3	National Foam Systems	Unknown	Unknown
Duponol EP	du Pont	Fatty alcohol aklylolamine sulfate	Anionic
Duponol RA	du Pont	Modified ether alcohol sulfate sodium salt	Anionic
Duponol WAQ	du Pont	Sodium lauryl alcohol sulfate	Anionic
Ethomid HT-60	Armour and Co.	Condensation of hydrogenated tallow amide and ethylene oxide	Nonionic
Hyonic FA-75	Nopco Chemical Co.	Modified fatty alkylolamide	Nonionic
Miranol HM Concentrate	Miranol Chem. Co.	Ethylene cyclomido 1-lauryl, 2-hydroxy ethylene Na alcoholate, methylene Na carboxylate	Amphoteric
Miranol MM Concentrate	Miranol Chem. Co.	Same as Miranol HM except myristyl group is substituted for lauryl group	Amphoteric
Nacconal NR	National Aniline Div., Allied Chemical & Dye Corp.	Alkyl aryl sulfonate	Anionic
Ninol AA62	Ninol Laboratories	Lauric diethanolamide	Nonionic
Ninol 1001	Ninol Laboratories	Fatty acid alkanolamide	Nonionic
Petrowet R	du Pont	Sodium alkyl sulfonate	Anionic
Pluronic L44	Wyandotte Chemical Corp.	Condensation product of ethylene oxide with propylene glycol	Nonionic
Product BCO	du Pont	C-cetyl betaine	Amphoteric
Renex 650	Atlas Powder Co.	Polyoxyethylene alkyl aryl ether	Nonionic
Sorbit AC	Geigy Chemical Corp.	Sodium alkyl naphthalene sulfonate	Anionic
Sulfanole FAF	Warwick Chem. Co., Div. Sun Chemical Corp.	Sodium salt of fatty alcohols, sulfated	Anionic
Triton AS-30	Rohm & Haas Co.	Sodium lauryl sulfate	Anionic
Triton X-100	Rohm & Haas Co.	Alkyl aryl polyether alcohol	Nonionic

The reader is referred to the reference list for discussion of the experimental aspects of the foam-drive process. The process should be capable of economical oil recovery where the process is carefully engineered and the field application is well supervised.

Novel Applications of the Gas-Drive Process

One of the novel and best applications of the gas-drive process for the recovery of oil has been in areas of complex geology, where updip oil reserves are nonproducible due to the lack of wells at the upper structural locations[15,16,17,18]. Figure 9-8 illustrates the

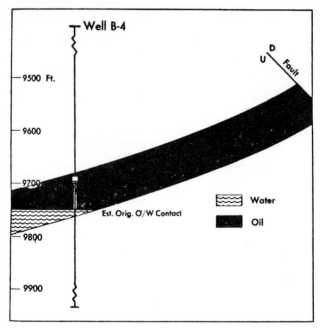

Figure 9-8. Cross-section of the 9,776-ft reservoir in the Ship Shoal Block 72 Field located offshore from Morgan City, Louisiana as the reservoir initially existed (from Franklin *et al.*[17]).

initial reservoir conditions which would be present. Figure 9-9 presents a structural interpretation of a reservoir where the ultimate oil recovery has been increased by the injection of gas. The present problem is one of complex geology where the updip configuration of the reservoir is not accurately known. The drilling of an additional well or wells is expensive, and affords no assurance that the exact top of the structure will be found. In this event, downdip gas injection in a reservoir having sufficient permeability and dip will result in the upstructure migration of the gas and the formation of a secondary gas cap. The injection of gas may be

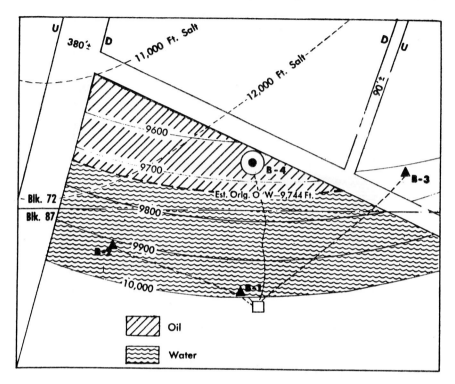

Figure 9-9. Structural interpretation of the 9,776-ft reservoir in the Ship Shoal Block 72 Field (from Franklin *et al.*[17]).

controlled on the basis of reservoir voidage, and the present wells used to recover most of the oil without the drilling of additional wells.

Figure 9-10 is a companion diagram to that presented as Figure 9-8, where the creation of a secondary gas cap has permitted the recovery of the "attic" oil. The authors [17] report that the estimated recovery by this oil recovery technique should approximate 200,000 to 300,000 stock tank barrels, in the case of the B-4 well in the Ship Shoal Block 72 field. The displacement efficiency of the secondary gas cap can be very good in steeply dipping reservoirs, as has already been discussed in this chapter. An estimation of the displacement efficiency can be obtained through use of the fractional flow and frontal advance equations.

Godbold [18] has reported the results of an exhaustive field test where inert gases were used in the place of natural gas for the re-

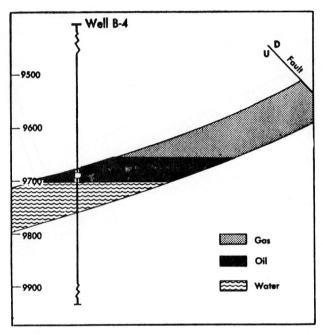

Figure 9-10. Cross-section of the 9,776-ft reservoir in the Ship Shoal Block 72 Field after gas injection (from Franklin *et al.*[17]).

covery of "attic" oil. The use of inert gas has merit where natural gas is not present in sufficient quantities, or where the natural gas has been committed to the market. Additional operating expenses are usually encountered in the use of inert gases over that where high pressure natural gas is available, since the gas must be compressed from atmospheric pressure to the required injection pressure. If engine exhaust gases are used, facilities will normally be required for the removal of nitric acid and for the dehydration of the gas. Godbold reports that the field trial yielded oil recoveries with the inert gas of similar magnitude to that which would have been obtained, had natural gas been used. In the field trial, 1 volume of the fuel gas yielded 9.4 volumes at standard conditions of inert gas for injection.

The use of alternate or continuous gas and water injection has been proposed, and is currently in the field test stage. In this case, the water serves to increase the areal, vertical and pore-to-pore sweep efficiency of the flood by improving the mobility ratio.

In a reservoir rock having an intergranular-type porosity, the injected gas will provide a trapped gas saturation, which results in a lowering of the residual oil saturation. This technique would be better termed a water flood with gas injection, since the best application would probably require that the water injection be the predominant flooding phase.

Crosby and Cochran[19] have reported success in the alternate repressuring and producing of a small oil-bearing Frio D-7 sand lens of the Rincon field in Starr County, Texas. The high gas saturation in the most permeable section of the reservoir prevented the use of gas injection, since a displacement front would probably not be formed. The presence of clays made the use of water inadvisable. The authors report that the success of the project was apparently attributable to the presence of oil reserves in the outlying areas and in sections of low permeability. The alternate injection and producing procedure permitted the economical recovery of some of this oil.

References

1. Muskat, M., "Physical Principles of Oil Production," Chap. 10, New York, McGraw-Hill Book Co., Inc., 1949.
2. Tarner, J., *Oil Weekly*, 32 (June 12, 1944).
3. Babson, E. C., *Trans. AIME* 155, 120 (1944).
4. Tracy, G. W., Petr. *Trans. AIME* 204, 243 (1955).
5. Pirson, S. J., "Oil Reservoir Engineering," Chap. 10, New York, McGraw-Hill Book Co., Inc., 1958.
6. Pirson, S. J., *World Oil*, 10 part series, 82 (September, 1962), 152 (October, 1962), 97 (November, 1962), 74 (December, 1962), 79 (January, 1963), 61 (February, 1963), 86 (March, 1963), 140 (April, 1963), 129 (May, 1963), 82 (June, 1963).
7. Nolan, W. E., and Locker, G. R., *Oil Gas J.*, 156 (October 7, 1957).
8. Reference 5, Chap. 11.
9. Bond, D. C., and Holbrook, O. C., U. S. Patent No. 2,866,507, 1958.
10. Fried, A. N., "The Foam-Drive Process for Increasing the Recovery of Oil," Report of Invest. No. 5866, USBM, 1961.
11. Raza, S. H., and Marsden, S. S., Jr., paper no. SPE 1205, presented fall Mtg. of *Soc. Petrol. Engrs.*, Denver, October, 1965.
12. Leonard, R. A., and Lemlich, R., *A.I.Ch.E. J.*, 18 (January, 1965).
13. Jamin, J. C., Compt. Rend. 50, 1860.
14. Bernard, G. C., and Holm, L. W., *Soc. Petrol. Engrs. J.*, 267 (September, 1964).
15. Morrow, R. M., *Petrol. Engr.*, B-28 (April, 1957).
16. Broom, J. C., and Dawsey, A. L., Jr., *Petrol. Engr.*, B-19 (February, 1959).

17. Franklin, L. O., Koederitz, W. A., and Walker, D., *Oil Gas J.*, 65 (July 24, 1961).
18. Godbold, F. S., *Oil Gas J.*, 133 (April 19, 1965).
19. Crosby, G. E., and Cochran, R. J., paper no. 1395-G, presented at the Oil Recovery Symp. of Southwest Texas, *Soc. Petrol. Engrs.*, Corpus Christi, October, 1959.

PROBLEMS

1. Derive the finite difference material balance equation for the depletion or solution gas drive reservoir case. Are solution gas drive reservoirs rate sensitive? Explain.

2. The following data have been obtained on an oil reservoir:

Sand thickness (oil)	37 feet
Porosity	20.7 per cent
Water saturation (irreducible)	26 per cent
Average absolute permeability	96 md
Original reservoir pressure	2354 psia
Bubble point pressure	2354 psia
Original solution gas-oil ratio	610 SCF/bbl
Gas specific gravity	0.7
Oil gravity	38° API
Oil well spacing	80 acres
Formation temperature	158°F

(a) Prepare the data necessary for a depletion drive calculation from information available in the petroleum industry literature:
   (i) Gas-oil relative permeability ratio: Wahl *et al.*, *J. Petr. Technol.*, 132 (June, 1958).
   (ii) Oil formation volume factor and gas in solution data: Standing, "Volumetric and Phase Behavior of Oil Field Hydrocarbon Systems," Reinhold Publ. Corp., 1952, charts.
   (iii) Gas formation volume factor: natural gas law.
   (iv) Gas and oil viscosity: Beal, *AIME Trans.*, 94 (1946).
(b) Using the above prepared data, calculate the depletion drive performance using the finite difference calculation technique.
(c) For the above reservoir data, is it possible for a gas cap to be in contact with the oil column?

3. A porous medium is 500 feet long, and one square foot in cross-sectional area. The porous media and fluid properties are as follows:

Porosity	25 per cent
Absolute permeability	5000 md
Irreducible water saturation	12 per cent
Oil formation volume factor at average pressure of system	1.500 bbl/STB
Gas in solution	600 SCF/STB
Oil viscosity	2.26 cp
Gas viscosity	0.0186 cp

The gas-oil relative permeability data are represented by the following:

Oil Saturation, per cent	k_{rg}/k_{ro}	k_{ro}
90	0	
85	0.03	
80	0.072	0.71
75	0.14	0.60
70	0.27	0.50
60	1.00	0.32
50	2.3	0.19
40	10	0.12
30	50	0.05
20	large	0

The system temperature is $72°F$. Calculate the following if gas is injected at 2 cu ft/hr (measured at system temperature and pressure).

(a) Pressure differential in psi across the system at the beginning of gas injection.

(b) Time elapsed before breakthrough of the gas in hours.

(c) Gas-oil ratio history before and after gas breakthrough.

(d) Oil producing rate and cumulative oil production as a function of time to a limiting gas-oil ratio of 40,000 SCF/STB.

4. The performance of the gas-injection project of Problem 3 is for a linear system. How would you modify the resulting answers for the following field cases? Describe the necessary calculation technique.

(a) An enclosed five-spot well pattern?

(b) An unenclosed five-spot well pattern?

(c) Injection to the pre-existing gas cap of an oil-bearing anticline having relatively steep dip angles on the flanks of the structure?

(d) A fault block of linear configuration on the side of a salt dome having steep dip, but no pre-existing gas cap or connecting aquifer?

CHAPTER 10

Oil Recovery by Miscible
Fluid Displacement

It is well-known that complete or total recovery of oil from an oil-bearing reservoir is not possible where displacement is by immiscible fluids, such as water or low pressure gas. Capillary forces and interfacial tensions will result in the leaving behind of a fixed residual oil saturation, even though many pore volumes of the immiscible displacing fluid have been passed through the system. The obvious approach to total oil recovery on a pore-to-pore basis would be to reduce capillary forces to zero. This can be accomplished if the displacing fluid is miscible with the displaced fluid or the oil (or gas), and perhaps the connate water. In this usage, a miscible fluid displacement would be defined as a displacement process where no phase boundary or interface exists between the displaced and displacing fluids. The fact that the displacing fluid is miscible, or will mix in all proportions with the displaced fluid, does not by any means insure that 100 per cent displacement on a total reservoir basis will occur. Factors such as stratification, pattern coverage, reservoir heterogeneity, gravity forces and displacing-displaced fluid mobility ratios will usually act to result in a lower total recovery of oil.

The main miscible fluid displacement processes presently recognized are as follows:

(1) High pressure dry gas miscible displacement.
(2) Enriched gas miscible displacement.
(3) Miscible slug flooding, where the leading edge of the slug is miscible with the displaced fluid.

(4) Aqueous and oleic miscible slug flooding (such as several of the alcohols).

(5) Carbon dioxide, flue or inert gas displacements.

In addition, various variations of these basic processes have been proposed, many of which have been laboratory and field tested.

The basic differences in the hydrocarbon type miscible fluid-fluid displacements can be shown best by reference to Figure 10-1,

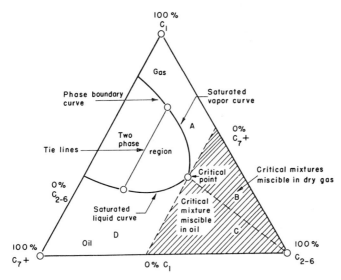

Figure 10-1. Ternary diagram for a hydrocarbon system (from Clark *et al.*[1]).

a ternary diagram for a hydrocarbon system. Although hydrocarbon mixtures are seldom approximately three component mixtures, it is convenient to use the ternary diagram to obtain a visual picture of phase behavior. The three-component system modeled consists of methane (C_1), the intermediates $(C_2$ through $C_6)$ and the heavier components $(C_7 +)$ Since the phase behavior of gases and liquids are a function of pressure, temperature and composition, it is necessary that pressure and temperature be constant for the diagram of compositions of Figure 10-1. A review of the critical temperatures and pressures of three groups of components making up the figure would show that methane (separately) is gaseous for reservoir pressures and temperatures of interest. The $C_7 +$ components would be liquids, while the state of the intermediate components will be determined by the prevailing temperature and pressure.

For the prevailing conditions, Figure 10-1 shows that a two-phase region containing gas and liquid exists. The tie line shown connects two points on the saturated vapor curve and the saturated liquid curve, respectively, the points representing the compositions of a liquid and a gas which are in equilibrium with each other at the prevailing temperature and pressure. If additional tie lines were constructed in the direction of increasing intermediate component concentration, a limiting tie line would result at the critical mixture of the components of the system. This would correspond to the critical point. A further increase in the intermediate components present results in a critical mixture, which is not strictly a liquid or a gas. Region D represents a 100 per cent liquid phase, while region A depicts a single-phase gas. The critical mixture region can be conveniently divided into two sections, B and C. Region B would show the range of compositions at the prevailing pressure and temperature which would be miscible with a dry gas of region A, while region C would contain mixtures which would mix in all proportions with oils in region D.

The engineer faced with the problem of designing a miscible displacement process will be concerned with the temperature of the reservoir, the prevailing pressure level and the pressure level at which the displacement will take place, the composition of the reservoir fluid and the possible compositions of the chosen displacing fluid. A diagram such as Figure 10-1 will help to determine the types of miscible displacement which are possible for the given reservoir. A high pressure dry-gas displacement will require that the dry gas, represented by a point in region A, be miscible with the oil to be displaced. This can only occur if the oil is rich in intermediate components—that is, having composition placing it in region C. In this instance, the dry gas will vaporize certain of the intermediates in the oil until a zone is created which is totally miscible, both in the injected gas and the reservoir oil. In an enriched gas drive, a gas having a composition in region B will be injected to displace an oil have a composition placing it in region D. Here, an absorption of the intermediates in the gas by the contacted oil will result in the formation of a buffer zone, miscible in both the original oil and the injected gas. This process has also been termed a condensing gas drive.

In view of the expense entailed in injecting a miscible fluid, it has become common practice to inject only sufficient material to form a bank miscible with the oil to be displaced, and perhaps miscible with the inexpensive fluid which follows. Such a system is termed "miscible slug flooding," and has a number of possible modifications. Considerable attention has also been accorded to the possible use of a solvent, miscible with both the oil and the connate water found in a reservoir. Several of the alcohols have this property, and have found limited use in Pennsylvania fields. Heightened interest in the use of carbon dioxide, or flue or inert gas has also occurred in recent years.

The various aspects of the main miscible and partially-miscible fluid displacement processes will be treated separately in the sections which follow.

High Pressure Dry-Gas Miscible Displacement

Much of the initial work and field application of the high-pressure dry gas or vaporizing gas miscible displacement was accomplished by the Atlantic Refining Company[2,3,4]. As pointed out in the introduction, a miscible displacement of oil with dry gas can be accomplished if the oil is rich in the intermediate components (C_2 through C_6). Figure 10-2 illustrates the phase conditions which are necessary for such a displacement to be feasible. The injected gas composition is represented by point A and the reservoir oil by point B. As the oil is contacted by the gas, the intermediates in the oil will be vaporized by the gas along the saturated vapor line, $V_1 - V_2 - V_3$, until the critical point, 0, is reached. The buffer zone of enriched gas thus created will be miscible with the oil having the composition represented by point B. Actually, a small volume of the oil in the immediate vicinity of the injection well(s) will be unrecoverable by this miscible flooding, since it will be stripped of some of its intermediate components and have a composition laying along the saturated liquid line, $L_1 - L_2 - L_3$. It is evident that the high pressure dry-gas miscible displacement process has application only to high pressure volatile (or high shrinkage) oil reservoirs. Fields meeting these requirements will normally be deep seated. The process should be instituted early in the primary producing history of the field, while pressure is still high.

It is interesting to consider the outcome of injecting a dry gas of composition A to a reservoir containing an oil having a compo-

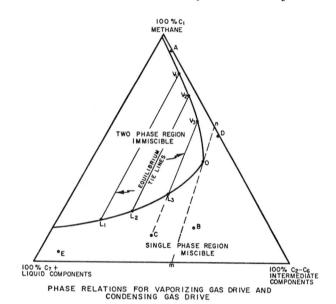

PHASE RELATIONS FOR VAPORIZING GAS DRIVE AND
CONDENSING GAS DRIVE

Figure 10-2. Ternary diagram illustrating conditions
necessary for a dry gas miscible displacement (from
Brigham *et al.*[5]).

sition represented by point C, in Figure 10-2. In this instance, a
mass transfer of components in the oil and gas will occur until, at
the contacting interface, a gas and oil of compositions V_2 and L_2,
respectively, will result. The connecting tie line will be defined
by the equilibrium ratios applicable to the two fluids, the compositions of the two fluids and the prevailing pressure and temperature.
Such a displacement process will be immiscible and a residual oil
saturation, due to capillary forces and interfacial tensions, will be
left behind the displacement front. Miscibility could be achieved
by altering the composition of the injected gas, or by increasing
the reservoir pressure. An increase in reservoir pressure will reduce the size of the two-phase envelope of Figure 10-2.

In order to define the PVT behavior of given reservoir fluid, a
representative sample of the fluid should be obtained in the field,
and a careful laboratory analysis made. While such an analysis
entails expense, the assurance of a miscible displacement which
such an analysis would provide is usually necessary to insure a
successful project. In general, a reservoir pressure in excess of
3000 psia will normally be necessary for a dry gas or vaporizing
gas miscible displacement process to be feasible.

Usually it will be found that any large increase of reservoir pressure to obtain miscible displacement with a dry gas will be uneconomic. In this case, it may be possible to inject a slug of enriched gas, which will serve as an artificially-induced buffer zone between the dry gas and reservoir oil. Such a system is termed an "enriched gas drive" (in part), and will be treated in a separate section in this chapter.

A factor which should be given careful consideration is reservoir dip. Displacement of an oil with a gas of lesser density can result in serious over-riding problems in a horizontal strata due to the effect of gravity. Steeply dipping reservoirs with gas being injected at the top of the reservoir will permit a beneficial segregation of the gas injected, the buffer zone and of the oil being displaced. The presence of continuous strata of differing permeability will usually have an adverse effect on the displacement, in both steeply dipping and horizontal beds that comprise an oil reservoir.

Determination of Miscibility Conditions. Miscibility conditions are most readily discerned by the laboratory analysis of the reservoir fluid. Construction of ternary diagrams, as illustrated in Figures 10-1 and 10-2 at the pressure and temperature at which the displacement would occur, will usually permit a meaningful determination of the conditions for miscibility as already outlined.

Deffrenne *et al.* outlines an experimental technique for determining the required pressure for miscibility for a reservoir fluid of given composition. In the actual field operation, the pressure at the displacement front will need to be maintained at the minimum pressure required for miscibility, from the point of injection and to the producing well as the front is moved forward, if miscibility is desired throughout the life of the project. Then too, it is wise to view estimates of miscibility from ternary diagrams with a conservative eye, due to lack of knowledge concerning the interrelationship between displacement behavior and phase behavior.

Rutherford[6] describes a procedure for the determination of miscibility using high-pressure experimental methods. Essentially, the approach utilizes a high-pressure sand pack mounted in a vertical position. Figure 10-3 provides a schematic diagram of the apparatus used. The technique has particular application to those systems which are not miscible on first contact of the displacing and displaced fluids. Since the primary interest in this type of experiment is whether the fluids are miscible, model scaling prob-

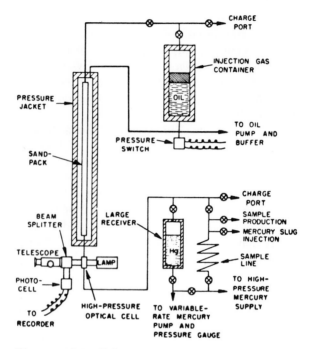

Figure 10-3. Schematic diagram of laboratory equipment for the determination of high pressure miscibility limits in porous media (from Rutherford[6]).

lems can be circumvented by restricting the use of apparatus to the role of bringing about multiple equilibrium contacts between the displacing and displaced fluids. The problems and complications which fingering of the fluids would cause can be avoided by restricting the flow rates in a vertically downward direction to that indicated by the following form of Darcy's equation:

$$q = \frac{kA(\rho_o - \rho_D)}{1,033(\mu_o - \mu_D)} \tag{1}$$

where

q = rate of flow vertically, cc/sec
k = permeability, darcys
A = cross-sectional area through which flow occurs, sq cm
ρ_o = displaced fluid density, gm/cu cm
ρ_D = displacing fluid density, gm/cu cm

μ_o = viscosity of displaced fluid, cp

μ_D = viscosity of displacing fluid, cp

$(\rho_o - \rho_D)/1033$ = potential gradient, in vertical direction, atm/cm

In practical field units, this equation becomes

$$q = \frac{0.1916 \times 10^6 \, kA(\rho_o - \rho_D)}{\mu_o - \mu_D} \qquad (1a)$$

where the units become barrels per day, and square feet for flow rate, q, and cross-sectional area, A, respectively. All other terms carry the same units as in equation (1). Since a vertically-mounted sand pack is not a reservoir model, this technique is not suitable for determining miscibility conditions for miscible slug processes. To avoid the possibility that the sand packed tube would expand under high pressure and establish permeability along the sides, an external pressure equal to the internal applied pressure is provided by a jacket containing oil as the pressure-applying medium. An additional jacketing provides a means to maintain the pack at the temperature of the reservoir being modelled. The reservoir fluid to be displaced is charged to the pack, and the displacing gas admitted at a rate less than that indicated either by equations (1) or (1a). The breakthrough of the displacing material in the apparatus shown in Figure 10-3 is detected by means of a combination high-pressure photometer and a visual observation cell in the bottom outlet line from the sand pack. A change in light intensity is recorded as a change in voltage on a meter or a recorder. Visual observations would also determine if the fluids are single or two phase. Information on the producing gas-oil ratios, densities, volumes and compositions may readily be obtained by standard laboratory analysis methods downstream of the sand pack.

Such laboratory tests as described are particularly valuable, since the economics of a miscible displacement process are heavily dependent upon the cost of the fluid which must be injected. Precise information on the minimum fraction of a given component in a high-pressure gas will permit the use of an injection fluid of minimum cost that will maintain miscibility with the fluid to be displaced in the reservoir.

Rutherford[6] found, as did Wilson[7], that the miscibility of a given light hydrocarbon with the reservoir fluid correlated very well with the pseudo-critical temperature of the injected phase. This means that the mixture of hydrocarbon components may vary as long as a minimum pseudo-critical temperature of the mixture is maintained that has been predetermined to be adequate for miscibility with a given reservoir fluid. Figure 10-4 presents data on a particular set

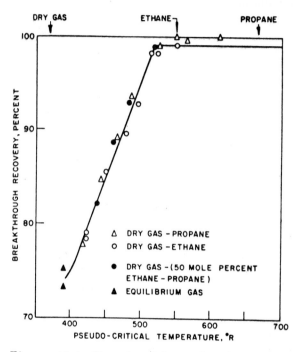

Figure 10-4. Experimental results for various light hydrocarbon mixtures displacing a reservoir fluid (from Rutherford[6]).

of laboratory experiments, where gases of different compositions were used to displace the same reservoir fluid. In this instance, an excellent correlation results, indicating that the displacing light hydrocarbon mixture must have a pseudo-critical temperature of approximately 520 °R for complete miscibility to be attained. Such a correlation would not be expected to apply to a displacing gas which is totally or partially composed of non-paraffin base hydrocarbons. The correlation suggests that the laboratory experiments only need to indicate the minimum pseudo-critical temperature which the displacing gas must have. It should be noted that pseudo-critical temperature of the displacing gas was in excess of

520 °R. This was apparently due to the precipitation of asphaltenes from the reservoir fluid.

In the laboratory, the amount of this precipitation can be approximated by flushing the core after the displacement with benezene, and removing the benezene from the high molecular weight semi-solid material by distillation. In most miscible displacements, it is doubtful that the deposition of asphaltene has any measurable detrimental effect upon the success of the project. If the quantity of the material present were high, the produced reservoir fluid would have been upgraded. Most of the volatile oils susceptible to high-pressure dry gas injection would have, at the most, only small quantities of the asphaltenes present.

Simon and Yarborough[8] have presented a correlation for predicting the critical pressure of a gas-solvent-reservoir oil system that is applicable to the dry gas miscible system, the enriched gas systems and to the miscible slug systems. The correlation applies where the mixture of the displaced and displacing fluids are in the range of 20 to 85 mole per cent methane, 5 to 65 mole per cent ethane-propane-butane fraction and 4 to 20 mole per cent $C_7 +$ fraction. The correlation is based upon the Dieterici equation of state and its first and second derivatives. The equation is:

$$p_c = \frac{a}{29.56 \ b^2} \qquad (2)$$

where

p_c = the critical pressure of a pure compound, psia
a = a measure of the cohesive forces between molecules
b = factor related to molecular volume

It can be observed that the critical pressure of a pure compound is a function of temperature, molecules in contact and molecular volume. In the case of a complex mixture, we would expect the critical pressure to be a function of composition, molecular volume, critical temperature and the energy of molecules in contact. With these functions in mind, the correlation parameters were chosen as:

(1) concentration of $C_2 - C_3 - C_4$ in the mixture.
(2) molecular weight of the $C_7 +$ fraction.
(3) K, the characterization factor of the $C_7 +$ fraction.

Figure 10-5 presents a means by which the characterization factor, K, may be determined from a knowledge of the kinematic vis-

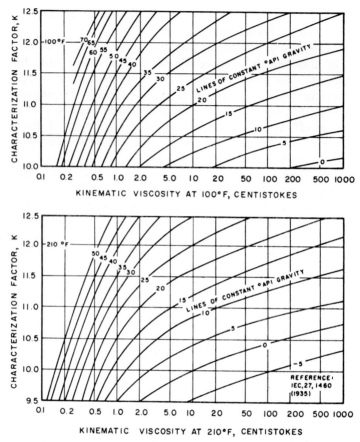

Figure 10-5. Characterization factor (K) for $C_7 +$ fraction of a crude oil viscosity and gravity (from Simon and Yarborough[8]).

cosity and the °API gravity of the reservoir oil to be displaced. When the characterization factor is larger than 11.9, Figure 10-6 may be used to determine the critical pressure of the mixture when the molecular weight of the $C_7 +$ fraction and the mole per cent of the intermediate fraction are known. Where the characterization factor, as determined from Figure 10-5, is less than 11.9, further correction is necessary according to the equation:

$$p_c = p_c^\circ \times (1 - g) \tag{3}$$

where

    g = aromaticity correction to the critical pressure as obtained from Figure 10-7.

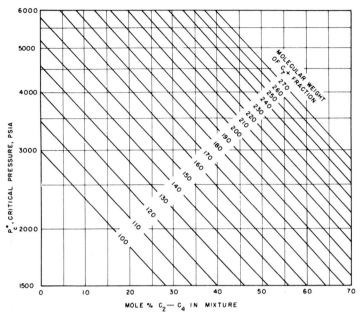

Figure 10-6. Correlation for critical pressure of a hydrocarbon mixture, $K = 11.9$ (from Simon and Yarborough[8]).

p_c^o = critical pressure of mixture as obtained from Figure 10-6, psia.

Simon and Yarborough found the correlation to be applicable where critical conditions were in the range of 100 to 280 °F, and 2,000 to 6,000 psia. For the hydrocarbon mixtures to which the correlation was applied, the average deviation was 5 per cent. The determination of the critical pressure, by the above-outlined technique, is valuable for separating the bubble-point and dew-point regions, and permitting a calculation of the physical properties of a given mixture of the displacing and displaced fluids by the principle of corresponding states.

Calculation of Reservoir Performance—High-Pressure, Dry-Gas Injection. Two basic approaches may be taken to the prediction of reservoir performance—a volumetric approach, or a compositional approach. The volumetric approach will be presented here.

If the displacement is to take place in a horizontal and thin stratum, the conventional approaches to the calculation of performance, as outlined in Chapter 7 for water flooding, may be used here. This assumes that gravity effects are neglected, and that

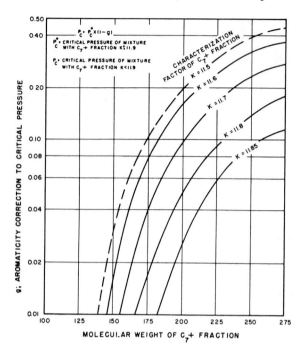

Figure 10-7. Correction for aromaticity to critical
pressure (after Simon and Yarborough[8]).

the more mobile dry gas will not segregate vertically. Then too,
the displacement front will be assumed to be narrow, and the flow
of the fluids will be controlled by the ratio of the mobilities of the
displaced and displacing fluids.   Since miscibility of the fluids
is a requirement, the mobility ratio will be merely the ratio of the
viscosities of the displaced and displacing fluids. Chapter 4 provides areal sweep efficiencies corresponding to the applicable mobility ratio, the pore volumes of displacing fluid injected and the
well pattern being used. A method such as that due to Stiles,
Higgins, or Dykstra and Parsons (see Chapter 7) will permit a calculation of the vertical sweep efficiency. The pore-to-pore sweep
efficiency should approach 100 per cent in those areas contacted
by the flood.

   Doepel and Sibley[9] have presented a technique for predicting
performance in a pattern flood in a multi-layer reservoir. This
method is similar to the method of Dykstra and Parsons for calculating water flood performance, as outlined in Chapter 7. It is well

known that the quantity of injected solvent or miscible material that flows into a given stratum is proportional to the injection capacity of that strata, as compared to the total injection capacity of the section exposed for fluid injection. In a pattern flood, this gives rise to three sweep efficiencies, the vertical sweep efficiency (C_V), the horizontal sweep efficiency (C_H), and the pore-to-pore or microscopic sweep efficiency (C_M). Total recovery from a given stratum would be the product of these three sweep efficiencies. In miscible flooding the microscopic sweep efficiency would depend upon the fraction of the reservoir fluid which might be left behind due primarily to the deposition of asphaltenes. This assumes that the flood is truly miscible throughout the life of the project. The horizontal sweep efficiency may be conveniently obtained from graphs, as presented in Chapter 4, for the applicable displaced-displacing fluid mobility ratio, the pore volumes of fluid injected and the appropriate enclosed well pattern. Information is also available in Chapter 4, for the case where the well-pattern is not enclosed, as would be the situation in pilot flooding operations. The vertical coverage in a layer system will be defined as being one in the layer of highest permeability. The vertical sweep efficiency will be proportionately less in the tighter layers, since the vertical efficiency will represent that fraction of area swept compared to that swept by the flood in the most permeable layer.

It is convenient to use the idea of a flow resistance, Ω, which is defined as being the denominator in Darcy's equation:

$$\Omega = \mu F_G \qquad (4)$$

where

F_G = geometric factor corresponding to the particular well pattern or geometry prevailing. (See Table 6-1, p. 170.)

μ = average fluid viscosity

It is apparent that the factors, μF_G, cannot be determined directly, since the average viscosity of the fluid in the system will depend upon the position of the displacement front. So long as the viscosity of the injected material and the reservoir fluid are not greatly different, the flow resistance will not vary significantly. Chapter 6 treats this problem, and provides graphs from which the flow resistance may be estimated from information on dimensionless injectivity, when the fluid mobility is markedly different from 1.

If it can be reasonably assumed that fluid flow is incompressible, that miscible conditions prevail, that gravity effects in the strata are negligible, that each individual layer has constant fluid and reservoir properties, that there is no cross-flow between layers, and that each layer is of the same thickness, the following equation can readily be developed[9]:

$$(I_f)_j = \frac{(k/\Omega)_j}{\sum\limits_{j=1}^{j=n} \left(\frac{k}{\Omega}\right)_j} \tag{5}$$

where

I_f = fractional solvent injectivity
k = effective permeability, darcys
j = index referring to a particular stratum

The calculation procedure[9] is as follows:

(1) Arrange the layers of the stratified reservoir in descending order of permeability, with the most permeable layer being layer 1. Determine the porosity and hydrocarbon saturation for each layer, calculate the hydrocarbon pore volume, and calculate the fractional hydrocarbon pore volume for each layer.

(2) Choose a hydrocarbon fraction of solvent injection for which a calculation will be made—perhaps 5 per cent.

(3) Using equation (5), calculate the amount of solvent which each layer accepts. Record the cumulative solvent injected to each layer, both as the hydrocarbon pore volume of the layer and of the total system.

(4) For a five-spot enclosed well network, Figure 4-17, p. 95, may be used to determine the areal coverage, C_A, for each layer, based upon the amount of solvent injected. Figure 4-24 could be used, if a line-drive well network were being employed. The microscopic coverage, C_M, will depend upon the amount of materials which are not moved in the miscible displacement, such as asphaltenes.

(5) Determine the volume of solvent in each layer as $C_A \times C_M$. The cumulative solvent production at reservoir conditions or pressure and temperature will be equal to the cumulative solvent injection, less the solvent present in each layer.

(6) The cumulative oil production from the system will be equal to the cumulative solvent injection, less the cumulative solvent

produced. If the solvent is a gas at reservoir and surface conditions (60°F and 14.7 psia), correct the value calculated in Step (5)
by means of the natural gas law. The cumulative oil production at
surface conditions may be determined from a knowledge of the oil
formation volume factor.

(7) Calculate the cumulative production of dry gas produced as
the sum of that produced in the oil (R_s × bbl of STO) and that dissolved in the solvent (cumulative solvent production × SCF of dry
gas per reservoir bbl of solvent). The last correction is not necessary if the solvent injected is totally dry gas.

(8) Calculate the incremental productions of oil, solvent and dry
gas by choosing successively higher fractions of solvent injection
as per Step (2), and repeat Steps (3) through (7) until economics
terminate the feasibility of further solvent injection.

(9) The incremental oil production for each assumed solvent injection step may now be calculated. Solvent cuts at reservoir
conditions, solvent-oil production ratios and dry gas-oil ratios at
reservoir and surface conditions as desired now may be determined.

(10) The areal coverage should be taken as the maximum value
calculated in Step (4). The vertical coverage, C_V, is obtained by
dividing the cumulative oil production by the product of C_A and C_M
from Step (4).

(11) The oil recovery factor is the product of C_A, C_M and C_V.

Figure 10-8 presents the results of a calculation on a fifty-layer
system where solvent was injected continuously, and the displacing-displaced phase mobility ratio was 17. The parameters of the
plot are permeability variation or variance and the total solvent injected as a fraction of the system pore volume.

The preceding performance calculation procedure will also be
applicable to the system where a buffer zone has been placed between the dry gas injected and the reservoir fluid. Miscibility in
the system would have to be maintained in this instance. Some
error could result if the injected buffer zone has a noticeably different mobility ratio to the reservoir fluid than did the dry gas.

Where the displacing fluid is injected into a steeply dipping
reservoir, the areal sweep efficiency may approach 100 per cent,
depending upon the location of the injection well(s). Where a
pattern well injection system is to be used on a dipping formation,
Chapter 4 provides information which will permit an estimation of
the influence of gravity on the areal sweep efficiencies which
would occur.

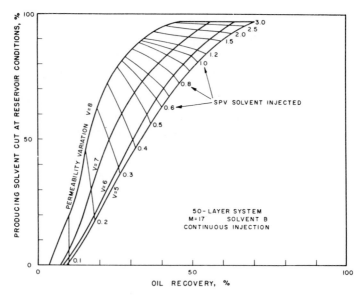

Figure 10-8. The effect of permeability variation and pore volumes of solvent injected on solvent cut and oil recovery for the conditions specified (from Doepel and Sibley[9]).

The performance of a high-pressure gas-injection project differs from that where water is injected, since the reservoir may be blown down. Techniques are available to predict such a performance[10].

Enriched Gas Miscible Displacement

The enriched gas miscible displacement is somewhat more complicated in application than that of high-pressure dry gas injection, since the enriching of gas is expensive. For this reason, it is common to inject a slug or buffer zone of enriched gas between the reservoir oil and the cheaper displacing fluid—usually dry gas, water, dry gas and water, or an inert gas. The calculation of the minimum slug size poses the largest problem.

The simplest form of enriched gas displacement would occur where only enriched gas is injected and the problem of slug size is eliminated. As explained in the introduction, the enriched gas drive process establishes miscibility by a mass transfer process, where the reservoir fluid is diluted with the intermediate-molecular-weight hydrocarbons (C_2 through C_6), condensed from the rich displacing gas. The general range of application of the process is to those reservoirs having a pressure of 2000 to 3000 psia. Where a slug or buffer zone of miscible material is injected to keep costs at a minimum, the slug material is usually composed of ethane, propane, butane, or mixtures of these components which will form a miscible transition zone, at a pressure of from 1200 to 2000 psia.

Figure 10-2 illustrates the phase conditions which are necessary for a miscible displacement to be attained. Since the reservoir pressure will be somewhat lower than that where a high-pressure gas drive would be used, the two-phase region will be proportionately larger. This has the effect of requiring a larger proportion of intermediates to be present, if a miscible displacement is to result. Figure 10-2 shows that an enriched gas of composition D may be used for the miscible displacement of an oil of composition C, or even of composition E. A larger volume of enriched gas of composition D, would have to be injected before miscibility with oil E would result, since more enriching of the reservoir oil would be necessary. The comments made in the discussion of the high-pressure dry-gas injection process, relative to the use of laboratory experiments to determine miscibility limits, to determine reservoir performance and to determine efficiencies, apply to the enriched gas process. The outlined performance calculation technique also applies. Kehn *et al.*[11] have shown that the recovery of oil by means of enriched gas injection can be simulated in the laboratory. Of course, the influence of gravity segregation, reservoir inhomogeneities and well patterns must be kept in mind when such data are applied to the field operation. Kehn also found that the presence of a free gas saturation was not detrimental to the formation of a displacing bank, and that relatively viscous oils could be recovered by the process. The sweep efficiency on a pattern well basis is usually low where the oil to be displaced is viscous, due to the adverse displacing-displaced phase mobility ratio which results.

Welge *et al.*[12] have presented a method for calculating the amount of oil which can be displaced from an inclined, linear, porous media by the injection of an enriched gas, for the case where the gas is only partially miscible with the reservoir oil. The calculation procedure is rigorous in approach and sufficiently lengthy that a solution using a high-speed computer is almost mandatory. Where it is not practical to maintain a completely miscible drive, the calculative technique of Welge *et al.* is recommended.

Barfield and Grinstead[13] have reported on the application of the enriched gas miscible displacement process to the Seeligson Zone 20B-07 Field of Kleberg County, Texas. A gas mixture of 50 per cent propane and 50 per cent separator gas was used to displace a 40° API gravity oil at a reservoir pressure of approximately 2650 psia. To determine whether the oil recovery was actually 100 per cent in the swept zone, a well was drilled and cored 100 feet downstructure from one of the injection wells. It was found that only a part of the core contained as much 1 per cent residual oil saturation. Subsequent workover of a well which had experienced a breakthrough of the enriched gas tended to show that viscous fingering, together with the effects of gravity segregation, had caused a premature breakthrough. It was concluded that ways should be devised to improve the volumetric sweep efficiency of the enriched gas drive process.

Miscible Slug Flooding—Oleic Miscibility Case

One of the chief drawbacks of the enriched gas miscible displacement process is the high cost of the large volumes of solvent that are needed. In high pressure dry-gas displacements, pressures in excess of 3000 psia are normally required. The miscible slug process[14,15] makes possible the injection of a cheaper follow-up fluid, and also application of miscible flooding to reservoirs having pressures as low as 1200 psia. The purpose of the miscible slug is to reduce the capillary pressure between the displacing phase and the oil, so that oil recoveries approaching 100 per cent may be obtained in those areas contacted by the flood. Unfortunately, the material comprising the miscible slug becomes diluted as it is pushed through the reservoir. Much of the following discussion will be concerned with the size of the slug which is necessary to maintain a miscibility with the reservoir oil.

In conventional miscible slug flooding, a slug or bank of propane or LPG, followed by dry gas, is injected to the reservoir. If the pressure is somewhat above 1100 psia, propane will be miscible with all reservoir oils and with nearly all dry gases for most reservoir temperatures. Ideally, the best application of this form of the technique would be to reservoirs containing oils with viscosities of less than 5 cp, with less than 25 feet of homogeneous pay section and low vertical permeability to offset segregation for the horizontal stratum case. Better application would be to a field having steep structural dip, and where the oil-producing rate is kept below the critical rate indicated by equation (1a).

For the case where ethane, propane, or butane are used in the slug, the possibility of precipitation of asphaltenes exists[16]. Indeed, propane has been used by oil refiners for the expressed purpose of precipitating asphalts. Actually, the precipitation of asphalt from crudes is the result of two separate mechanisms, the solubility of oil in propane is that of a true solution, while the precipitation of asphalt is a problem in colloid chemistry. In general, little problem as a result of asphalt precipitation would be expected, since the precipitation would occur throughout the sections of the reservoir contacted by the slug. This may result in a small reduction in the formation permeability, but would not be expected to cause total plugging. Where large amounts of asphaltic materials do exist in reservoir oil, and where the host reservoir is being considered for a miscible slug process, laboratory tests should be made to determine if local plugging of the porous media would be expected to occur. In general, the tendency of a slug of miscible material to precipitate asphalt decreases as the molecular weight of the injected material increases. The injection of ethane would normally result in the largest precipitation of asphaltic materials from a given crude.

Determination of Slug Size. The size of the miscible slug which should be used is a subject of considerable controversy. Due to the cost of solvent, the injection of an excess amount has serious economic consequences, while the injection of an insufficient quantity, with a losing of miscibility in the system, has equaled serious results.

The first theoretical work on the microscopic displacement phenomena, and resulting mixing between miscible fluids of equal density and viscosity, was reported by Taylor[17], and extended by Aris[18]. Taylor's work showed that the mixing of the two fluids was a function of injection rate and distance traveled by the displacing front. von Rosenberg[19] interpreted experiments on miscible displacement in porous media as essentially verifying the extension of the conclusions of Taylor to the reservoir system. Koch and Slobod[20] report that their laboratory experiments indicate a dependency upon system length, but did not find any significant rate sensitivity. These last authors concluded that the injection of a slug amounting to 2 to 3 per cent of the pore space should be adequate to maintain miscibility between the reservoir oil and the following reservoir gas.

Hall and Geffen[21] also found that the displacement rate and the porous media type (homogeneous) did not materially affect the length of the mixing zone after the zone had stabilized for the linear flow geometry case. The mechanism controlling the size of the mixing zone was stated as being primarily that due to dispersion. Here, dispersion is taken to be that due to the changes in velocity imparted to particular fluid elements, as the result of differing local pore geometry in the porous media. Some of the fluid elements would be expected to slow up and others to speed ahead, thus becoming spread out, both to the side and in the direction of flow.

Other investigators[22,23,24] have concluded that the slug size would need to be considerably larger than 2 or 3 per cent of the pore volume. Some of the experiments have resulted in the conclusion that the length of the mixing zone is a function of the square root of the linear distance traversed, while others have found mixing to be directly proportional to distance. Perrine[25,26] has presented a stability theory which provides a theoretical basis for reconciling the various reported results. Essentially, Perrine concludes that the great variance in flow behavior, necessary for the differing experimental results, is due to different mechanisms at the displacement front. When the densities and the viscosities of the reservoir oil solvent slug displacing gas are the same or the mobility ratio is favorable, the solvent bank spreads by the mechanism of dispersion. Where the mobility ratio is adverse (due to differing fluid viscosities), viscous fingering occurs. Extreme viscous fingering in laboratory flow models has been observed by Lacey et al.[27] and by Habermann[28]. Figure 10-9 illustrates the viscous fingers which can be seen in a laboratory model, representing one-quarter of a five-spot well pattern when the mobility ratio is adverse. The subscripts o, s, and d refer to the oil, slug, and displacing fluids, respectively. It can be observed that the miscible slug is concentrated at the ends of the fingers, and that an immiscible displacement is taking place in those places where the driving fluid is in direct contact with the reservoir oil. Habermann states that at adverse driving-to-driven fluid mobility ratios, viscous fingering will occur, even if the reservoir is homogeneous as to rock properties. If the sand contains permeable streaks, fingers will develop and follow these zones. Additional fingers will form in the tight sections. He further concludes that the fingers will not stabilize, but will continue to lengthen with dis-

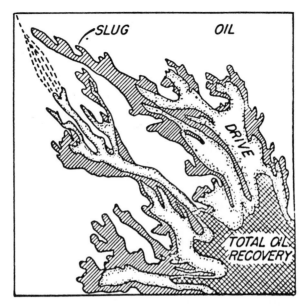

Figure 10-9. Configuration of the miscible slug and displacing fluid at adverse mobility ratio, slug size = 10% of pore volume, μ_o/μ_s = 27.1, μ_s/μ_d = 1.72, μ_o/μ_d = 46.4 (from Habermann[28]).

tance from the point of injection. It is then evident that two separate mechanisms control the size of the mixing zone, and in turn, the size of the miscible slug required, depending upon whether the mobility ratio is less than or equal to one, or greater than one.

If the mobility ratio is less than or equal to one, the longitudinal dispersion will be controlled by the following equation[17]:

$$\frac{\partial C}{\partial t} = K \frac{\partial^2 C}{\partial x_1^2} \tag{6}$$

where

C = concentration, volume fraction
t = time, sec
K = dispersion coefficient, cm^2/sec
x_1 = distance from the center of the flood front, cm

Equation (6) is the well-known diffusion equation, with the exception that the diffusion coefficient has been replaced by the dispersion coefficient, K. The variable, x_1, can be related to x, the distance from the inlet to the porous medium, by the following expression:

$$x_1 = x - vt = x - \frac{Lt}{T} \qquad (7)$$

where

    L = total length of the linear system, cm
    T = time for injection of one pore volume of fluid, sec
    t = injection time, sec

The solution of equation (6) is as follows:

$$C = 0.5 \left[1 - \mathrm{erf} \left(\frac{x_1}{2\sqrt{Kt}} \right) \right] \qquad (8)$$

Values for the $\mathrm{erf}(u)$ are available in Table 12-8 (p. 462), since $\mathrm{erf}(u)$ is equivalent to $(1 - \mathrm{erfc}(u))$. Equation (8) indicates that the spread of the mixing zone will be proportional to the square root of the dispersion coefficient, K. Brigham *et al.*[29] have tested the applicability of the equation to miscible displacements where the mobility ratio is equal to, or less than one, by noting that:

$$x_1 = L \left(\frac{V_p - V}{V_p} \right) \qquad (9)$$

and

$$t = T \left(\frac{V}{V_p} \right) \qquad (9a)$$

where

    V_p = pore volume of the linear system, cm^3.
    V = volume of displacing fluid injected during time t, cm^3.

The substitution of equations (9) and (9a) into equation (8) results in a defining of the concentration profile in porous medium at a given point in the system, rather than at a given time:

$$C = 0.5 \left[1 - \mathrm{erf} \left(\frac{L(V_p - V)}{2\sqrt{TKV_pV}} \right) \right] = 0.5 \left[1 - \mathrm{erf} \left(\frac{L}{2\sqrt{KTV_p}}(U) \right) \right] \qquad (10)$$

where $U = \dfrac{V_p - V}{V}$, a parameter accounting for the predicted growth of the mixing zone.

Figure 10-10 presents a plot of the effect of system length upon the error function parameter. It can be observed that a linear relationship exists as the displacing fluid concentration varies from 5 to 80 per cent for laboratory displacements in glass bead packs. This supports the validity of a square root law for the growth of the mixing zone in a linear system, when the displacing-displaced phase mobility ratio is equal to, or less than one over most of the concentration range.

Taylor[17] has shown that the dispersion coefficient may be defined in the mixing zone between any two fluid compositions:

$$K = \frac{1}{t}\left[\frac{x_{90} - x_{10}}{3.625}\right]^2 \tag{11}$$

where x_{90} and x_{10} are the distances from the point of injection to the point of 90 and 10 per cent displacing fluid concentrations, respectively. Brigham *et al.* have altered equation (11) to express the dispersion coefficient in terms of the error function parameter, U:

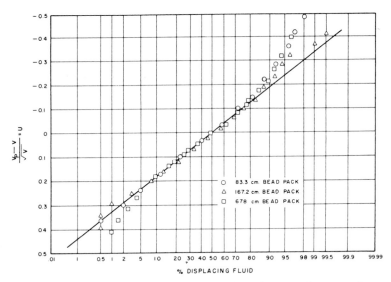

Figure 10-10. Effect of linear system length upon the error function parameter, U (after Brigham *et al.*[29]).

$$K = \frac{1}{V_p T} \left[\frac{L(U_{90} - U_{10})}{3.625} \right]^2 \tag{12}$$

The values for U_{90} and U_{10} are read from a straight line drawn through data points from tests on the porous medium in question, with the experimental results plotted as in Figure 10-10. The constant 3.625 is read from any table of error integrals. Had the concentration range been from 80 to 20 per cent, the constant would have been 2.380.

Figure 10-11 shows the effect of viscosity ratio (the mobility ratio in a miscible system) on the displacing phase concentrations observed at the outlet end of a 83.3-cm long glass bead pack. The

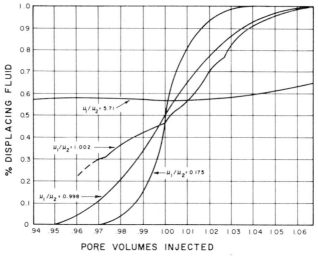

Figure 10-11. Effect of viscosity ratio on the displacing phase concentration for the described bead pack (after Brigham *et al.*[29]).

average bead diameter was 0.100 mm, the porosity 34.6 per cent and the permeability 7.1 darcys. It is evident that the mixing zone becomes longer as the mobility ratio is increased to a limit of one. Above one, viscous fingering phenomena prevail, and the effluent curve departs from the S-shaped curve that the error function curve would predict. Brigham *et al.*[29] have also studied the effect of flow velocity on the dispersion coefficient in glass bead packs and in Torpedo and Berea sandstone, and have proposed the following equation for both high and low flooding rates:

$$\frac{K}{D} = \frac{1}{F\phi} + \alpha \left(\frac{\overline{r}v}{D}\right)^{1.20}$$ (13)

where

D = Fick diffusion coefficient, cm^2/sec

F = formation resistivity factor

α = mixing coefficient, function of rock inhomogeneities under dynamic conditions

$\overline{r}$ = average pore radius obtained from Kozeny's equation, cm

v = average pore velocity, cm/sec

The values for α have been calculated as 0.69, 0.49, 0.30, 23.2, and 53 for 0.044 mm beads, 0.100 mm beads, 0.470 mm beads, Torpedo sandstone and Berea sandstone, respectively, for miscible fluids having a mobility ratio of 0.175. An equation has also been developed which indicates the pore space velocity which will result in the minimum mixing zone length[29]:

$$v = \frac{D}{\overline{r}} \left(\frac{1}{0.2\alpha F\phi}\right)^{1/1.20}$$ (14)

where units are the same as those of equation (13).

Prepared figures, such as Figure 10-11, may be used to estimate the quantity of injected solvent that would be required to maintain miscibility to the outlet end of a linear system. Preferably, these determinations should be done in the laboratory at controlled injection rates, and the observation that the amount of mixing is proportional to the square root of the distance traveled, used to extrapolate the experimental results obtained. Preferably, the porous media considered for flooding and the fluids in place and to be injected should be utilized. The typical S-shaped concentration curve would be obtained only when the mobility ratio of the displacing-displaced fluids is equal to, or less than one. Where a slug is to be injected, it should be remembered that a mixing zone will result at both, and the leading and trailing edges must be taken into account. Again, the mobility ratios must be favorable for the preceding theory to be valid, at both the leading and trailing edge of the slug. Laboratory miscibility studies will permit the determination of the minimum slug concentration for miscibility to be maintained.

Agan and Fernandes[30] have rewritten equation (8) in a convenient form for determining the volume of solvent necessary for maintaining a miscible displacement in a radial flow geometry, for a reservoir of homogeneous properties, and where the mobility ratio is favorable or unity. The equation has the form:

$$C = \text{erf} \left(\sqrt{\frac{3R^4}{64Kr^3}} \right) \qquad (15)$$

where

R = radius of the slug volume, ft
r = radial distance slug travels and remains miscible, ft
K = coefficient of longitudinal dispersion, ft

Notice that in equation (15) the units on the longitudinal dispersion term, K, have become feet, instead of cm/sec. Very little data are available for K, as defined. Agan et al. used 0.1327 feet for K for a miscible slug displacement for the pessimistic case of dispersion between the dry gas propelling the slug and the reservoir oil, although the dispersion is between the slug and the oil, and the dry gas and the slug. To use equation (15), it is necessary to determine the minimum slug concentration, C, to maintain system miscibility by laboratory experiments, by available information on a similar crude system, or by theoretical equilibrium phase considerations. Knowing the radial distance, r, over which the miscible displacement is required, it is a simple matter to determine the slug radius R. This value may be readily converted to reservoir barrels of miscible slug required knowing the pay thickness, the porosity and the water saturation. If the system were stratified, the resulting answer, as a per cent of the floodable volume, would be applicable only to separate strata. A following section will detail how the performance of a stratified system may be predicted.

Fitch and Griffith[31] have presented an equation for a calculation of the length of the mixing zone for the homogeneous linear reservoir case, where the mobility ratio of displacing-displaced fluids is favorable:

$$\text{Mixing zone length, per cent of total} = \left(\frac{1000}{L} \frac{\Delta\mu}{\mu_i/\mu_d} \right)^{0.5} \qquad (16)$$

Equation (16) is based upon the experimental data of Hall *et al.*[21]. The subscripts i and d refer to the injected and displaced fluid viscosities, respectively. The system length is L. Equation (16) may be applied to the miscible slug problem by calculating the mixing at both the leading and trailing edges of the slug. An assumption would have to be made as to the slug concentration necessary to maintain miscibility. With this decided, it is then possible to make an estimation of the amount of solvent which would need to be injected. Fitch *et al.* suggest that the mixing zone volume could be considered as being 50 per cent solvent at the limit of miscibility.

Where the mobility ratio is adverse between the slug and the reservoir oil, between the driving fluid and the slug, or between the driving fluid and the reservoir oil, as in the case of a miscible slug dry-gas process, the process could be expected to have only limited economic field application. The process has distinct possibilities in reservoirs of steep dip, where the rates at the front are sufficiently slow so that gravity effects dominate and prevent the formation of viscous fingering. Extensive stratification, in either a horizontal or dipping formation, would be expected to increase the slug requirements if miscibility is to be maintained over much of the project life.

It is apparent that the mobility ratio between the displacing and displaced fluids has to be improved if the miscible slug process is to find wide and successful application in the field. The mobility ratio can be improved[32] by the following of the dry gas behind the slug, with simultaneous water and gas injection. In a pattern flood, this procedure will increase the areal coverage of the flood. Unfortunately, this method will not prevent the deterioration of the miscible slug, since the mechanism controlling viscous fingering will be unchanged. It has also been proposed to inject a buffer zone at the front and back of the miscible slug to permit a gradation of the viscosities, and in turn the mobility ratios between the displaced and displacing fluids. This should be beneficial, but does, of course, require that the buffer zone remain intact during the displacement.

A number of miscible displacement projects have been initiated[33-38] during the past eight years, with various degrees of success reported in the literature. Premature breakthrough of the miscible slug to the producing wells has been a common problem in many of these projects. Apparently, confusion has existed as to the role of reservoir stratification, of mobility ratio, and of slug orientation. Laue *et al.*[39] have recently reported the preliminary results of a propane slug-miscible displacement project in a reservoir having a 2° regional dip. The careful drilling and testing of an observation well indicate that the propane slug is horizontal. It was also found that viscous fingering apparently was occurring when the producing rate was above that indicated by equation (2). Laboratory tests on a long core indicated that propane injection would result in a deposition of asphalt and 16 per cent reduction of the reservoir permeability. This apparently has created no formation plugging problem in the field operation. It should be evident that the orientation of the miscible slug will have a significant bearing on the quantity of miscible slug which would be required for a given project. Gardner *et al.*[40] and Slobod *et al.*[41] have presented laboratory and analytical studies of gravity segregation of miscible fluids in linear models, for the case when the viscosity ratio of the displacing-displaced fluids is small, or the mobility ratio is favorable.

Calculation of Performance—Miscible Slug Flooding. The method of Doepel and Sibley, described in a preceding section of this chapter, may be used with only slight modifications for the prediction of reservoir performance in a stratified reservoir where a miscible slug is used. The volume of slug necessary to maintain miscibility in a given strata, for all or a part of the history of the flood, can be estimated from information in the immediately preceding section.

Agan and Fernandes[30] have presented a prediction technique which has application to the prediction of performance of a miscible slug process in a highly stratified reservoir. The method can also be used to calculate performance when the mobility ratio is unfavorable, if it may be assumed that the areal sweep efficiencies of Figures 4-12, 4-17, 4-18, and 4-24, for example, for the various well patterns, is indicative of the performance which occurs in an enclosed field-wide flood—especially when the mobility ratios are adverse.

The following equation may be used to determine the relative amounts of injected fluids which flow into each layer:

$$G_{ix} = G_{in} \frac{k_x}{k_n} \tag{17}$$

where
  G_{ix} = pore volume injected into layer x
  G_{in} = pore volume injected into layer n

The subscripts x and n refer to layers x and n, respectively. The permeabilities used should be the mean value for each given layer in millidarcys. If, for the mobility of a given system, the breakthrough of the miscible slug will result when a set pore volume has been injected to the most permeable layer, the amount of injection which has occurred to each of the other layers can be readily calculated from equation (17). On the other hand, the pore volume that has been injected to slug breakthrough in layer 2 can then be used to determine the pore volumes injected to layer 1, etc. The pore volumes injected to any layer at the breakthrough of any other layer may then be generated.

Table 10-1 presents a summary of the pore volumes required for breakthrough of a reservoir having three layers, a mobility ratio of 5, and an enclosed five-spot well pattern. The mean permeabilities of layers 1, 2, and 3 are 100, 50 and 10 md, respectively. Figure 4-17 of Chapter 4 was used to obtain values for the area swept for each layer of Table 10-1.

TABLE 10-1. Injection Pore Volumes Required to Obtain
Breakthrough in Each Layer*.

Layer	Layer 1		Layer 2		Layer 3	
	Injection	Swept	Injection	Swept	Injection	Swept
1	0.1667	0.1667	0.3333	0.2250	1.6667	0.3167
2	0.0834	0.0834	0.1667	0.1667	0.8334	0.2867
3	0.0167	0.0167	0.0555	0.0555	0.1667	0.1667
	0.2668	0.2668	0.5555	0.4472	2.6668	0.7701

*Initial breakthrough at 50 per cent of layer volume.

It can be reasonably assumed that the production and injection rates for each layer would be a direct function of the capacity (kh) of that layer. Then too, as the best approximation, it could be reasoned that the oil production and gas-oil ratio from each layer would remain constant during the fill-up period (due to an initial gas-saturation), and that the gas-oil ratio would remain constant from fill-up until breakthrough of the slug. If steady-state conditions prevail, the producing rate should equal the injection rate at reservoir conditions after fill-up. The producing rate in reservoir bbls/day for layer x after fill-up would then be: ,

$$q_{tx} = i_g \frac{(kh)_x}{(kh)_t} \qquad (18)$$

where

i_g = total injection rate, reservoir bbl/day

t = subscript indicating a property of the total system

If the layer has already experienced slug breakthrough:

$$q_{tx} = q_x \text{ unswept } + q_x \text{ swept} \qquad (19)$$

The oil production from layer x is thus taken to be equal to the production, from areas contacted and not contacted by the displacement front. In this instance the production, q_x, may be estimated from the observation that:

$$q_x = \psi_{sx} q_{tx} \qquad (20)$$

The term ψ_{sx} is the fraction of the layer x production which is from the swept area, and may be ascertained from Figure 4-12, p. 87, for the five-spot well pattern. For those layers which have not had fill-up:

$$q_{tx} = q_{oi} \frac{(kh)_x}{(kh)_t} \qquad (21)$$

where q_{oi} is the oil producing rate in reservoir bbl/day before injection has begun. Equations (17) through (21) are sufficient to calculate the performance of the project, provided that an independent determination of the necessary slug volume has already been made.

The prediction of project behavior is accomplished by making a calculation of the performance for various values of the cumulative pore volumes of slug and displacing fluid injected. A review of the information for the particular pore volume calculation step, in a tabulation such as Table 10-1, would indicate which layers have had breakthrough, and which have or have not had fill-up. A simple summing of the producing rates from each layer, as calculated by means of equations (18) through (21), results in the total oil producing rate. The producing rate of the slug material would be from those layers having breakthrough. Since the miscible slug would be produced from the swept area and would ideally be

100 per cent slug material, the producing rate at standard conditions would be the sum of the q_x values from the broken through layers, as calculated by equation (20) for each layer, corrected by the appropriate gas formation volume factor in reservoir bbl/scf. The gas producing rate is obtained from the barrels of oil produced, times the constant solution gas-oil ratio prevailing at the beginning of the project. The stock tank oil producing rate would be the reservoir barrels corrected by the oil formation volume factor.

Figures 4-20 and 4-22 (pp. 98, 99) permit a calculation of the performance of a miscible slug process in a stratified reservoir to be made for the direct-line and staggered-line drive enclosed well patterns, by the method just outlined for the five-spot well pattern. The assumptions made in the calculation of the performance must be kept in mind, especially when the mobility ratio is adverse. Doubt still exists as to the applicability of figures such as Figures 4-12, 4-20 and 4-22 (pp. 87, 98, 99) to the field case, due to the severe viscous fingering which was seen in laboratory models by Habermann[28]. The method does provide a calculation which yields reasonable answers for project performance from data available to the engineer.

Greenkorn *et al.*[42] have reported the results of extensive field tests of miscible displacements for a range of mobility values. It was concluded that the viscous fingering observed in laboratory models occurs in the field, and that laboratory models can be constructed to reproduce or predict field behavior, even at unfavorable mobility ratios. Good information must be available on the degree and location of permeability variations, so that this can be reproduced in the model. Then too, the operation procedure apparently should be the same. Unfortunately, the use of scaled models also has many drawbacks, such as the expense and the inability to properly model phase behavior. However, these models can be very useful for semi-quantitative prediction.

One of the drawbacks to a miscible slug process using dry gas as the scavenging fluid is the poor areal coverage and the viscous fingering which results at the displacement front. Fitch and Griffith[31] have presented a calculation technique for an alternate gas-water injection behind the miscible front, designed to improve the areal sweep efficiency. It was also reported that laboratory tests indicate that a small volume of water injected ahead of the miscible slug will reduce the amount of fingering in the following miscible displacement. It is not known if such a procedure would be beneficial in the field application.

Two-phase solvent floods offer advantages in the control of solvent fingering into the reservoir oil at the displacement front. Handy[43] describes the use of a solvent slug where the driving gas is essentially insoluble in the miscible slug. Such a slug can be propelled through a porous media, at pressures somewhat in excess of the vapor pressure of the solvent, thus permitting the use of the miscible slug process at less than the previously mentioned 1200 psia reservoir pressure for a totally miscible system. In this case, the gas and solvent are partially miscible. The solvent and driving-gas form two equilibrium phases where the slug contains a quantity of the driving gas, and the driving gas, in turn, contains certain components taken from the slug. The main difference between the totally miscible slug and two-phase displacement processes is in the quantity of driving gas required to move the slug through the reservoir. In general, the quantity of gas required would be higher for the two-phase displacement process.

Oil Recovery by Water-Driven Miscible Slugs

One of the major drawbacks to the use of a miscible slug driven by dry gas for the recovery of oil is the poor areal coverage which can result. The use of a water-solvent flooding technique may have the high sweep efficiency of the conventional water flood and the high pore-to-pore sweep efficiency of the miscible displacement. Caudle and Dyes[32] have suggested the simultaneous injection of water with the miscible fluid. Blackwell *et al.*[44] have pointed out that the efficiency of the water-driven miscible slug process depends upon the oil-solvent viscosity ratio, on the positioning and maintaining of a miscible slug ahead of the water and upon gravity segregation due to differences in fluid densities. In addition, an immiscible displacement will occur between the solvent and the injected water.

Water-Driven Miscible Slugs—Horizontal Reservoirs. Laboratory studies of the displacement with visual models indicate that the solvent and water will usually segregate in the reservoir into two distinct layers due to the differences in fluid densities. This being the case, it is possible to develop a simple mathematical model for horizontal sands as depicted in Figure 10-12.

The following equation, developed by Blackwell *et al.*, permits a calculation of the amount of solvent which should be used to

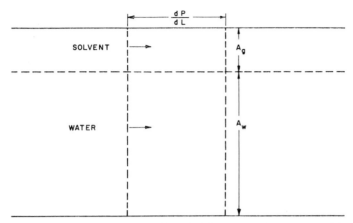

Figure 10-12. Mathematical model for water-driven miscible slugs in horizontal beds (from Blackwell *et al.*[44]).

cause the solvent and water fronts to advance at the same rate:

$$\frac{q_w}{q_g} = \frac{(S_{w,rg} - S_{wi}) - (1 - S_{wi}) \dfrac{k_{w,rg}\,\mu_g}{k_{g,wi}\,\mu_w}}{1 - S_{w,rg}} \tag{22}$$

where

$S_{w,rg}$ = water saturation at the residual hydrocarbon saturation
$k_{w,rg}$ = permeability to water at the residual hydrocarbon saturation, md
$k_{g,wi}$ = permeability to solvent at the initial water saturation, md

Equation (22) is developed from material balance considerations and pressure gradients in the solvent and water layers. Simultaneous injection of solvent and water, at a ratio less than that given by equation (22), should have the effect of the solvent layer slightly preceding the water which would be desirable.

For the water-solvent mathematical model described by Figure 10-12, it should be evident that the total effective mobility of the water and solvent would be:

$$\gamma_{\text{total}} = \frac{k_{g,wi}}{\mu_g} A_g + \frac{k_{w,rg}}{\mu_w} A_w \tag{23}$$

where A_g and A_w refer to the cross-sectional areas available to the flow of solvent and water, respectively. Since the total area through which the displaced oil flows amounts to a summing of the areas through which the solvent and water flows, the following equation for the displacing-displaced fluid mobility ratio can be readily developed[44]:

$$M = \left[\frac{1 + q_w/q_g}{\dfrac{k_{w,rg}\,\mu_g}{k_{g,wi}\,\mu_w} + \dfrac{q_w}{q_g}} \right] \frac{k_{w,rg}\,\mu_o}{k_{o,wi}\,\mu_w} \tag{24}$$

where $k_{w,ro}$ = permeability to water at the residual oil saturation, md.

Manipulation of equation (24) will show that the indicated mobility ratio for this segregated water-solvent type of displacement will be only slightly different than would be calculated assuming a uniform mixture of the water and solvent. In general, equation (24) shows that the mobility ratio will be somewhat more adverse than would be the case for an equivalent water-oil displacement without the solvent. The areal sweep efficiency will then approach that obtained in a conventional water flood. The advantage of the combined water-solvent injection will be a higher total oil recovery due to an improved pore-to-pore sweep efficiency. The actual benefits to be derived by the oil producer will depend upon the cost to inject the solvent. It should be noted that since the displacement front between the water and solvent is immiscible, in those areas already swept by the solvent, the following water will trap a fraction of the solvent behind the front as the residual hydrocarbon saturation.

Water-Driven Miscible Slugs—Dipping Reservoirs. Since gravity segregation plays a dominant role in reservoir processes where fluids have differing densities, certain benefits in the displacing mechanism could result if the rates and points of injection and withdrawal were carefully controlled. Unfortunately, the miscible slug is of lower density than the reservoir oil, and should be placed at the top of the structure with displacement downward, while the water has the reverse property, and should be injected at a low point with displacement upward.

Figure 10-13 shows the type of profile which could be expected when a water-solvent mixture is displacing a lower viscosity oil downward in a reservoir of shallow dip angle, at rates where the solvent-water and the solvent-oil fronts are essentially stable. The following form of Darcy's equation can be used to calculate

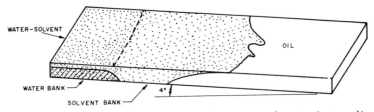

Figure 10-13. Schematic drawing of a water-solvent mixture displacing oil downward in a reservoir of small dip angle (from Blackwell *et al.*[44]).

the maximum solvent-oil front advance for a stable displacement in a homogeneous reservoir:

$$\frac{q}{A} = \frac{k(\rho_o - \rho_g) \sin \alpha}{\mu_o - \mu_g} \qquad (25)$$

where ρ_o and ρ_g are the oil and solvent densities, and α is the reservoir dip angle.

Since the injected water will tend to underrun the solvent and the reservoir oil, the minimum rate for a stable displacement of the water-solvent front must also be known:

$$\frac{q}{A} = \frac{(\rho_w - \rho_g) \sin \alpha}{\dfrac{\mu_w}{k_{w,rg}} - \dfrac{\mu_g}{k_{g,wi}}} \qquad (26)$$

Manipulation of equations (25) and (26) will show that the range of injection and producing rates may be very low if the reservoir dip or formation permeability is low. Then too, if the oil is two or three times as viscous as the injected water, the minimum rate calculated by equation (26) for a stable water-solvent front will exceed the maximum rate for the maintenance of a stable solvent-oil, as calculated by equation (25). A stable displacement would not be possible, and severe fingering could be expected. From their experiments, Blackwell *et al.* concluded that in a homogeneous reservoir containing an oil of 0.3 cp or less, total recovery of oil from the pattern area can be attained. If the reservoir is stratified or contains oil of viscosity higher than 0.3 cp, some residual oil will be left in those parts of the pattern swept by the water-solvent mixture.

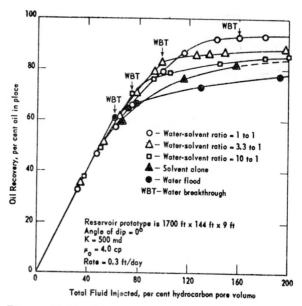

Figure 10-14. Effect of the water-to-solvent ratio on oil recovery in a linear system (from Blackwell et al.[44]).

Figure 10-14 presents laboratory results where the water to solvent ratio was varied from one to one to infinity (the water flood case). Since the oil viscosity exceeds 0.3 cp, total recovery was not attained in this linear system experiment.

Water-Driven Miscible Slugs—Application to Watered-Out Reservoirs. Since total oil recovery is the ultimate goal of the reservoir engineer, one of the challenges he may encounter is the large quantity of oil which remains in an oil reservoir after water flooding, either by natural or artificial means. The possibility of using a miscible slug driven by water to obtain the remaining oil has been investigated by Csaszar and Holm[45], for the case where reservoir pressure is between 200 and 1000 psia. This range of pressures is of particular interest, since many water floods have been done in shallow fields, or at rather low reservoir pressures. Propane is miscible with oil at low pressures. However, essentially all of the various driving gases and water would not be miscible with the propane. As the flood progresses, the solvent injected could be methane, which would normally be of lower cost than many other solvents.

Csaszar and Holm report that laboratory studies indicate that oil can be recovered from both uniform and non-uniform watered-out reservoirs by the injection of a propane slug amounting to 4 to 5 per cent of the pore space. Recovery was increased further by the injection of approximately 4 per cent of the pore space of dry natural gas or nitrogen. Oil recovered amounted to two to three times the liquid volume of propane injected. The method worked best if the reservoir oil had a viscosity of 3 cp or higher, since in this instance the residual oil saturation would be higher. The authors felt that the oil recovery was due to: (1) the higher residual oil left by oils of higher viscosity, (2) recovery of oil in tight sections by-passed previously by the water, (3) the formation of a trapped gas-phase on a hydrocarbon replacement basis, and (4) lower oil saturation remaining due to the swelling of the original residual oil by the propane.

Oil Recovery by Alcohol Flooding

The use of alcohol for the recovery of reservoir oil is not a new idea[46,47], but has received detailed attention only during the last six years or so. Gatlin and Slobod[48] were the first to show that despite the high cost of alcohol, the process could have commercial application. At the present time, the application of the technique appears limited to those fields where the price of crude is high. As with many other oil recovery techniques, an increase in the selling price of crude oil would result in a wider application of the method.

The alcohol slug process differs from the miscible displacement processes, discussed earlier in the chapter, in that both the reservoir oil and the connate water will be displaced if the alcohol concentration in the slug is sufficiently high. If the alcohol concentration drops below a certain level, miscibility is lost, and the system will revert to a water flood when water is used as the fluid to move the slug through the formation. Water would normally be used to displace the alcohol, due to its low cost and favorable effect on areal sweep efficiency in a pattern flood.

Types of Alcohols. The type(s) of alcohol(s) which would be used in a particular field application will depend upon the efficiency with which it displaces oil, relative to the alcohol and water volumes required. Alcohols which have been used during laboratory studies of alcohol flooding include isopropyl alcohol

(IPA), methyl alcohol (MA), normal amyl alcohol (nAA), normal butyl alcohol (nBA), secondary butyl alcohol (SBA) and tertiary butyl alcohol (TBA). The use of nBA in a commercial application would be limited due to its high cost. Each of the alcohols have varying degrees of solubility in water and reservoir oil, depending upon the relative concentration of each phase at the displacement front.

A ternary diagram is useful for the understanding of the alcohol slug process. Figure 10-15 is such a diagram, with the apexes A, B, and C representing the three pure components which can be plotted. At a set temperature and pressure, the ternary diagram

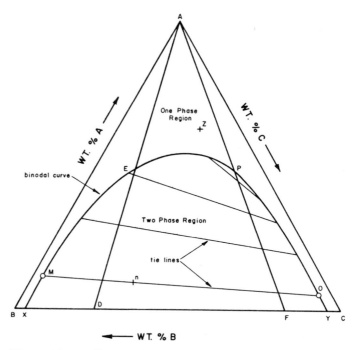

Figure 10-15. Ternary diagram describing the phase behavior of an alcohol-brine-reservoir oil system at constant pressure and temperature (after Burcik[49]).

can be used to plot the volume per cent, mol per cent, or weight per cent of each of the pure components, and also describe those concentrations where miscibility is or is not obtained. It should be evident from Figure 10-15 that A is miscible in all proportions with B and C, but that B and C are only miscible in the presence of a substantial concentration of A. If the system concentration were to change along the line from A to D as components B and C were added, complete miscibility would be maintained until the point E was reached. At this point, components B and C would not be miscible, although component A would be present in both B and C. At point D, the concentration of A would be zero; the composition of the two coexisting phases, B and C, would be represented by points X and Y. The phase represented by X would be rich in component C. The line connecting the phases is termed the "tie line". Considering a mixture of fluids of composition n on Figure 10-15, the quantity of the oleic phase is proportional to the line segment mn, and the quantity of the aqueous phase is proportional to the line segment no. The ratio of the amount of the B-rich phase to that of the C-rich phase, on a weight basis, would then be given by the ratio no/mn. As the amount of component A in the system increases, the length of the tie lines decreases but does not necessarily remain parallel to the base of the ternary diagram, since the amount of A which enters the B-rich and C-rich phases does not have to remain the same.

In the limiting case, the tie line shrinks to a point known as the "plait point," the point where the two coexisting phases become identical. In this regard, the plait point is similar to the critical point of a three-component hydrocarbon mixture. The two-phase region, traced by the ends of the tie lines, is called the "binodial curve." The distance from the peak of the binodial curve to component A indicates the mutual solubility of the solvent in components B and C. The greater the distance, the greater the indicated mutual solubility.

Figure 10-15 is typical of the type of diagram which results for the alcohol-brine-crude oil system. Point A corresponds to the alcohol, point B to the brine and point C to the reservoir oil. In this instance, the alcohol is preferentially water-soluble, since the plait point is to the left of the binoidal curve peak and nearer to the oil corner. IPA usually performs in this manner. TBA normally results in a plait point to the left of the binodial curve peak, since it is preferentially oil soluble. The displacement mechanism, which results with the injection of an alcohol slug followed by water to an oil reservoir, will depend to a large extent upon the location of the plait point[50,51,52]. Holm and Csaszar[52] found that the quantity of total solvent required for total oil recovery in a linear system was less when an oil soluble alcohol slug was followed by a water soluble slug driven by water. When a single slug was used, a smaller amount of a preferentially oil soluble alcohol was required, as compared to a preferentially water-soluble slug. This means that the process should be controlled wherever possible by devising a system where the plait point is to the left of the binodial curve peak. Ideally, the alcohols used should result in the maximum distance between the peak of the binodial curve and point *A* of Figure 10-15, since this would decrease the alcohol concentration necessary for system miscibility.

Displacement Mechanisms with Alcohol Flooding. The first logical step in the investigation of alcohol flooding for a given field application would be to determine the phase diagram, complete with plait point and tie lines, along the lines of Figure 10-15. The plot could be made on the basis of volume per cent, mol per cent, or weight per cent. One of the major problems would be to choose the proper alcohol, or combination of alcohols for the slug, so that the phase diagram could be prepared in the laboratory. Burcik[49] makes brief mention of the laboratory technique which can be used to prepare a phase diagram.

The choice of the solvent would be a compromise between cost and availability of the solvent, and the solvent which performs the most efficiently in the displacing of the reservoir oil. A number of authors have found that the displacement would be more efficient if the plait point is to the left of the peak of the binodial curve[51-54]. In addition, the composition path followed by the system as it achieves miscibility would also be influenced by the type of displacement which occurs. Then too, the displacement is less efficient as the path followed by the system to achieve miscibility is moved away from the plait point[51]. Figure 10-16 gives the results of a laboratory study where a 6 cp oil was displaced by 27 per cent of pore volume of a given alcohol (n-amyl, s-butyl, isopropyl, or ethyl), followed by an 18 per cent ethyl alcohol slug, which was in turn followed by water. The oil recovery by water flood is also plotted. The n-amyl alcohol was found to perform the best due to its preferential oil-solubility.

One of the problems in maintaining an efficient displacement of the "piston" type is the relative velocity of the components, which make up the displacement front between the alcohol(s) and the reservoir oil and water of the reservoir. Where the alcohol is preferentially soluble in the oil, the oil phase flows at a higher velocity than that of the water phase. Figure 10-17 illustrates this

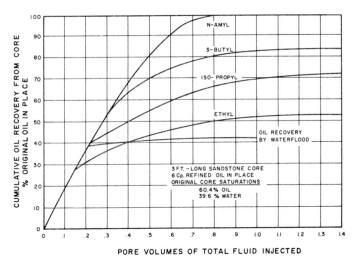

PORE VOLUMES OF TOTAL FLUID INJECTED

Figure 10-16. Oil recovery from sandstone cores using water-driven alcohol slugs (after Holm and Csaszar[52]).

phenomena as it occurs when TBA is injected. Where the alcohol is preferentially soluble in the water, the water phase is the faster, as shown in Figure 10-18. Figure 10-19 shows the saturation distribution which would be expected in a linear system where two solvents are used. In this instance, solvent 1 would be preferentially soluble in the reservoir oil. Butyl or amyl alcohol would exhibit this behavior. The second solvent, S_2, would be miscible in both water and oil, but in this application has preferential solubility in the first solvent. The careful tailoring of the dual solvents or buffer zones can equalize the velocity of the water and oil phases at the displacement front. This will result in piston-like

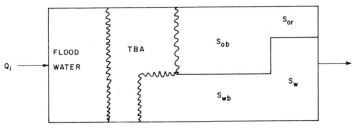

Figure 10-17. Preferentially oil-soluble alcohol slug driven by water in a linear system (after Holm and Csaszar[52]).

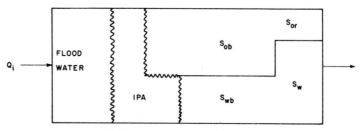

Figure 10-18. Preferentially water-soluble alcohol slug driven by water in a linear system (after Holm and Csaszar[52]).

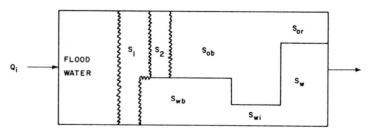

Figure 10-19. Dual alcohol slugs driven by water in a linear system (after Holm and Csaszar[52]).

behavior, and minimize the dilution of the alcohol slug which would occur if either the oil or water phases were to lag[51,53].

Taber and Meyer[51] have suggested the premixing of alcohol and reservoir oil to control the displacement mechanism, with the purpose of approximating a piston-like behavior of the displacement front. Figure 10-20 illustrates the injection sequence which would be followed for a laboratory system of $IPA-CaCl_2$ brine-isooctane. This system has a plait point to the right of the binodial curve peak; that is, the alcohol is preferentially soluble in the water. If alcohol were added to the system at the displacement front along the line *ABC*, at point *B* the relative oil phase volume shrinks to zero, as evidenced by the tie lines. This will result in unequal velocities of the oil and water phases. This can be adjusted by the injection of an alcohol and oil mixture so that a path is toward the plait point. Tabor *et al.* report that the path should be to the right of the plait point, or to point *E*, for the best system performance. In this case, the displacement at the front is with an alcohol-oil mixture that is preferentially oil miscible to a small degree. The oil-alcohol section of the slug should be followed by an alco-

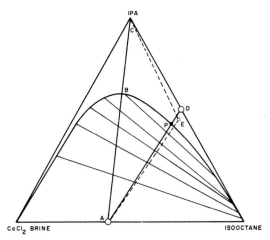

Figure 10-20. Ternary diagram to illustrate the adjustment of phase volumes (from Taber and Meyer[51]).

hol slug to prevent the system from reverting to a two-phase or immiscible system, as the flood front moves forward.

Sandrea and Stahl[50] have studied the effect of composite solvent slugs on the recovery of crude oil. It was found that propane, or LPG, works well as a buffer zone between IPA and a reservoir oil that are incompatible. IPA would often be used in the field application of alcohol flooding due to its lower cost and availability in quantity. It is preferentially water soluble, as illustrated in Figure 10-16. Propane, or LPG, is at the present time one of the least expensive of the oil soluble solvents. Despite the adverse mobility ratio between the propane and the reservoir oil, the propane, or LPG, yielded higher displacement efficiencies during laboratory tests.

Performance Calculations in Alcohol Flooding. A number of authors have presented alcohol flooding performance calculation techniques[48,51,53,55]. Most calculation methods have succeeded in reproducing the alcohol displacement histories that have been observed. Unfortunately, no field data are available to verify the validity of the various approaches taken. For this reason, no attempt will be made to present a calculation technique here. The reader is encouraged to read the original references. The approach of Wachmann[55] is of particular interest, since the theory is essentially an extension of the Buckley-Leverett construction for two immiscible fluids to systems of two slightly miscible fluids, and also to the three component system. The assumptions of the calculation technique are no viscous fingering, chemical equilibrium between the phases and no longitudinal mixing.

As a first approximation, the size of the alcohol slug may be determined in the same way as for miscible slug flooding, as discussed in a previous section of this chapter.

Carbon Dioxide, Flue or Inert Gas Displacements

As early as 1941, Pirson[56] suggested the high pressure injection of carbon dioxide for the recovery of the residual oil saturation of reservoirs by vaporization. This phase of carbon dioxide repressuring of reservoirs for the recovery of oil has been studied by Menzie and Nielsen[57]. Carbon dioxide has also been proposed and used in conjunction with water for carbonated water floods, with the benefit of oil viscosity reduction and oil swelling[58,59,60,61]. In addition, carbon dioxide has application as a water driven slug for the recovery of oil in a miscible displacement[62,63].

A variation of the processes using carbon dioxide is possible, through the use of flue or engine exhaust gases, which contain 87 per cent nitrogen, 11 to 13 per cent carbon dioxide, and inerts[64]. The technique is based upon the observation that the carbon dioxide will be dissolved in the oil and reduce the oil viscosity, while the nitrogen will boost the reservoir pressure. Present applications of the technique are similar to that of the steam "huff-and-puff" well stimulation technique, where the same well is used for injection and production on a cyclic basis.

Correlations for Carbon Dioxide—Crude Oil Systems. Simon and Graue[65] have presented correlations for predicting the solubility, swelling, and viscosity of carbon dioxide—crude oil systems. The authors have compiled all the available data on the subject, and then extended these data experimentally as necessary. The CO_2 solubility data cover a range of temperatures from 110 to 250°F for pressures up to 2300 psia. For correlation purposes, the "UOP" or "Watson" characterization factor, K, is used. The factor is commonly used in refinery technology to indicate the type of hydrocarbon, such as paraffin, naphthene, or aromatic, which is present. Its values range from 12.5 to 13.0 for purely paraffinic-based crudes, to 10 or less for aromatic materials[66]. By definition, the K factor is the ratio of the cubic root of the average boiling point of the crude in degrees Rankine, to the specific gravity of the crude at 60°F. It has been found[66] that viscosity and specific gravity determine the characterization factor, K, to sufficient accuracy in most hydrocarbon mixtures. Figure 10-5 permits an estimation to be made of K, from API gravity and kinematic viscosity of the oil. The kinematic viscosity may be determined from published correlations[67] if a laboratory determination is not at hand.

Figure 10-21 presents the solubility of CO_2 as a mol fraction in an oil having a K factor of 11.7, as a function of saturation pressure and the prevailing temperature. Figure 10-22 provides a means for converting the solubility of CO_2 value obtained from Fig-

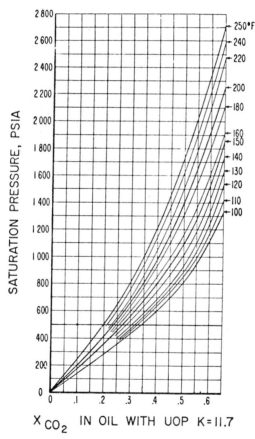

Figure 10-21. Carbon dioxide solubility in oil (from Simon and Graue[65]).

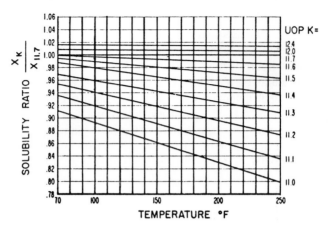

Figure 10-22. Carbon dioxide solubility in oil correction relationship (from Simon and Graue[65]).

ure 10-21, to oils having K values different from 11.7. Knowing the CO_2 solubility in the oil, Figure 10-23 may be used to determine the swelling factor, with molecular weight of the oil divided by oil density in gm/cc, at 60°F as a parameter. The following equation can be used to estimate the molecular weight of a crude oil[68]:

$$M_o = \frac{6084}{°\text{API} - 5.9} \tag{27}$$

The swelling factor is defined as the volume of the oil and dissolved CO_2 at saturation pressure and system temperature, divided by the CO_2-free volume of the oil at the same temperature, and 14.7 psia.

Figure 10-24 provides a means by which the viscosity of a CO_2-crude oil mixture, at saturation pressure and 120°F, can be determined from the CO_2-free crude oil viscosity. Since Figure 10-24 is for the specific system temperature of 120°F, Figures 10-25 and 10-26 are provided so that the oil viscosity can be adjusted to this temperature and the saturation temperatures altered according to a change in the saturation temperature. The utility of Figures 10-21 through 10-26 is best illustrated by an example problem:

EXAMPLE 1:
    Calculate the viscosity, the swelling factor and the CO_2 solubility in oil of a CO_2- 25° API crude oil mixture, at a CO_2 saturation pressure of 800 psia and 140°F.

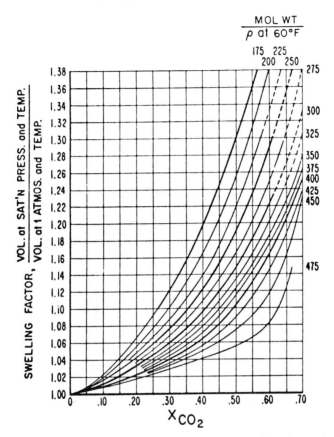

Figure 10-23. Oil swelling factor as a function of CO_2 mol fraction and oil characteristics (from Simon and Graue[65]).

(1) Determine the viscosity of gas-free crude oil at $140°F$ to be 12.5 cp, and at $120°F$ to be 16 cp from Beal[67].

(2) Determine the saturation pressure of the CO_2-crude oil mixture at $120°F$ by paralleling the curves on Figure 10-26, beginning with $140°F$ and 800 psia. The indicated saturation pressure is 730 psia.

(3) Determine the viscosity of the CO_2-crude oil mixture using Figure 10-24. Since the saturation pressure is 730 psia at $120°F$, and the oil viscosity is 16 cp, the value of the viscosity ratio, μ_m/μ_o, is 0.228. This yields a value of 3.7 cp for μ_m.

(4) Determine the viscosity of the CO_2 mixture at $140°F$ by paralleling the lines of Figure 10-25. The indicated viscosity is 2.4 cp.

(5) Determine the kinematic viscosity of the 25 °API crude. Since the density of the oil is 0.905, the kinematic viscosity is equal to 12.5 divided by 0.905, or 13.8 centistokes.

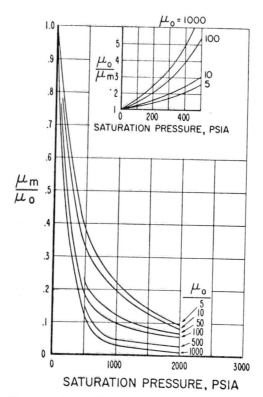

Figure 10-24. Viscosity of CO_2-crude oil mixtures at $120°F$ (from Simon and Graue[65]).

(6) Determine the characterization factor of the oil. Figure 10-5 yields a K value of 12.1. Since the K value is larger than 11.9, no correction for aromaticity is required.

(7) Determine the mol fraction of CO_2 in a CO_2-crude oil mixture at a saturation pressure of 800 psia. Figure 10-21 yields a value of 0.425 for x_{CO_2}. Figure 10-22 shows that the ratio $x_K/x_{11.7}$ is 1.012. The mol fraction of CO_2 is calculated as 0.430, the CO_2 solubility in oil. This value could be readily converted to SCF/bbl of 25 °API oil by means of the natural gas law.

(8) Determine the molecular weight of the 25 °API oil to be 318 lb/mole from equation (27).

(9) Determine the swelling factor from Figure 10-23. The M_o/ρ value is .352 with a CO_2 mole fraction of 0.430. The swelling factor is 1.008.

Discussion of Displacement by High Pressure Non-Hydrocarbon Gases. Koch and Hutchinson[69] and others have found that the composition of the injected gas is of little relative importance in

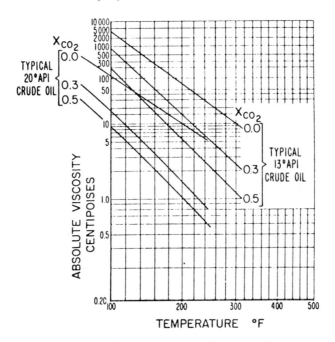

Figure 10-25. Oil viscosity as a function of temperature and CO_2 solubility (from Simon and Graue[65]).

determining the miscibility pressure for a particular reservoir fluid.
The high pressure gas process finds application where the reservoir contains a large concentration of the intermediates (C_2 through
C_6). A mass transfer of these intermediates, from the reservoir fluid
to the displacing gas, will result in miscibility being achieved at
the displacement front. Reservoir pressures have to be at least
3000 psi for the process to be technically feasible when a lean or
dry hydrocarbon gas is used as the displacing fluid.

Laboratory work accomplished by Koch *et al.*, used nitrogen
rather than flue gas, since it was reasoned that the 12 mol per cent
CO_2 normally present in a flue gas would be stripped from the injected gas in the reservoir by going into solution in the connate
water saturation. It was found that in excess of the miscibility
pressure, the displacement was 100 per cent as results when a dry
hydrocarbon gas is used. For one particular reservoir fluid, the
miscibility pressure using nitrogen was 3870 psia and 3500 psia,
using a dry hydrocarbon gas. This suggests that the pressure necessary for miscibility will be lower if a small amount of dry gas is

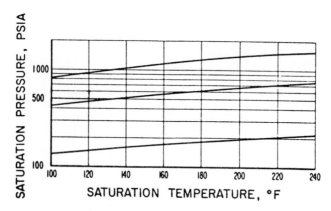

Figure 10-26. Saturation pressure of CO_2-crude oil mixtures versus saturation temperature (from Simon and Graue[65]).

added to the nitrogen that is to be injected. With other reservoir fluids, the pressures for miscibility could be anywhere from approximately 3000 to 4500 psia.

The calculation of performance would be accomplished in a similar fashion to that described earlier in this chapter for the high pressure dry gas process.

Discussion of Displacement by Cyclic Carbon Dioxide Repressuring. Menzie and Nielsen[57] have described laboratory experiments where it was possible to vaporize and produce a 35° API crude oil by cyclic injections of carbon dioxide. The lighter ends of the oil were produced in the first cycles, with the remaining oil becoming progressively of lower API gravity. It was found that about 30 cycles were necessary to vaporize and recover most of the oil, in the case of the sand-filled laboratory cell.

Recently, Clark *et al.*[64] have described a similar oil recovery technique, termed the CO_2 condensing gas drive process. Here, flue gas containing approximately 12 mol per cent CO_2 is injected in a cyclic fashion for the recovery of oil. The swelling of the oil by the CO_2, with a resulting reduction in viscosity at relatively low pressures, permits the application of the process to oils of moderate to low gravities at shallow depths. The nitrogen in the injected flue gas increases the reservoir pressure and provides a driving force for the recovery of oil on the producing part of the cycle. The method has the advantage that the response in terms of produced fluids occurs very soon after the initiation of the project. If it is found that the process is not applicable to a particular reservoir, the exhaust gas generating unit can be easily moved to a location more favorable for the process. The method does, of course, have application as a well stimulation technique, during both the primary and secondary recovery phases of a field producing history.

The flue gas is generated in the field by a natural gas, or LPG, fueled 4-cycle internal combustion engine. The resulting hot exhaust gases are treated to remove corrosive oxygen, nitrogen oxides and water vapor before entering the compression cylinders driven by the engine. Each standard cubic foot of natural gas burned will yield approximately 10 standard cubic feet of flue gas. If 1 standard cubic foot of propane gas is burned, roughly 24 standard cubic feet of flue gas will be formed. The process has been field tested[64].

Ramsey and Small[70] have reported the use of carbon dioxide for the improvement of water injectivity of wells being used for conventional water flooding. The methods of treatment have normally included the CO_2, water, a non-ionic surfactant, and LPG. The treatment has been in the form of highly carbonated water, low level carbonated water and slugs of CO_2. The method apparently has best application to a formation with swelling clays. Improvement of injectivity is due to the shrinking of clays with the weak carbonic acid which the CO_2 forms in the presence of water. The CO_2 is usually supplied by a service company.

Partially-Miscible Gas-Solvent Systems. The high pressure requirements for a miscible displacement, with either a dry or inert gas, limits the application of the technique to high pressure reservoirs at considerable depth. If the field had been produced for a time, reservoir pressure might be too low for consideration of such a process. Handy[43] has described a process where a partially-miscible gas solvent system would be used. Such a system would result when a solvent slug, which is miscible in the reservoir fluid, is displaced through the reservoir by a gas which is partially miscible in the slug. Partial miscibility would require that the solvent and the driving gas form two equilibrium phases where the gas phase contains a sizeable quantity of the solvent, and the solvent in turn contains an appreciable quantity of the driving gas.

The main advantage other than lower system pressure would be due to improved fluid mobility ratios, resulting in less solvent fingering. This would permit the use of smaller solvent slugs for a given reservoir system. Partial miscibility would require that the pressure at the displacement front be between the vapor pressure of the solvent and the critical pressure of the gas-solvent mixture. Handy[43] has presented a simple method for estimating the gas requirements for such a system.

The Water-Driven Carbon Dioxide Process. Laboratory investigations on linear flow systems[71,72] indicate that substantial benefits can be obtained by the use of a carbon dioxide slug driven by water or carbonated water over that obtained by ordinary water flooding procedures. Since the CO_2 is miscible with the reservoir oil at pressures in excess of 700 psia, it is possible to achieve 100 per cent recovery in those areas swept by the CO_2. Since CO_2 is also highly soluble in water at elevated pressures, water, or carbonated water, would be the logical choice as the fluid to displace the CO_2 slug through the reservoir. A previous section of this chapter outlines the benefits of oil swelling and reduction of oil viscosity due to solubility in CO_2, and provides a means for predicting these factors numerically.

The method has application even to those reservoirs containing a rather viscous oil, which might respond poorly to a conventional water flood due to adverse mobility ratios. A reduction in the viscosity of the oil at the displacement front, and the use of water as a displacing fluid, would serve to reduce the seriousness of the viscous fingering problem. The application of the method would, of course, require access to a carbon dioxide source, with economics determining whether the technique could be used. The method has the advantage that the carbon dioxide also swells the residual oil saturation. Since the size of the residual oil saturation is a function of the pore configurations, the wettability of the system and the oil surface tension, the leaving behind of a lower viscosity oil-CO_2 mixture could only result in a lower residual oil saturation. The injection of uncarbonated water would result in a stripping of the CO_2 from the residual oil. Additional oil recovery is also obtained during the blow-down phase of the flood.

Carbonated Water Flooding. Carbonated water flooding has been proposed and tested in the field[61]. The method increases the oil recovery over that obtained by conventional water flooding in two ways:

(1) By reducing the viscosity of the oil, thereby improving the mobility ratio.

(2) By swelling the oil. Plain water following the carbonated

water, or the period of blow-down will result in a release of CO_2 in the residual oil, resulting in a lower residual oil saturation at the end of the project life.

If the field to which the process is applied has a sensitivity to other injection waters, the carbonated water may result in an increased water injectivity, due to the shrinking of water-sensitive clays. This would have the desirable effect of shortening the project life, since the water injection rate will often determine the oil producing rate which can be sustained. Holm[73] indicates that the method will result in less ultimate oil recovery than the water-driven CO_2 process. The advantage, however, is less pronounced as the reservoir oil becomes more viscous—at least in laboratory studies of the processes.

de Nevers[74] has presented a calculation technique for predicting recoveries by carbonated water flooding. Essentially, the approach is a modification of the Buckley-Leverett frontal displacement theory. It provides a solution technique for the problem where a carbonated water slug is followed by plain water as the scavenging fluid. The reader is referred to the original article for details of this specialized calculation.

Limitations of Equations and Errors

Techniques have been presented in the chapter for predicting the reservoir performance of both the continuous solvent injection and the solvent slug injection cases, based upon dividing the formation into layers of different permeability. The amount of solvent injected into each layer during a time interval is calculated, and the areal sweepout within a layer estimated from sweepout curves, provided in Chapter 4, with mobility and well configuration as parameters. For such a treatment to be strictly valid, the displaced and displacing fluids must have the same density and viscosity, and the layers must be noncommunicating, or the strata under consideration must have sufficient dip to allow the displacement to proceed at a rate where gravity forces prevent the initiation of viscous fingers. These conditions have already been stressed.

It should be recognized that in displacements with purely hydrocarbon systems, the density and viscosity conditions are seldom met, the condition of noncommunicating layers may only be approximated, and reservoirs with sufficient dip as well as fluid and rock properties allowing a satisfactory critical displacement rate are only occasionally found. The reservoir engineer, using the techniques outlined, should recognize that reservoir and fluid properties always play a controlling role in the success or failure of a miscible displacement project. Failing to recognize the controlling mechanisms may result in the multilayer calculation technique yielding only a reasonable first approximation to reservoir behavior in situations that differ too markedly from conditions that have been discussed. This would be particularly true for slug operations.

With regard to slug operations, it was recommended that the minimum slug size be determined through phase behavior knowledge and laboratory experiments in short porous media, which are extrapolated by the square-root-of-length relationship to reservoir dimensions. This is an unscaled experiment, and is only strictly valid for equal density-viscosity fluids. In view of the fingering behavior that has been observed in scaled laboratory experiments, this procedure may not be adequate in some cases. Even for the ideal equal density, equal viscosity case, serious objection could be made to a performance prediction procedure that utilizes a square-root-of-length slug size relationship, if the different layers are in sufficient communication for transverse dispersion to occur. Reference is made to the work of Koonce and Blackwell[75], which shows that required slug size, on a pore volume basis, might increase linearly with distance traveled under special conditions.

In short, no generally recognized or accepted procedure has appeared in the literature which will properly model the performance of all reservoirs subjected to miscible fluid displacement. Each field application will be different. Special modifications to the calculation procedure will normally be necessary to temper the analytical results obtained through calculation procedures. Certain parts of the large card house of technology, which has been constructed in recent years, has been inadequately checked by field applications (through lack of application) at the time of this writing.

References

1. Clark, N. J., Shearin, H. M., Schultz, W. P., Garms, K., and Moore, J. L., *J.Petrol. Technol.*, 11 (June, 1958).
2. Slobod, R. L., and Koch, H. A., Jr., *Oil Gas J.*, 84 (April 6, 1953).
3. Brownscombe, E. R., *Oil Gas J.*, 133 (June 14, 1954).

4. Kern, L. R., Kimbler, O. K., and Wilson, R., *J. Petrol. Technol.*, 16 (May, 1958).
5. Brigham, W. E., and Dew, J. N., Proc. of the 1st Annual Adv. Petr. Engr'g. Sem., Univ. of Okla., February 9-10, 1960, p. 56.
6. Rutherford, W. M., *Soc. Petrol. Engrs. J.*, 340 (December, 1962).
7. Wilson, J. F., *Trans. AIME* **219**, 223 (1960).
8. Simon, R., and Yarborough, L., *J. Petrol. Technol.*, 556 (May, 1963).
9. Doepel, G. W., and Sibley, W. P., paper no. SPE-137, presented fall mtg. of *Soc. of Petrol. Engrs.* of **AIME**, Dallas, October, 1961.
10. Katz, D. L., *et al.*, "Handbook of Natural Gas Engineering," p. 403, New York, McGraw-Hill Book Co., Inc., 1959.
11. Kehn, D. M., Pyndus, G. T., and Gaskell, M. H., *J. Petrol. Technol.*, 45 (June, 1958).
12. Welge, H. J., Johnson, E. J., Ewing, S. P., Jr., and Brinkman, F. H., paper no. 1525-G, presented fall mtg. of *Soc. Petrol. Engrs.* of **AIME**, Denver, October, 1960.
13. Barfield, E. C., Grinstead, W. C., Jr., paper no. SPE-136, presented fall mtg. of *Soc. Petrol. Engrs.* of **AIME**, Dallas, October, 1961.
14. Kennedy, H. T., *Oil Gas J.*, 58 (June 30, 1952).
15. Koch, H. A., Jr., and Slobod, R. L., *Trans. AIME* **210**, 40 (1957).
16. Bossler, R. B., and Crawford, P. B., Proc. of 11th Oil Recovery Conference, Texas Petr. Res. Committee, Bull. No. 67, 210 (May, 1958).
17. Taylor, G., *Proc. Roy. Soc.* **A 219**, 186 (1953), **223**, 446 (1954), and **225**, 473 (1954).
18. Aris, R., *Proc. Roy. Soc.* **A 235**, 67 (1956).
19. von Rosenberg, D. U., *A.I.Ch.E. Jour.* 2, 55 (1956).
20. Koch, H. A., Jr., and Slobod, R. L., *Trans. AIME* **210**, 40 (1957).
21. Hall, H. N., and Geffen, T. M., *Trans. AIME* **210**, 48 (1957).
22. Lacey, J. W., Draper, A. L., and Binder, G. G., *Trans. AIME* **213**, 76 (1958).
23. Griffith, J. D., and Nielsen, R. F., Bull. No. 58, Min. Ind. Exp. Sta., Penn. State Univ., 71.
24. Everett, J. P., Gooch, F. W., Jr., and Calhoun, J. C., Jr., *Trans. AIME* **189**, 215 (1950).
25. Perrine, R. L., *Soc. Petrol. Engrs. J.*, 17 (March, 1961).
26. Perrine, R. L., *Soc. Petrol. Engrs. J.*, 9 (March, 1961).
27. Lacey, J. W., Faris, J. E., and Brinkman, F. H., *J. Petrol. Technol.*, 805 (August, 1961).
28. Habermann, B., *J. Petrol. Technol.*, 264 (November, 1960).
29. Brigham, W. E., Reed, P. W., and Dew, J. N., *Soc. Petrol. Engrs. J.*, 1 (March, 1961).
30. Agan, J. B., and Fernandes, R. J., *J. Petrol. Technol.*, 81 (January, 1962).
31. Fitch, R. A., and Griffith, J. D., *J. Petrol. Technol.*, 1289 (November, 1964).
32. Caudle, B. H., and Dyes, A. B., *Trans. AIME* **213**, 281 (1958).
33. Gibbon, A., *World Oil* **144**, 6, 92 (May, 1957).
34. Jenks, L. H., Campbell, J. B., and Binder, G. G., Jr., *Trans. AIME* **210**, 34 (1957).
35. Marrs, Doyle G., *J. Petrol. Technol.*, 327 (April, 1961).

36. Relph, J. L., *J. Petrol., Technol.*, 24 (May, 1958).
37. Justen, J. J., Hoenmans, P. J., Groeneveld, H., Connally, C. A., Jr., and Mason, W. L., *Trans. AIME* **219**, 38 (1960).
38. Sessions, R. E., *J. Petrol. Technol.*, 32 (January, 1963).
39. Laue, L. C., Teubner, W. G., and Campbell, A. W., *J. Petrol. Technol.*, 661 (June, 1961).
40. Gardner, G. H. F., Downie, J., and Kendall, H. A., *Soc. Petrol. Engrs. J.*, 95 (June, 1962).
41. Slobod, R. L., and Howlett, W. E., Cir. No. 63, Min. Ind. Exp. Sta., Penn. State Univ., 60 (1962).
42. Greenkorn, R. A., Johnson, C. R., and Haring, R. E., paper no. SPE 1232, presented fall mtg. *Soc. Petrol. Engrs.*, **AIME**, Denver, October, 1965.
43. Handy, L. L., paper no. SPE 446, presented fall mtg. *Soc. Petrol. Engrs.*, **AIME**, Denver, October, 1962.
44. Blackwell, R. J., Terry, W. M., Rayne, J. R., Lindley, D. C., and Henderson, J. R., *Trans. AIME* **219**, 293 (1960).
45. Csaszar, A. K., and Holm, L. W., *J. Petrol. Technol.*, 643 (June, 1963).
46. Morse, R. A., German Patent No. 849,534 (July, 1952).
47. Paulsell, W. G., "The Effect of a Mutually Miscible Intermediate Phase on Immiscible Fluid Displacement in a Porous Medium," M. S. Thesis, U. of Okla., 1953.
48. Gatlin, C., and Slobod, *Trans. AIME* **219**, 46 (1960).
49. Burcik, E. J., Bull. No. 61, Min. Ind. Exp. Sta., Penn. State Univ., 156.
50. Sandrea, R., and Stahl, C. D., *Soc. Petrol. Engrs. J.*, 45 (March, 1965).
51. Taber, J. J., and Meyer, W. K., *Soc. Petrol. Engrs. J.*, 37 (March, 1964).
52. Holm, L. W., and Csaszar, A. K., *Soc. Petrol. Engrs. J.*, 129 (June, 1962).
53. Taber, J. J., Kamath, I. S. K., and Reed, R. L., *Soc. Petrol. Engrs. J.*, 195 (September, 1961).
54. Farouq Ali, S. M., and Stahl, C. D., *Producers Monthly* **27**, 1, 2 (1963).
55. Wachmann, C., *Soc. Petrol. Engrs. J.*, 250 (September, 1964).
56. Pirson, S. J., "Tertiary Recovery of Oil," paper presented before the Central Appalachian Section, **AIME** (June 26, 1941).
57. Menzie, D. E., and Nielsen, R. F., *J. Petrol. Technol.*, 1247 (November, 1963).
58. Weber, G., *Oil Gas J.* **49**, 38, 171 (January 25, 1951).
59. Johnson, W. E., Macfarlane, R. M., Breston, J. N., and Neil, D. C., *Producers Monthly* **17**, 1, 15 (November, 1952).
60. Martin, J. W., U. S. Patent 2,875,833.
61. Anon., *Oil Gas J.* **60**, 29, 106 (July 16, 1962).
62. Beeson, D. M., and Ortloff, G. D., *J. Petrol. Technol.*, 63 (April, 1959).
63. Holm, L. W., *Trans. AIME* **216**, 225 (1959).

64. Clark, N. J., Roberts, T. G., and Lindner, J. D., *Petrol. Engr.*, 43 (August, 1964).
65. Simon, R., and Graue, D. J., *J. Petrol. Technol.*, 102 (January, 1965).
66. Watson, K. M., Nelson, E. F., and Murphy, G. B., *Ind. Eng. Chem.* **27**, 1460 (1935).
67. Beal, C., *Trans. AIME* **165**, 94 (1946).
68. Cragoe, C. E., "Thermodynamic Properties of Petroleum Products," Vol. 1, Monograph 10, Bureau of Mines, (N.Y.: Amer. Gas Assoc.) 1957.
69. Koch, H. A., Jr., and Hutchinson, C. A., Jr., *Trans. AIME* **213**, 7 (1958).
70. Ramsay, H. J., Jr., and Small, F. R., paper no. SPE 595, presented Rocky Mtn. Joint Regional Mtg., *Soc. Petrol. Engrs.*, **AIME**, Denver, October 27-28, 1963.
71. Beeson, D. M., and Ortloff, G. D., *J. Petrol. Technol.*, 63 (April, 1959).
72. Holm, L. W., *Trans. AIME* **216**, 225 (1959).
73. Holm, L. W., *Producers Monthly*, 6 (September, 1963).
74. de Nevers, N., *Soc. Petrol. Engrs. J.*, 9 (March, 1964).
75. Koonce, K. T., and Blackwell, R. J., paper no. SPE 1231, presented at fall mtg. of *Soc. Petrol. Engrs.* of **AIME**, Denver, October, 1965.

PROBLEMS

1. Sketch ternary diagrams to illustrate the effect upon the two-phase region where the system pressure is increased. Assume that the system contains a hydrocarbon, such as illustrated in Figure 10-1. What can be said about the tie-lines and the critical point?

2. Discuss the basic differences between the ternary diagrams for a hydrocarbon system and an alcohol-brine-reservoir oil system.

3. A $35°$ API crude oil has a viscosity of 4.6 cp. If the reservoir temperature is $175°F$, what would the critical pressure of a gas-solvent reservoir oil system? What value does the resulting answer have in a miscible flooding problem?

4. An oil reservoir being considered for high pressure dry gas injection has the following reservoir and fluid data available:

Irreducible water saturation (all layers)	18 per cent
Porosity (all layers)	24 per cent
Oil saturation (all layers)	82 per cent
Well spacing	40 acres
Net oil pay thickness	26 ft
Wellbore radius	0.375 ft
Gas in solution (initially)	850 SCF/STB
Oil viscosity at reservoir conditions	0.85 cp
Oil formation volume factor	1.80 bbl/STB
Solvent viscosity at reservoir conditions	0.30 cp

Specific gravity of gas	0.70
Gas viscosity (60°F and 14.7 psia)	0.02 cp
Reservoir pressure	3800 psia
Reservoir temperature	240°F

The following permeability profile information is thought to be representative of the reservoir as a whole:

Depth, ft	Permeability, md
10,018–20	36
10,020–22	176
10,022–24	194
10,024–26	178
10,026–28	256
10,028–30	305
10,030–32	208
10,032–34	276
10,034–36	145
10,036–38	76
10,038–40	24
10,040–42	56
10,042–44	18

Assuming that miscibility conditions prevail, and that no gravity segregation or crossflow between the strata occurs, calculate the performance of a 40-acre segment of the reservoir for a steady-state reservoir injection and producing rate of 200 bbl/day. The plot should consist of oil producing rates, gas producing rates, cumulative oil produced, and cumulative gas produced as a function of time. The economic limit occurs at 40,000 SCF gas per STB of oil. What fraction of the oil is recovered at the economic limit?

5. Use the reservoir and fluids data of Problem 4 and the method of Doepel and Sibley to calculate performance by miscible slug flooding. Assume that the miscible slug is propane, and that the economic limit will be reached at 20,000 SCF/STB.

6. Determine the viscosity, the swelling factor, and the CO_2 solubility in oil, where CO_2 is injected to a reservoir having a pressure of 1125 psia and reservoir temperature of 156°F. The reservoir crude is of 27° API gravity.

CHAPTER 11

Oil Recovery by
In Situ Combustion

In situ combustion for the secondary recovery of oil consists of igniting the oil in the porous rock by means of an igniter, and driving a combustion zone through the reservoir rock by means of air and/or combustion gases. The heat generated serves to increase the mobility of the oil, permitting its displacement to producing wells.

In all processes designed to recover oil from porous media, an examination of the factors in Darcy's equation for flow will show the effect of a change in one of the variables upon the producing rate. Since producing rate is inversely proportional to oil viscosity, heat could be expected to produce rather dramatic results. The idea is not new. Lewis[1], in 1917, stated that: "Of the oil retained in the sand after a well has reached economic exhaustion under customary production methods, part is retained so tightly by capillarity in the finest pores as well as by adhesion to the surface of the sand grains that its removal seems possible only by using heat or solution." The in situ combustion principle was set out by Howard[2], in 1923. Actual in situ combustion, though not recognized as such, apparently occurred even earlier in some of the air injection projects, due to spontaneous ignition as evidenced in cores taken from new wells in the reservoirs, in more recent years. The idea of in situ combustion received attention in Russia as early as 1934[3,4,5]. Unsuccessful attempts to initiate and maintain a combustion front during field tests were reported in the Russian technical journals during 1939[6,7]. It was not until the early 1950's that extensive laboratory studies and field testing of in situ combustion showed the technique to have considerable promise for the recovery of secondary oil from certain types of oil reservoirs[8,9,10].

Some of the basic in situ combustion processes which have received attention are:

(1) *Forward Combustion.* The flame front moves from the injection well toward the producing well(s), or that is, in the same direction as the injected air. This process has been varied slightly, and termed the "heat wave process," where gases of limited oxygen content have been injected to control the rate of advance of the burning front[11].

(2) *Reverse Combustion.* The flame front moves from the producing well toward the injection well(s), or that is, in the opposite direction as that of the injected air.

(3) *Variations.* Injection of slugs of water to transfer heat to improve the areal sweep efficiency has been suggested[12]. Produced gases can be cycled, in order to halt the combustion part way between the producing and injection wells, and propel the heat wave toward the producing wells[13].

For the present, discussion will be limited to that of the forward combustion process as the variations incorporate many of the same general principles.

Reservoir Properties Necessary for In situ Combustion

The following upper and lower limits on reservoir properties for a forward combustion injection project are those which are indicated by laboratory data and field results to this date, and may change as our understanding of the mechanism of in situ combustion becomes more complete:

(1) *Oil Content.* Since as much as 300 barrels per acre foot of hydrocarbons may be consumed by the combustion front, at least 600 barrels should be present in the reservoir initially. In other words, the oil saturation should be high, and the water saturation low.

(2) *Pay Thickness.* The pay thickness should be at least five feet thick. Thinner pay sections are indicated to have high vertical heat losses, with subsequent danger of the temperature dropping below that necessary to maintain a combustion front. Preferably, the pay thickness should not exceed 50 feet as with

increased thickness, air requirements sufficient to maintain a combustion front movement rate of at least 0.25 feet per day become excessive, with respect to the practical limitation imposed by compression equipment.

(3) *Depth.* Depth should be greater than 200 feet. In general, depths less than 200 feet would severely limit the pressure at which air could be injected. Operating in deep reservoirs results in high well costs, as well as substantial air compression expense, and economic considerations will impose a practical depth limit. This may be on the order of 2500 to 4500 feet.

(4) *Oil Gravity and Viscosity.* In general, oils of gravity greater than 40 degrees API do not deposit sufficient coke (fuel) to maintain a combustion front. Then too, oils of this high gravity can usually be water-flooded at lesser cost and higher recovery. Oils of gravity less than 10 degrees API are usually too viscous to flow ahead of the combustion front, where a low reservoir temperature prevails under the influence of forward combustion. Recovery of such extremely low gravity oils may be possible by reverse combustion, where the produced oil flows through the heated zone, and the oil composition has been structurally altered.

(5) *Reservoir Permeability.* Where oil viscosity is high (a reservoir containing 10 degrees API oil), a permeability greater than 100 millidarcys would be necessary, especially if the reservoir is of shallow depth and injection pressures were correspondingly limited. A 30 to 35 API gravity oil at 2500 feet might respond to in situ combustion, with permeabilities as low as 25 to 50 millidarcys.

(6) *Size of Reservoir.* A reservoir should be sufficiently large so that if a small scale pilot flood is successful, an economically successful full scale operation can be instituted. Depending upon sand pay thicknesses, the minimum size of such a project might be 100 acres.

(7) *Reservoir Confinement.* Where possible, the reservoir chosen for in situ combustion operations should have no gas cap or water zone within the area of operation. Presence of zones, where fluid mobility is larger than that in the oil zone to be flooded, would permit by-passing of the injected air. Hopefully, the wells in the field would be in good condition, their location would be known, and the spacing would be regular. This would avoid the expense of abandoning oil wells and completing new ones which would help insure the economic success of the operation.

Qualitative Description of Forward Combustion

Figure 11-1 illustrates the saturation and temperature distributions which could be expected to exist in a forward combustion, or "heat wave" process. Notice that the temperature profile is wavelike. Szasz[11] has divided the system into five distinct zones:

(1) A constant, low temperature zone at the injection end of the system which may be approximately the temperature of the injected air.

(2) A zone of increasing temperature as the zone of combustion is approached.

(3) A narrow zone of combustion, of relatively high but constant temperature, where the combustion reactions take place.

(4) Temperature decreasing in the direction of flow.

(5) A zone of relatively low and constant temperature which for all practical purposes is the original formation temperature.

The flowing fluids, air, gas, water, steam and oil transfer heat from zone to zone. The temperature in zone 2 is reduced, while the temperature in zone 4 is increased at given fixed points in the

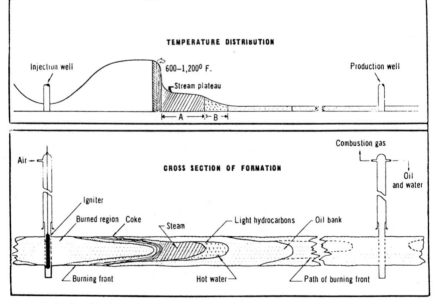

Figure 11-1. Temperature and saturation distribution in the reservoir during forward combustion (after Nelson and McNiel[14]).

reservoir, as time passes. Since new heat is generated in zone 3, this results in the movement of the heat wave forward as the process continues. No heat movement occurs in zones 1 and 5.

In forward combustion, air is usually first injected into the injection well(s), and produced from the producing wells, to ensure that the formation has permeability to air and a flue system for the combustion gases. Experience in a given area will dictate the advisability of this step. The oil in the formation is then ignited at the injection well. A variety of ignition devices are available [15,16,17]. In general, ignition of the oil may be accomplished with any type [18] of heating of the oil in the presence of oxygen, and indeed may result from spontaneous ignition [19]. Electrical resistance heaters, or gas or liquid-fueled heaters are widely used, and have the advantage that temperatures can be maintained at desired levels for sufficient time over the entire interval to be ignited. This would insure ignition, if ignition is possible. Surface heat injection and chemical methods also have been used. The injection of approximately one-half barrel of linseed oil per foot of zone to be lighted, followed by air, will result in auto-ignition. Burning charcoal can also be dumped from the surface, and followed by air injection in those wells where the bottom hole equipment is sufficiently simple so that damage will not result. Usually, the reaction between the oxygen in the injected air and the crude oil increases rapidly. Indications are that at about 500°F, a portion of the hydrogen in the oil burns to form water, and coke is layed down [20]. Ignition of the coke occurs at approximately 700°F, with the evolution of heat and the normal combustion products. The burning front temperature is in the range of 700 to 1200°F, with extinction occurring at approximately 600°F.

With the movement of the burning front away from the air injection well, a number of distinct zones may be identified. If the formation contains water at the immobile saturation, and gas and oil at saturations where finite values for oil and gas relative permeability values exist, then gas and oil will flow under an imposed pressure gradient due to the injection of air. Then too, the combustion front will burn the sand clean as long as the temperature remains above 700°F, forming a heat wave which will vaporize the water and oil saturations present. The "steam plateau," ahead of the combustion zone, will contain these vaporized materials, plus the combustion gases, water of combustion and any air which might have bypassed the combustion front. Since the vaporized oil and water and the products of combustion are moving at a faster rate than the combustion front, colder areas of the reservoir are soon reached which contain reservoir oil. The reservoir oil is subjected to heat, to miscible displacement by the vaporized and recondensed light components and to hot water drive. This is a zone of changing temperature with resulting changes in oil, gas and water viscosities, and fluid mobilities. This is depicted as zone B of Figure 11-1. This zone is the most difficult to analyze, due to the phase changes taking place.   Ahead of zone B, no phase changes are taking place, and no fractionation of the oil is occurring.   Displacement here is due to an immiscible displacement, controlled by the relative permeabilities of the oil, water and gas.

The narrow zone closest to the combustion front is characterized by a temperature sufficiently high to crack the oil—perhaps 500 to 700°F, resulting in three products: a volatile fraction which condenses further downstream, a lighter gaseous fraction which imparts a certain (usually quite low) heating value to the combustion gases, and a non-volatile residue of coke, tar, and/or pitch on the sand grains which provides the fuel for combustion. In addition, temperatures may be high enough in this zone to result in chemical and phase changes in the solid matrix. This zone, as well as the combustion zone, is also influenced in a significant manner by the heat flowing vertically to the colder formations overlaying and underlaying the section being burned.

In situ Combustion Performance Calculations— Nelson-McNiel Technique

Nelson and McNiel[21] have published a procedure for the setting up of a full scale in situ combustion project. The method is relatively simple, and can be accomplished in a relatively short time by using a slide rule. It does, however, require experimental data where the particular crude oil and reservoir conditions of porosity and saturation are duplicated as nearly as possible. Where this is done in a burning tube in the laboratory, observations are made of rates and amounts of air injected, the amounts of oil and water produced, and the amount and composition of the gases produced. The following discussion will follow very closely the procedure laid out by Nelson and McNiel.

Calculation of Air Requirements. The following laboratory data will be required:

D = internal diameter of the burning tube, ft
L = length of the pack burned, ft
ϕ = porosity, fraction
V_g = produced gas volume on dry basis, std cu ft
N_{2a} = nitrogen in air injected, fractional volume
O_{2a} = oxygen in air injected, fractional volume
N_{2g} = nitrogen in produced gas, fractional volume
O_{2g} = oxygen in produced gas, fractional volume
CO_{2g} = carbon dioxide in produced gas, fractional volume
CO_g = carbon monoxide in produced gas, fractional volume

The combustion equation is as follows, where no attempt has been made to balance the equation since the carbon-hydrogen ratio will not be known:

Air(nitrogen and oxygen) + fuel(carbon and hydrogen) $\longrightarrow$

Nitrogen + carbon dioxide + carbon monoxide + water +

unreacted oxygen   (1)

It has been observed that the amount of nitrogen injected equals the amount of nitrogen produced, since nitrogen does not enter into the reaction. The volume of nitrogen injected and produced in the laboratory then is:

$$N_2 \text{ injected} = N_2 \text{ produced} = V_g \times N_{2g} \qquad (2)$$

The quantities of other materials, either injected or produced according to equation (1), then become:

$$O_2 \text{ injected} = (V_g \times N_{2g}) \times (O_{2a}/N_{2a}) \qquad (3)$$

$$CO_2 \text{ produced} = V_g \times CO_{2g} \qquad (4)$$

$$CO \text{ produced} = V_g \times CO_g \qquad (5)$$

$$\text{Unreacted } O_2 \text{ produced} = V_g \times O_{2g} \qquad (6)$$

Carbon in fuel burned = W_C, lb = $[(V_g \times CO_{2g}) +$

$(V_g \times CO_g)] \times (12/379)$ (7)

Water formed by combustion = W_w, lb =

$2[(V_g \times N_{2g} \times O_{2a}/N_{2a}) - (V_g \times O_{2g}) - (V_g \times CO_{2g}) -$

$0.5(V_g \times CO_g)] \times (18/379)$ (8)

Hydrogen in the fuel burned $= W_H$, lb $=$

$$2\left[(V_g \times N_{2g} \times O_{2a}/N_{2a}) - (V_g \times O_{2g}) - (V_g \times CO_{2g}) - 0.5(V_g \times CO_g)\right] \times (2/379) \quad (9)$$

Since the fuel consumed at the flame front is essentially 100 per cent carbon and hydrogen, the total weight of the fuel used is:

$$W_F, \text{ lb} = W_C + W_H \qquad (10)$$

The pounds of fuel used per cubic foot of sand burned in the laboratory tube then is:

$$W = 4 W_F/\pi D^2 L \qquad (11)$$

On a reservoir basis, the pounds of fuel used per acre-foot of sand burned would be:

$$W_R = 43,560 \; WF \qquad (12)$$

where

$$F = (1 - \phi_R)/(1 - \phi_P)$$

and the subscripts R and P refer to properties of the reservoir and sand pack, respectively. The factor F corrects for the density difference which might exist between the reservoir rock and the laboratory sand pack, due to porosity difference. Notice that if the porosity of the system is low, a large fraction of the oil present may be required for fuel.

It is now possible to calculate the relationships between the fuel burned, the total air injected, and the volume of sand burned:

$$\text{Total air injection} = N_2 \text{ injected} + O_2 \text{ injected}$$

or

$$V_a = V_g \times N_{2g} + (V_g \times N_{2g})(O_{2a}/N_{2a}) \qquad (13)$$

The air injected per cu ft of reservoir burned in scf then is:

$$A = V_a/W_F \times WF = 4 V_a F/\pi D^2 L \qquad (14)$$

The actual volume of air per acre-foot that will be required to burn a given pattern in the field will be determined by the product of the volumetric sweep efficiency and the unit air consumption. Nelson and McNiel have used an areal sweep efficiency of 62.6 per cent for the five-spot pattern. Reference to Table 4-1 (p. 85) and Figure 4-10 (p. 86) indicates that there is a lack of agreement as to what the areal sweep efficiency should really be. Since the shadowgraph work used miscible fluids and a porous media, some basis exists for setting the areal sweep efficiency at approximately 50 per cent. The movement of injected air and combustion gases, relative to the movement of the viscous oil usually found in an in situ combustion project, justifies the use of an infinite mobility ratio in choosing the areal sweep efficiency. The volumetric sweep efficiency will be a product of the areal and vertical sweep efficiencies. This differs from that where immiscible fluids are displaced, since in this instance a pore-to-pore sweep efficiency is also used. In fire flooding, the sand is burned clean so that the pore-to-pore sweep efficiency term has no significance.

Data from tests of in situ combustion in the field which have been published, indicate that the vertical sweep efficiency is very sensitive to stratification, and can vary over a range from 35 to 90 per cent. Stratification of the reservoir also contributes to the problem of air bypassing the combustion zone, and to the poor utilization of injected oxygen. A safety factor can be imposed upon the calculations so that the air requirements will not be underestimated by assuming a 100 per cent vertical efficiency. The areal sweep efficiency of 62.6 per cent will be used, thus resulting in an overall, or volumetric sweep efficiency, of 62.6 per cent. The air required in MMSCF per acre-ft in a five-spot flooding pattern then becomes:

$$\text{Air injected/acre-ft of pattern} = 0.626 \times 43,560 \times A/10^6 \quad (15)$$

Notice that experience in a given area would permit the adjustment of the volumetric sweep efficiency for a more accurate prediction of the air requirements. Then too, the areal sweep efficiencies for patterns other than the five-spot can be estimated from Chapter 4. The estimation of air requirements will be found to be of considerable importance, since the cost of compressing air is one of the largest expenses of insitu combustion. Good oxygen utilization will be largely responsible for the economic success of such a project. The total air requirement for a given pattern will be merely the results of equation (15), multiplied by the total acre-ft of sand to be burned, irrespective of the area to be contacted by the flame front, i.e. V_T as needed for equation (22).

Calculation of Injection Rates and Pressures. The reservoir properties and operating economics of the project will determine the rate of advance of the combustion front, which in turn specifies the air injection rates and pressures during the life of the project. For the maximum rate of return on investment, the rate of advance of the combustion front should be as high as possible. On the other hand, this rate should not be so high that the producing wells cannot produce all of the displaced oil. The lower limit is set by the minimum rate at which a flood front must move before heat transfer results in a temperature at which combustion cannot occur. Practical considerations usually dictate a burning rate of 0.125 to 0.5 feet per day, where the formation is approximately 20 to 30 feet in thickness. Thinner sand sections would require a somewhat higher minimum rate, due to the proportionately higher heat loss from the narrow flame front.

It is convenient for rate calculations to convert air requirement values from that required to burn a cubic foot of formation to the "air flux," u, or the volume of air required per square foot of burning front per day:

$$u = Av \qquad (16)$$

where A is determined from equation (14), and v is the rate of burning front advance in feet per day.

It should be evident from considerations outlined in Chapter 4, that the air flux will be different from point-to-point along the combustion front, depending upon the relative location of the producing and injection wells, and the position of the combustion front. Where the air flux is insufficient to support combustion, the flame goes out and only the heat wave is displaced forward until it is dissipated. The limited sweep efficiency which results can be related to a dimensionless flow term, i_D, which is calculated from the maximum air rate to be used, i_a, and the minimum air flux required to sustain combustion, u_{min}. The equation defining the dimensionless flow rate is as follows:

$$i_D = i_a/(u_{min}\,ah) \qquad (17)$$

where a is the length of the shortest streamline in feet between the injection and producing well, and h is the average pay thickness between wells in feet.

Nelson and McNiel provide a brief table showing the sweep efficiencies corresponding to several dimensionless flow rates. Since equation (17) uses the shortest streamline along which the air would be moving at a maximum rate, the dimensionless flow rate calculated is the limiting value at which a combustion front can exist, for the minimum air flux assumed to apply to the system. In a five-spot well system, the areal sweep efficiency at breakthrough, corresponding to i_D values of 3.39, 4.77, 6.06 and infinity, are 50.0 per cent, 55.0 per cent, 57.5 per cent, and 62.6 per cent, respectively. These results are based upon potentiometric studies, and yield higher sweep efficiency values than do the shadowgraph techniques as discussed previously. It is evident from equation (17) that the only way an areal sweep efficiency of 62.6 per cent could be reached would be infinitely high air injection rates. Practical considerations will usually limit the areal sweep efficiency to approximately 55 per cent, and a corresponding i_D value of 4.77.

The maximum air rate for one enclosed well pattern may be calculated from the following equation:

$$i_a = i_D u_{min} \, ah \tag{18}$$

If an enclosed five-spot well pattern was being considered, the maximum air rate would be based upon a 55 per cent areal sweep efficiency, and on the minimum air flux corresponding to a selected minimum flame front advance rate. For formations on the order of 20 to 30 feet thickness, the minimum rate is on the order of 0.125 feet per day. After the in situ combustion project has been ignited, a period of increasing air flux will be necessary. Assuming that the maximum rate of flood advance for efficient displacement, v_1, is to be 0.5 feet per day, the air flux rate in scf per day, during the radial phase of the displacement, can be simply calculated from the following equation:

$$i_a = 2\pi r_f h A v_1 \qquad \text{for} \quad r_f \geqq r_1 \tag{19}$$

where r_1 is the radial distance in feet to the burning front at the end of the increasing air injection rate period, and r_f is the radial distance to the burning front. Usually about 10 per cent of the pattern area will have been swept during the radial displacement phase of the flood front. Calculations of air injection rates, for a range of values of r_f, are calculated until the maximum air rate specified by equation (18) has been reached. Beyond this point, the air flux rates will begin to decline, and the burning front advance will drop below 0.5 feet per day for the first time. Then too, as the burning front becomes distorted due to the influence of the producing wells, the air flux along the shortest streamline will tend to increase, while the air flux along the longer streamlines will decrease.

The time in days required for the increasing rate period of the fire flood may be calculated as follows:

$$t_1 = r_1/v_1 = i_a/2\pi h A v_1^2 \tag{20}$$

Since the increase in air injection is essentially linear during this phase, the volume of air injected to this point in MMSCF is:

$$V_1 = t_1 i_a/2 \tag{21}$$

Figure 11-2 illustrates the air requirements for a confined five-spot well pattern. If, for the purpose of balancing a burning operation where a number of five-spot patterns are involved, a similar

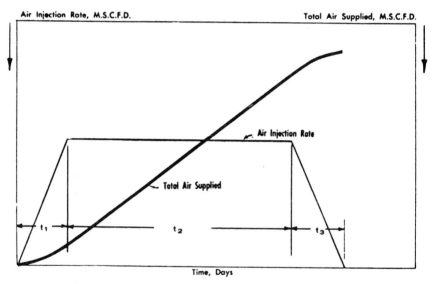

Figure 11-2. Air requirements of a confined five-spot well pattern (after Nelson and McNiel[21]).

period exists at the end of the flood where decreasing air injection rates are used, then a volume of air, V_3, will be injected over time, t_3. In this instance, V_1 will equal V_3, and t_1 will equal t_3. The volume of air injected at the constant rate i_a in MMSCF will be:

$$V_2 = V_T - V_1 - V_3 \tag{22}$$

the time in days for this part of the project being:

$$t_2 = V_2 \times 10^6 / i_a \tag{23}$$

The time for the overall burning project in the pattern would be the sum of the times during the three phases. Figure 11-2 illustrates this.

Since the cost to compress air for injection is one of the major expenses of secondary recovery by in situ combustion, it is apparent that the simultaneous operation of a number of patterns would permit a programming of the operation, such that something approaching a 100 per cent compressor loading could be maintained. Figure 11-3 illustrates this concept. As in all secondary recovery projects, operating economics are usually improved in large-scale operations.

Where a pilot in situ combustion project is planned before deciding whether a full-scale operation is warranted, the air injection pressures required to inject air at calculated flux rates will be observed directly. Where a pilot operation is to be accomplished, a calculation of the approximate injection pressures required will be necessary in designing the compression facilities for the project.

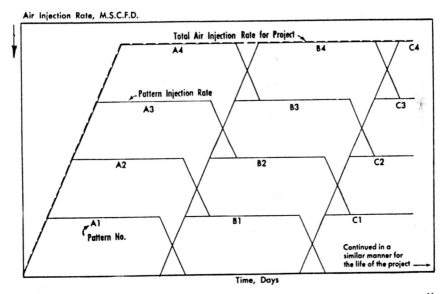

Figure 11-3. Programming air injection where a number of five-spot well patterns are to be fire flooded (after Nelson and McNiel[21]).

For the five-spot well network, whether confined or not, the following adaptation of Darcy's law for compressible flow during the radial flow period, should indicate the maximum pressure required:

$$p_{iw}^2 = p_w^2 + (i_a \mu_a T_f / 0.703 \, k_g h) \times [\ln (a^2 / r_w v_1 t_1) - 1.238] \quad (24)$$

where

p_{iw} = sand face pressure at the injection well, psia
p_w = sand face pressure at the producing well(s), psia
μ_a = viscosity of air at injection pressure and temperature, cp
T_f = formation temperature, °R
a = well-to-well spacing, ft

and where the remaining terms have been defined previously.

In equation (24), p_w will be determined by pump submergence at the producing well(s), and will usually be a low value for maximum oil production. The effective permeability to gas in millidarcys will normally not be known. It will usually suffice to assume a value of 5 per cent of the absolute permeability. Additional pressure drops beyond that indicated by equation (24) will occur in the injection wellbore and in the injection system, extending from the compressor to the well. Knowing the maximum air injection rate and the pressure required, the size of the compressor can be determined. The number of compression stages required will then allow a choice of compressor type to be made, consistent with engine fuels available. This phase of the design should be carefully done, since the costs associated with air injection will largely control the economics of the project.

Calculation of Oil Recovery. As pointed out earlier, the mechanism of oil displacement by a combustion front is complex. Laboratory and field data do indicate that in addition to the oil displaced by the front to the producing wells, a substantial amount of oil is recovered due to heating of the formation, with subsequent reduction of oil gravity, which then can move under a potential gradient to the producing wells. The actual oil recovery, as a fraction of the oil displaced due to the sweep of the burned zone, may be rather small, since the volumetric efficiency is a product of the areal and vertical sweep efficiencies, E_A and E_1 respectively:

$$E_v = E_A \times E_1 \quad (25)$$

As is often the case, the vertical sweep efficiency might approximate 55 per cent, while horizontal sweep efficiency might be in the range of 55 per cent, for an overall efficiency of 30 per cent. Then, the actual oil recovered from the swept zone will be the difference between the original oil content of the region swept by the combustion zone, and the oil consumed as fuel. Since coke, tar, and/or pitch used as fuel will have a specific gravity of approximately one, the following equation may be readily developed for the barrels of oil displaced per acre-ft of reservoir burned:

$$N_1 = 43,560 \ [(S_o \ \phi_R/5.615) - (WF/350)] \tag{26}$$

where 350 is the weight in pounds of one barrel of liquid having a specific gravity of one.

Cores, taken from regions where a combustion front had passed through a section of a reservoir in the field, indicate that roughly half of the oil in the regions not contacted by the front may have been produced by a combination of gravity drainage and gas drive. Nelson and McNiel[21] indicate that for preliminary design purposes, a recovery of 40 per cent of the oil from the regions contiguous to, but not swept by the flame front, may be assumed. Present technology does not permit a direct calculation of this recovery factor, since the amount of oil displaced appears to be due to reservoir permeability, gravity drainage, proximity to the flame front, oil type and viscosity and degree of heterogeneity of the formation. Results of a pilot in situ combustion operation would permit a valid determination of the recovery factor. The equation for the oil displaced in barrels per acre-ft of unburned reservoir is:

$$N_2 = 43,560 \times 0.40 \times S_o \times \phi_R/5.61 \tag{27}$$

Total recovery from the burned and unburned zones would be:

$$N_3 = E_v \times N_1 + (1 - E_v) \times N_2 \tag{28}$$

with the overall recovery efficiency as a per cent being:

$$E_R = \frac{N_3 \times 5.615}{43,560 \times S_o \times \phi_R} \tag{29}$$

Assuming that the water saturation in the unburned zone is immobile and unchanged during the flood, the total water production due to in situ combustion can be calculated as the sum of the water originally present in the burned zone, plus the water formed during combustion. The total water produced in barrels per acre-ft of reservoir rock in the well pattern is:

$$W_p = 43,560 \times E_v \times [(4W_w F/350\pi D^2 L) + (S_w \phi_R/5.615)] \tag{30}$$

Calculation of Producing Rates. Published oil and water production curves for in situ combustion projects of a number of different reservoirs[22,23,24] indicate that the production-rate history will vary widely from one reservoir to another. Usually, oil production rates will be small during the air injection phase, prior to ignition of the oil at the injection well. Production rates will remain small during the increasing air injection phase, unless well spacing is very close. Production increases rapidly as the heat of the combustion front reaches the vicinity of the producing well, since the oil viscosity will be substantially reduced. As the displacing gas-to-oil mobility ratio improves, the gas drive producing mechanism will be increasingly effective in moving oil to the producing wells. As the combustion front advances, the gas driving mechanism is complemented by the presence of a cold and finally a hot water displacement of the oil.

For the purposes of design, it will usually suffice to assume that the oil production will be proportional to the air-injection rate. If a number of well patterns are being flooded in a staggered operation schedule, such an assumption should not be seriously in error. The oil recovered in barrels per MMSCF of air injected may then be calculated:

$$N_p' = \frac{10^6}{43,560 \times 0.626 \times A} [E_v N_1 + (1 - E_v) N_2] \qquad (31)$$

The oil producing rate curve will have the same shape as the air injection rate curve.

The water producing rate could be expected to be zero during the initial air injection phase and ignition phases of the insitu combustion project, if the initial water saturation is immobile. As the combustion front advances, a water bank will be formed which will result in increased water producing rates after the flush production from the oil bank has been obtained. To facilitate a determination of a water producing rate, it will be assumed, however, that the water-oil ratio is constant. In view of equation (31), this means that the barrels of water produced per MMSCF of air injected will be a constant:

$$W_p' = \frac{E_v \times 10^6}{0.626 A} \left[\frac{4 W_w F}{\pi (350) D^2 L} + \frac{S_w \phi_R}{5.615} \right] \qquad (32)$$

Discussion of In situ Combustion Performance

Much time and effort has been brought to bear on the many facets of in situ combustion performance during the past 15 years. Many questions have been answered, and many more questions have been raised. Some of the topics which have great importance to the successful economic fire flooding of a reservoir are the effect of pressure and oil type, effect of flooding patterns, the role of stratification, the influence of porosity and permeability, the role of rock thermal characteristics, the influence of oil viscosity, air requirements and minimum burning rates for the sustaining of a combustion front.

Effect of Pressure. Wilson *et al.*[25] have performed laboratory experiments in linear adiabatic systems to determine the effect of pressure during both forward and reverse combustion. The procedure involved the packing of the combustion tube with unconsolidated sand, water and oil, in the desired proportions.

Tests made under the conditions of forward combustion indicate that the process was fuel-dominated, so far as the peak temperature and combustion zone velocity were concerned. Figure 11-4

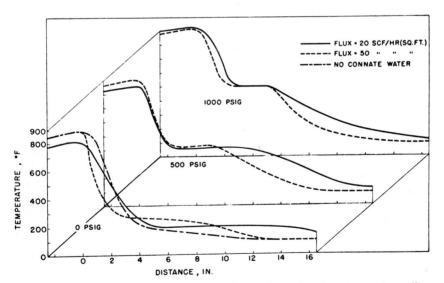

Figure 11-4. Temperature profiles in forward combustion for various flux rates and operating pressures (after Wilson *et al.*[25]).

shows the temperature profiles that were observed for two different flux rates and operating pressures of 0, 500, and 1000 psig. In general, it can be observed that the combustion zone and steam plateau regions were short, and that the temperature distribution between the steam plateau and combustion zone was independent of time for the rates used. Figure 11-4 shows that even when the initial water saturation is zero, a water bank will form ahead of the combustion zone.

Figure 11-5 shows the very small increase in air requirements that were noted during laboratory tests as the operating pressure was increased. Increasing the pressure does increase the peak temperature when the air flux is low. At higher flux rates, both peak temperature and rate of combustion front movement are found to be independent of the pressure, provided that no bypassing of the air occurs.

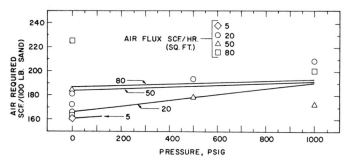

Figure 11-5. Air requirements in forward combustion at various operating pressures (after Wilson *et al.*[25]).

In general, oil recovery and the air-to-oil ratio required during in situ combustion operations are essentially independent of pressure, temperature and air flux rates that would normally be encountered in field operations.

Effect of Heat Conduction and Convection. The in situ combustion process depends upon the formation of a heat wave due to the movement of an expanding heat source through the formation from the injection well. Initially, the heat source would be cylindrical in form until the influence of the producing well(s) causes a cusping of the combustion front. In any analytical treatment of such a system, the main problem is one of properly determining the transient temperature distributions with position and time for varied conditions of heat conduction and convection. Ramey[26] and Bailey and Larkin[27] have independently published solutions to the heat transfer problem where the effects of heat conduction alone have been treated. In a later publication, Bailey and Larkin[28] include the effect of convection for the linear and radial flow cases. Selig and Couch[29] have also presented analytical solutions for heat transfer in the radial model where vertical heat losses were at least partially accounted for by setting the temperature on the formation boundaries at the ambient value. This would give an upper limit for the vertical heat flow. Chu[30] has provided a solution for the problem by considering that the bed being heat-wave flooded is bounded by an impermeable layer, where heat transfer is by conduction only. In the flooded bed, heat transfer was considered to be due to both heat convection and heat conduction.

Thomas[31] has presented the results of a mathematical model, where heat generation occurs over only a fraction of the thickness of the formation being heat-wave flooded. This is a reasonable representation of the field case, since cooling at the formation boundaries and stratification of the reservoir will usually result in an incomplete vertical coverage by the combustion front or heat source. To permit an analytical solution for the temperature distribution in the system, Thomas[31] made the following assumptions:

(1) Heat capacities, thermal conductivities and thermal diffusivities do not change with changing temperature and pressure, and are the same in the reservoir and the bounding beds.

(2) Oxygen utilization is 100 per cent.

(3) The combustion zone is very narrow as is observed in the laboratory and in the field.

(4) The injection rate of oxygen-bearing gas is constant.

(5) The fuel concentration is constant, and is completely consumed at the combustion front.

(6) Ignition is limited to a vertical thickness of h of the total formation thickness.

(7) The gas mass rate of flow remains constant.

(8) Heat radiation is neglected.

(9) Vaporization and condensation effects in the various fluid phases are neglected.

(10) Liquid phase flow through the combustion zone is zero.

(11) All heat transfer by convection is by the flow of gas.

The consequence of assumptions (4) and (5) is that the propagation rate of the moving combustion front is inversely proportional

to its radial position. This specifies the areal coverage of the
flame front since, as the radius to the front increases, the frontal
velocity in a part of the pattern will eventually drop to the point
where heat losses due to conduction and convection will create a
temperature drop, sufficient to cause the extinction of the flame.

The following development for temperature distribution, caused
by the radial movement of a combustion front, is due to Thomas[31].
It can be shown that the equation for conservation of energy for the
radial flow of gas in cylindrical coordinates is:

$$\frac{1}{r} \frac{\partial}{\partial r} (r q_T) - k \frac{\partial^2 T}{\partial z^2} + \frac{\partial}{\partial t} (H) - S(r, z, t) = 0 \qquad (33)$$

where

r = radial space coordinate, ft
q_T = heat flux, Btu/hr-sq ft
k = thermal conductivity, Btu/hr-ft-°F
T = temperature increase, °F
t = time, hours
H = enthalpy, Btu/cu ft
z = vertical space coordinate, ft

and where $S(r, z, t)$ is a source density function, with the units of
Btu/hr-ft^3, representing the heat source that moves radially through
the system. The enthalpy for a porous media containing gas and
liquid can be expressed as follows:

$$H = [\phi \rho_g c_g + (1 - \phi_g) \rho_m c_m] T \qquad (34)$$

where the subscripts g and m refer to the gas and the liquid-rock
matrix, respectively. The specific heat in Btu/lb-°F is represented by c. Equation (34) can be simplified to the following form:

$$H \cong [(1 - \phi_g) \rho_m c_m] T \qquad (35)$$

since the term $(1 - \phi_g) \rho_m c_m$ is much larger than the term $\rho_g \phi_g c_g$.

Since heat flux, due to radiation, is assumed to be zero by radiation, Fourier's law yields:

$$q_T = -k \frac{\partial T}{\partial r} + \rho_g c_g u_g T \qquad (36)$$

where the second term on the right-hand side has been added to account for heat convection in the gas phase. The variable, u_g, is

the gas flux in scf/hr-sq ft which, in a radial system, is equal to $V/2\pi r$ where V is the injection rate in scf/hr per foot of formation thickness. In view of equations (35) and (36), equation (33) takes the following form:

$$\frac{\partial^2 T}{\partial r^2} + (1 - 2\nu) \frac{1}{r} \frac{\partial T}{\partial r} + \frac{\partial^2 T}{\partial z^2} - \frac{1}{\alpha} \frac{\partial T}{\partial t} = -\frac{S(r, z, t)}{K} \qquad (37)$$

where

$\alpha = k/(1 - \phi_g)\rho_m c_m$, the thermal diffusivity constant
$\nu = \rho_g c_g V/4\pi k$, a dimensionless constant

The solution of the above differential equation can be solved where the boundary conditions required that the temperature of the heat source is initially zero degrees above ambient, and that the temperature distribution throughout the system is continuous and bounded.

Bailey[32] has shown that the heat source can be of the form:

$$S(r, z, t) = \Omega(z, h/2) H_m v \, \delta(r - r_f) \qquad (38)$$

where

$$\Omega(z, h/2) = \begin{cases} 1, & |z| \leq h/2 \\ 0, & |z| > h/2 \end{cases}$$

$\delta(r - r_f)$ = the Dirac delta function
r_f = radial distance to the combustion front, ft
H_m = heat generated per unit volume of rock, Btu/cu ft

Since air flux is constant, and a radial system is assumed, the rate of movement of the combustion front will vary inversely with distance from the injection well; therefore:

$$v = \frac{dr_f}{dt} = \frac{\lambda}{2 r_f}, \text{ from which } r_f^2 = \lambda t \qquad (39)$$

where λ = proportionality constant.

Thomas[31] has presented a solution for the temperature distribution of the system, as modelled by equations (33) through (39), by means of the appropriate Green's function, G:

$$T(r, z, t) = \int_0^t \int_0^\infty \int_0^\pi \int_{-\infty}^\infty S(r_0, z_0, t_0) G \, dt_0 \, r_0 \, dr_0 \, d\theta_0 \, dz_0 \qquad (40)$$

where the Green's function is

$$G(r, \theta, z, t \mid r_0, \theta_0, z_0, t_0) =$$

$$\frac{\left(\dfrac{r^2}{4\,\alpha\,(t-t_0)}\right)^{\nu} e^{-\dfrac{R^2+(z-z_0)^2}{4\,\alpha\,(t-t_0)}} \sin^{2\nu}(\theta-\theta_0)}{4\,\pi\,\Gamma\left(\nu+\frac{1}{2}\right)[\alpha\,(t-t_0)]^{3/2}\,\rho_m\,c_m} \tag{41}$$

where

$$R^2 = r^2 + r_0^2 - 2\,rr_0\,\cos(\theta-\theta_0)$$
$$t_0, z_0, \theta_0 = \text{variables of integration}$$

Equation (40) may now be manipulated to obtain the following form:

$$T(r, z, t) = \frac{H_m}{8\,k}\,\lambda\,r^{\nu}\int_0^t \frac{e^{-\dfrac{r^2+\lambda t_0}{4\,\alpha(t-t_0)}}}{(t-t_0)(\lambda t_0)^{\nu/2}}\,I_{\nu}\left(\frac{r\sqrt{\lambda t_0}}{2\,\alpha(t-t_0)}\right)\times$$

$$\left[\text{erf}\left(\frac{h/2+z}{2\sqrt{\alpha(t-t_0)}}\right)+\text{erf}\left(\frac{h/2-z}{2\sqrt{\alpha(t-t_0)}}\right)\right]dt_0 \tag{42}$$

where I_{ν} refers to a modified Bessel function of the first kind of order ν, and:

$\lambda = 4\,kvC_0\,\omega/F_0\,c_g$, sq ft/hr
C_0 = oxygen concentration, lb/lb
ω = fuel-oxygen ratio, lb/lb
$H_m = HF_0/\omega$, heat generated per unit volume of rock, Btu/ft^3
ΔH = heating value of oxygen, Btu/lb
F_0 = fuel concentration, lb/ft^3

Equation (42) may be numerically integrated after assigning numerical values to the constants. Thomas has published solutions to the equation for the following fixed values of the thermal constants:

k	1 Btu/hr-ft-°F
$(1-\phi_g)\rho_m\,c_m$	27.5 Btu/ft^3-°F
ρ_g	0.0765 lb/ft^3
c_g	0.24 Btu/lb-°F
α	0.0364 ft^2/hr
ω	0.331 lb/lb
H	5780 Btu/lb

Figure 11-6 shows the results of a numerical integration where a gas injection rate of 941 scf/ft-hr ($v = 3/2$) was used, and typical radial temperature profiles for various positions of the combustion zone were plotted. Three temperature profiles are presented for each position of the flame zone, showing the substantial differences in temperatures between that observed at the center line of the formation ($2\,z/h = 0$), and the formation boundary ($2\,z/h = 1$).

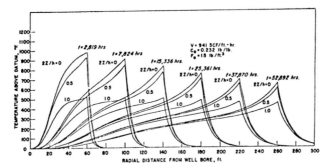

Figure 11-6. Temperature profiles in a forward combustion project for a number of radial distances from the injection well (after Thomas[31]).

As could be expected, the temperature of the system drops as the combustion front moves farther radially into the formation, since the air flux is constant and the areal extent of the flame front is larger. Eventually, for a given air flux and crude and formation properties, the propagation rate would drop to the point where heat transfer would be sufficient to lower temperature to the extinction point of the combustion zone. Laboratory experiments indicate that this occurs at approximately 600°F. It should be remembered that radial movement of the combustion front occurs over only a part of the displacement between the injection and producing wells. Nevertheless, Figure 11-6 provides an insight into the reasons for choosing limited areal and vertical sweep efficiencies, as explained in the Nelson-McNiel[21] method for calculating fire flooding performance.

Figure 11-7 presents typical isotherms as obtained by using equation (42), and the same data as that used in the construction of Figure 11-6, where the combustion zone is 100 feet into the formation. Notice that there is a considerable loss of heat vertically and behind the combustion front, due to convection and conduction of heat. The temperature distribution through the system

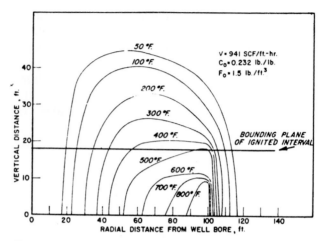

Figure 11-7. Isotherms for a combustion zone in forward combustion for specified data (after Thomas[31]).

is of prime importance, since this determines whether a flame front can be sustained over a given distance. Since the gas injection is assumed to be constant in the development of equation (42), it would be expected that the radial distance a combustion front could move from the injection well before extinction, would be directly influenced by the oxygen concentration. Figure 11-8 shows that for the data used, the relationship is essentially a linear one. Similarly, the combustion front could be maintained over a greater distance by either increasing the gas injection rate, or the fuel concentration, or both.

Kunii and Smith[32] have published information on thermal conductivities of porous rocks filled with stagnant fluid. These conductivities were measured in the laboratory by taking data from which values could be calculated from Fourier's law for heat conduction:

$$k = \frac{q_T}{2 \pi L} \frac{d(\ln r)}{dT} \tag{43}$$

where

k = thermal conductivity, Btu/hr-ft-°F
q_T = heat flux, Btu/hr-ft^2
L = system height, ft
r = radial distance from heat source, ft
T = temperature, °R

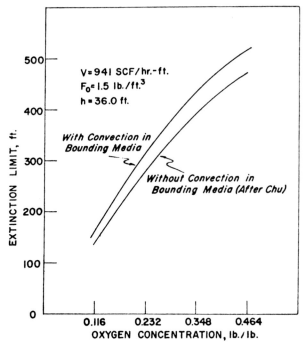

Figure 11-8. Effect of oxygen concentration upon distance a combustion front can move radially before extinction in forward combustion (after Thomas[31]).

Figure 11-9 shows the dependency of thermal conductivity upon the fluid in the rock, the reservoir pressure, and the rock type. Kunii and Smith also propose a technique for predicting stagnant conductivity analytically from information based upon a consolidation parameter. Where information on thermal conductivity is needed for a solution of equation (42) for a specific rock type, the recommended calculation procedure would have merit.

Somerton and Boozer[33] have studied the thermal characteristics of rocks at elevated temperatures. Table 11-1 presents a description of the rocks, while Table 11-2 summarizes the pertinent properties determined by laboratory analysis. A shale sample was also tested, since sand beds subjected to in situ combustion are often bounded by such a material. Heat losses do occur in these layers. In general, it was found that the thermal conductivity, k, decreased with increasing temperature. Then too, repeat runs on the samples,

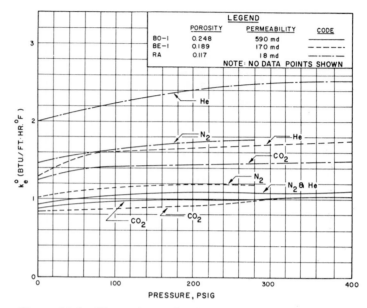

Figure 11-9. Thermal conductivities of three sandstones saturated with three different fluids (after Kunii and Smith[32]).

TABLE 11-1. Description of Rock Samples
(after Somerton and Boozer[33]).

Sample	Description	Principal Minerals		
		Quartz Per cent	Felds. Per cent	Other
Bandera sandstone	Well-cons., very fine grained	35	25	Calcite, clay
Berea sandstone	Well-cons., fine grained	65	10	Calcite, sericite, clay
Boise sandstone	Well-cons., med. grained	40	35	Clay, sericite
Limestone	Small vugs, med.-coarse grained			Calcium carbonate
Shale	Hard, laminated, very fine grained	50		Clay, iron oxides, biotite
Tuffaceous sandstone	Well-cons., large to very small grains	10	60	Clay, pumice lapilli, calcite
Rock salt	Crystalline			Halite

TABLE 11-2. Properties of Rock Samples
(after Somerton and Boozer[33]).

Sample	Porosity	Bulk Density (original) lb/cu ft.	Bulk Density (after testing) lb/cu ft.
Bandera sandstone	0.200	134.2	131.8
Berea sandstone	0.205	134.8	126.0
Boise sandstone	0.265	118.9	116.0
Limestone	0.186	140.2	78.5*
Shale	0.170	137.1	128.8
Tuffaceous sandstone	0.280	115.3	107.2
Rock salt	0.010	135.0	128.0
*After conversion to calcium oxide.			

after being subjected to 1500°F, resulted in lower conductivities on the sandstone and shale samples. This would indicate that the thermal conductivity of rock behind the combustion front (disregarding the presence of fluid) could be expected to be less than that ahead of the front. Thermal reactions will be discussed in a following section.

Asaad[34] has developed an approximate correlation between rock and fluid properties and thermal conductivity, which may be expressed as follows:

$$k = k_1 \left(\frac{k_2}{k_1} \right)^{\phi c} \tag{44}$$

where

k = thermal conductivity of fluid-saturated rocks, Btu/hr-ft-°F

k_1 = thermal conductivity of rock solids, Btu/hr-ft-°F

k_2 = thermal conductivity of saturating fluid, Btu/hr-ft-°F

c = correlating factor

Somerton[35] has pointed out that k_1 is difficult to evaluate, since it is apparently influenced by the degree of grain contact which occurs between the crystals or grains making up the porous media. Randomly oriented quartz crystals, in good thermal contact, have a k_1 value[36] of 5.0 Btu/hr-ft-°F, while feldspar has a k_1 value of 1.2 Btu/hr-ft-°F. The accuracy of equation (44) at high temperatures and pressures has not been tested[33]. Table 11-3 presents data applicable to equation (44) for several rock types:

TABLE 11-3. Calculated Correlation Values[33].

Sample	k_1	ϕ	c
Sandstone	5.7	0.196	2.3
Sandstone	2.2	0.40	1.0
Silty sand	2.3	0.43	0.9
Siltstone	2.2	0.36	1.1
Limestone	4.8	0.186	1.7
Sand (fine)	5.4	0.38	1.2
Sand (coarse)	5.4	0.34	1.2

It can be observed that the correlating factor c has a value of about unity for unconsolidated samples, but a noticeable deviation from unity for the consolidated samples. This may be due to the structural characteristics of the rock matrix.

Lange[37] presents information on thermal conductivity for a variety of insulating materials, solids, and fluids.

Adivarahan *et al.*[38] have presented data and calculation methods for determining effective thermal conductivities in porous media through which gases or salt water are flowing—both parallel and counter-current to the flow of heat. It was found that the effective thermal conductivity was not measurably influenced by a change in temperature or pressure over a modest range. Effective conductivities were found to increase with fluid velocity due to fluid mixing within the pores of the system. Figure 11-10 illustrates the changes in effective conductivity observed for five different rock samples as the flow rate of a 10 weight per cent sodium chloride solution was varied. Where the fluid movement rates in an in situ combustion project are relatively high, corrections to the heat conductivity value for the media, k, should be made for the specific fluid and porous media being treated.

The choosing of the proper heat conductivity value, for a specific problem in heat transfer and temperature distributions in in situ combustion, takes careful consideration of the dynamics of the system. Most of the present published literature on the topic of heat conduction in porous media is for the specialized cases of either stagnant fluids or for the flow of a few select fluids under single-phase flow conditions. The problems on heat transfer and temperature distributions in in situ combustion engineering are seldom this simple since a number of regions are involved; usually more than one phase will be present, and heat conduction will be radial over only a portion of the projected project life.

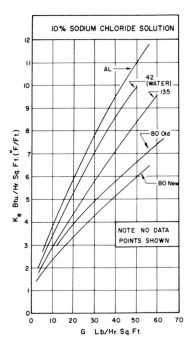

Figure 11-10. Effect of a 10 weight percent aqueous salt solution upon effective conductivity for 5 different porous medias (after Adivarahan *et al.*[38]).

Thermal Effects on Porous Media. Since in situ combustion for the recovery of oil takes place within the porous media, it might be expected that a number of temperature-induced reactions with the formation could occur. Bailly and Mangold[39] have identified the possible reactions as being loss of water, oxidation, crystal structure breakdown, crystal inversion, decomposition and melting with possible accompanying changes in volume, pressure and composition. Since a number of authors[18,21] have reported that maximum observed temperatures in the laboratory and the field during in situ combustion approximate 1200°F, those reactions which take place at higher temperatures are not normally of interest in the engineering of such a project.

Reference to Table 11-4 shows that for a variety of rock samples, the thermal reactions observed in the laboratory were an α-β quartz crystal inversion, and clay and salt dehydration. Since the α-β quartz conversion occurs at approximately 1030°F and is reversible, very little lasting effect in terms of retardation or augmentation of the fire-flood should occur. The clay reaction in the shale sample was apparently due to the loss of hydroxyl water from kao-

TABLE 11-4. Heats of Reaction for Minerals in Reservoir Rocks[40]

Mineral	Temp. Range, °F	Heat of Reaction, Btu/lb	Reaction
Ca-Montmorillonite	75–430	228	Desorption
Mg-Montmorillonite	75–430	243	Desorption
Mg-Illite	750–1280	115	Decomposition
Kaolinite	850–1190	455	Decomposition
Ca-Montmorillonite	1030–1330	121	Decomposition
Quartz	1065	8.67	α-β inversion

linite at approximately 800°F, and was endothermic. Table 11-4 is
a summary of heats of reaction for some minerals that are found in
reservoir rocks, and which do react at temperatures less than 1200°F.
The reactions with the clay minerals are endothermic over the temperature range of interest to in situ combustion secondary recovery
of oil. In some of these reactions, the heat energy required for the
reaction may be nearly as great as that required to raise the temperature of the rock for the purposes of in situ combustion. The
heat of reaction values given in Table 11-4 are those observed at
atmospheric pressure, and could be expected to be somewhat different at elevated pressures encountered in a flooding project.
Brindley and Nakahira[41] report that the temperature at which the
kaolinite reaction begins is 180 degrees higher under a water vapor
pressure of 88 psia.

Where the crude type results in the deposition of a rather small
quantity of coke to sustain the combustion front, the detrimental
effect of clays in the reservoir rock is evident. So much of the
available heat may be lost to the thermal reactions of the rock
minerals, that premature flame extinction results. Pilot flooding
operations would permit an evaluation of mineral reaction problems, which could result in failure of a project.

Somerton and Selim[42] have presented information showing that
the thermal expansion of rocks due to the high temperatures of
in situ combustion probably causes structural damage, as a result
of the differential expansion of the mineral grains. It is unlikely
that this would cause any difficulty during in situ combustion operations, with the possible exception being those sands where such
treatment could instigate sand sloughing at the wellbores.

Effect of Oil Viscosity. In situ combustion operations have their
best application in those oil fields where oil viscosity is high, and
recovery by primary producing means has been low or none at all.

Where viscosity is high, the time-honored water flooding method of secondary recovery is of little value due to the adverse water-to-oil mobility ratio. Figure 11-11 shows how the viscosity of oil changes with temperature. It can readily be seen that oils that would not flow at economic rates at low reservoir temperatures can become quite fluid with the addition of heat.

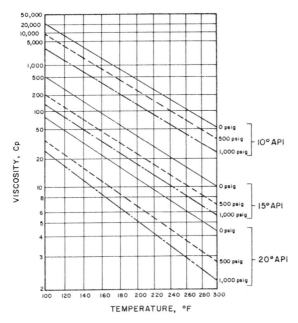

Figure 11-11. Temperature effect on oil viscosity (after Owens and Suter[43]).

In the in situ combustion process, the displacement mechanism is quite complicated. However, the temperature reached in the process does result in a complete vaporization of the oil and water, such that the interface between the water and oil would disappear. A residue of coke is left behind to support the combustion front. Ahead of the combustion zone, the water and the oil condenses, and a hot-water drive of the oil results. Then too, heat conduction and convection results in a gravity drainage of oil from the unburned zones, due to decreased oil viscosity.

Air Requirements and Fuel Availability. Dew and Martin[44] have presented an equation for air requirements in forward combustion, which is based upon the heat capacity of the rock:

$$A_s = \frac{\rho_s \, c_s \, (1 - \phi)(T_2 - T_1)}{Y \, H_a (1 - L)} \tag{45}$$

where

A_s = air required per unit volume of sand, scf/cu ft
ρ_s = sand grain density, lb/cu ft
c_s = heat capacity of the sand, Btu/lb-°F
T_2 = final temperature, °F
T_1 = initial temperature, °F
Y = fraction of injected oxygen that is consumed
H_a = heat from consumption of air, Btu/SCF
L = heat loss to system, fraction

and where the heat capacity of the reservoir rock can be approximated by:

$$c_s = \frac{T_1 + 2000}{10,000} + \frac{T_2 - T_1}{20,000} \tag{46}$$

Equation (45) is simple in concept but difficult in application, since the factors Y and L will seldom be known with any accuracy in an area where in situ combustion has not been tried before. Where laboratory data or field results from a pilot operation are available, a reasonable assumption of these values can usually be made.

The previous sections outline the ins and outs of heat loss. This is perhaps one of the most unsatisfactory aspects of the calculation of secondary recovery performance under in situ combustion. Figure 11-12 shows the dependency of air requirements upon heat loss and combustion efficiency in the generation of a given temperature rise, as given by equation (45), where the porosity of the rock is 30 per cent. Figure 11-13 will permit the estimation of fuel requirements to heat 1 acre-foot of reservoir rock under the condition of no heat loss. For the actual project or in the designing of a pilot flood, the fuel requirement, as given by Figure 11-13, should be upgraded to account for the heat losses to the formation, and the possibility of thermal reactions within the rock.

Dew and Martin[44] state that coke burned at the combustion front will have a heating value of 15,000 to 21,000 Btu/lb, with a good average being 18,000 Btu/lb. Since a fraction of the reservoir oil

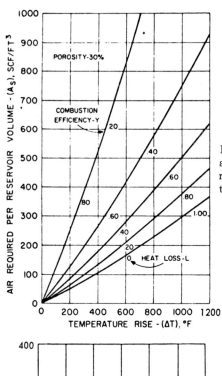

Figure 11-12. Effect of heat loss and combustion efficiency upon air requirements (after Dew and Martin[44]).

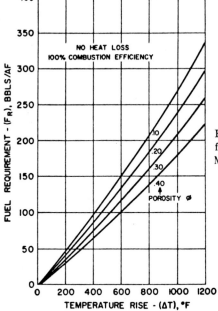

Figure 11-13. Fuel requirements for forward combustion (after Dew and Martin[44]).

will be left behind in the form of coke for the combustion front, the quantity of fuel available is important for the maintaining of adequate temperatures at the front. Figure 11-14 can be used for an initial estimate of fuel availability, as related to porosity and crude oil gravity. For the development of the graph, a fuel gravity of 1.00 and a sand grain density of 2.63 was used. Since fuel requirements in the field have been in the range of 150 to 300 barrels per acre-foot, it is doubtful that crudes lighter than 30 degrees API can be successfully flooded. Then too, a lower limit of oil gravity for the forward combustion recovery method is perhaps 10° API, since fuel availability becomes excessively large with little oil remaining for displacement to the producing wells. Figure 11-14 is of particular interest since it permits an estimation of the oil volume not consumed to support the combustion front, and also a measure of the air requirements, since the air-fuel ratio is normally insensitive to fuel composition. Since a barrel of coke as fuel weighs approximately 350 pounds, and each pound of fuel can generate an average of about 18,000 Btu/lb, the air requirement can be calculated from the observation that 100 Btu of heat is released from each scf of air. This means that 180 scf of air are required per lb of coke consumed, or some 350 times 180 scf per bar-

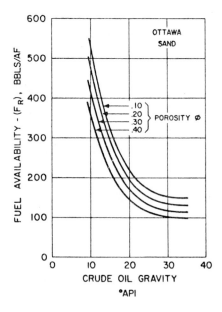

Figure 11-14. Effect of oil gravity and formation porosity upon fuel availability (after Dew and Martin[44]).

rel of coke consumed. Total air requirements can then be estimated. A correction should also be included for oxygen utilization. Field tests indicate that this value can range between 60 and 100 per cent.

Variations of the Basic Forward Combustion Process

A number of possibilities exist for varying the forward combustion process, as outlined in the preceding sections, for the purposes of improving economics, for improving sweep efficiency and for controlling temperatures at the combustion front. Some of the methods proposed, or that have been tried in the field and reported in the literature, are discussed in the following sections.

Controlled Temperatures at the Front. Since total combustion of the coke deposits at the front is considered to occur at any temperature above 600°F, the expense of heating the sand above this temperature probably serves only to increase the cost of air injection. Where thick formations are present, high temperatures could be tolerated, since this would serve to heat the unburned zones with beneficial gravity drainage of the oil. In thin zones where good vertical sweep is occurring, economies can sometimes be effected by reinjecting some or all of the produced gases with air[45], to reduce overall oxygen concentration[46]. Since the produced gases are available under pressure at the producing wells, compression costs can be reduced. The optimum oxygen concentration is a function of the type and amount of coke to be consumed at the front. Field experience to date would indicate that proper combustion front performance requires straight air injection—at least for the crudes heavier than 25° API.

Smith and Watson[45] have suggested the alternate injection of air and produced gases in those reservoirs where heavy oils have left quantities of coke in excess of that required to maintain the minimum combustion zone temperature. The procedure would involve the injection of an ample quantity of produced gases to move the "heat wave" sufficiently ahead of the frontal position where combustion had last occurred, so that temperature would drop enough that the coke would be below combustion temperature. One of the complicating features of this system would be the low temperature oxidation of the coke behind the front, or even reignition of the coke with no resulting gain in injection economies, due to the burning of a lesser amount of the residual fuel.

One of the problems in fire flooding reservoirs containing light oils is the problem of insufficient residual fuel, in the form of coke or tar, to support frontal temperatures at a level necessary to maintain combustion. Szasz[11] reports that a mixture of recycle gas containing combustible components, with the injection air outside of explosion limits, will allow the propagation of a combustion front in a porous medium, without residual fuel (coke or tar) being present. Again, it should be remembered that only the excess oxygen beyond that required to consume the injected gaseous fuel is available for burning whatever residual fuel that might be present.

Gates and Ramey[19] have suggested that good oil recoveries may be obtained from those reservoirs where the vertical coverage by the combustion zone is poor. In this instance, nearby sections of the oil reservoir are heated, and oil recovered by a combination of the gravity drainage and gas drive oil-producing mechanisms. Such a condition could be accomplished intentionally by lighting only a section of the reservoir at the injection well, or by injection of produced gases rather than air, to reduce burning rates in some parts of the flooded area.

Walther[47] has published information on a project where combustion occurred above ground, and the resulting hot gases and steam were injected into the reservoir. Since some oxygen was still present in the injected gases, a limited amount of burning apparently occurred in the formation. Such a system would appear to be somewhat inefficient, due to the considerable expense for surface equipment and for the maintenance of high temperatures throughout the reservoir, from the injection sand face to the combustion front. Then too, large amounts of condensed water, beyond that required for the displacement of oil, could be expected ahead of the combustion zone.

Improvement of Sweep Efficiencies. One of the problems associated with in situ combustion is the poor areal sweep efficiency which occurs when the mobility ratio is infinite (of the order of 50 per cent), and when the poor vertical sweep occurs in a stratified reservoir. Sweep efficiencies for such cases could be improved if an injection fluid of lower mobility could be used, rather than air or recycled gases. Water has a much lower mobility ratio than injection gases. However, the water-to-oil mobility ratio is still adverse, even when the oil temperature has been increased due to the in situ combustion phase of the project. Langnes and Beeson[48] have reported the results of a laboratory study where in situ combustion was followed by water flooding. In general, it was concluded that such a combination of recoveries was possible, but that the recovery by the hot water flooding of the reservoir which results is less than that by combustion alone. In some instances, economics may be improved by the injection of water as a final phase of the project. Again, the tests were in a laboratory tube, where sweep efficiency by the in situ combustion mechanism should have been quite complete. In a well pattern, sufficient benefit from improved areal and vertical sweep efficiencies might result, making the process potentially attractive.

Oil Recovery by Reverse Combustion

Reverse combustion differs from forward combustion in that the combustion zone moves in a direction opposite to that of the air flow[49]. Figure 11-15 illustrates the reverse combustion process. In most instances, forward combustion will prove to be the more attractive economically. Forward combustion does depend, however, upon the oil having sufficient mobility ahead of the combustion front, so that permeability to the injected gases and gases of combustion will be maintained. Where oils are extremely viscous, or even semi-solid, and/or reservoir temperatures are low, difficulty in moving the oil to the producing wells can result. Then too, the

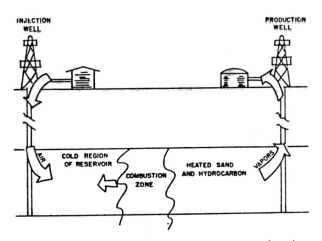

Figure 11-15. Illustration of the reverse combustion process (after Berry and Parrish[50]).

resulting reduction in air permeability which occurs smothers the combustion[51]. Reverse combustion does not suffer from this drawback, since the vaporized oil and water move in the direction of flow which is through the combustion zone and the burned sand which is already hot. So long as the sand remains hot, there is no practical upper limit on the original oil viscosity. Then too, the produced oil will have been significantly upgraded with a resulting increase in API gravity, and decrease in viscosity.

The process is less efficient than forward combustion, since the fuel used will be an intermediate fraction with coke remaining on the sand. This coke is neither consumed nor used in the process for the recovery of oil, and therefore represents a significant energy loss. On the other hand, the upgrading of the oil to a more valuable product may offset to some extent the thermal inefficiency which the coke residue represents. Reed *et al.*[52] report that the produced oil, after fractionation in the reservoir, characteristically is of 25 API gravity, with a viscosity of 15 cp at 100°F. The process depends upon heat transport by conduction countercurrent to the air flow, exceeding the heat transported forward by the moving gas[11], in the direction of air flow. This difference is vital to the movement of the combustion front.

Several papers have been published[52,53] which present theoretical considerations of reverse combustion based upon laboratory and mathematical models. Current understanding of the reverse combustion process remains incomplete, due to a lack of definitive field tests performed to date.

References

1. Lewis, James O., *U. S. Bur. Mines Bull.* 148, 128 (1917).
2. Howard, F. A., U. S. Patent 1,473,348, November 6, 1923.
3. Lapuk, B. B., *Azerbaidzhanskoe Neftyanoe Khoz.*, No. 2, 31 (1939).
4. Sheinman, A. B., and Dubrovai, K. K., "Underground Gasification of Oil Formations and Thermal Methods of Oil Production," p. 26, Moscow, 1934.
5. Sheinman, A. B., Dubrovai, K. K., Sorokin, N. A., Charuigin, M. M., Zaks, S. L., and Zinchenko, K. E., *Neftyanoe Khoz.* 28, 4, 48 (April, 1935). English translation by A. A. Boehtlingk, *Petrol. Engr.* 10, 3, 27 (December, 1938); No. 5, 91 (February, 1939).
6. Lapuk, B. B., *Azerbaidzhanshkoe Neftyanoe Khoz.*, No. 12, 10 (1939).
7. Sheinman, A. B., *Novosti Tekhniki*, No. 15, 30 (1939).
8. Kuhn, C. S., and Koch, R. L., *Oil Gas J.* 52, 14, 92 (August 10, 1953).
9. Grant, B. F., and Szasz, S. E., *Trans. AIME* 201, 108 (1954).

10. Buffum, F. G., *Petrol. Engr.* **26**, 6, B-99 (June, 1954).
11. Szasz, S. E., Proc. of 1st Adv. Petr. Engr. Seminar, Univ. of Okla., 3 (February 9-10, 1960).
12. Langnes, G. L., and Beeson, C. M., *Petrol. Engr.*, 92 (July, 1965).
13. Nelson, T. W., and McNiel, J. S., Jr., "Fundamentals of Thermal Oil Recovery," p. 27, Dallas, Petrol. Engr. Publ. Co., 1965.
14. Nelson, T. W., and McNiel, J. S., Jr., *Oil Gas J.*, 86 (January 19, 1959).
15. Moss, Jon T., *Petrol. Engr.*, 72 (May, 1965).
16. Strange, L. K., *Petrol. Engr.*, 105 (November-December, 1964).
17. Champion, F. E., Glass, E. D., Kirkpatrick, J. W., and Vaughn, J. C., U. S. Patent 2,997,105, August 22, 1961.
18. Gates, C. F., and Ramey, H. J., Jr., "Thermal Recovery—1965," p. 51, Tulsa, *Oil Gas J.*, 1965.
19. Gates, C. F., and Ramey, H. J., Jr., *Trans. AIME* **213**, 236 (1958).
20. Tadema, H. J., Proc. of Fifth World Petr. Congress, Sec. II, paper 22.
21. Nelson, T. W., and McNiel, J. S., Jr., *Oil Gas J.*, 58 (June 5, 1961).
22. Clark, G. A., Jones, R. G., Kinney, W. L., Schilson, R. E., Surkalo, H., and Wilson, R. S., *Soc. Petrol. Engrs.* paper no. SPE 956, **AIME**, presented Houston, October, 1964.
23. Parrish, D. R., Rausch, R. W., Jr., Beaver, K. W., and Wood, H. W., *Trans. AIME* **225**, I-197 (1962).
24. Emery, L. W., *Trans. AIME* **225**, I-671 (1962).
25. Wilson, L. A., Reed, R. L., Reed, D. W., Clay, R. R., and Harrison, N. H., *Soc. Petrol. Engrs. J. of* **AIME**, 127 (June, 1963).
26. Ramey, H. J., *Trans. AIME* **216**, 115 (1959).
27. Bailey, H. R., and Larkin, B. K., *Trans. AIME* **216**, 123 (1959).
28. Bailey, H. R., and Larkin, B. K., *Trans. AIME* **219**, 321 (1960).
29. Selig, F., and Couch, E. J., *Österr, Ingenieur-Archiv*, **Bd. xv**, Heft 1-4 (1961).
30. Chu, C., *J. Petrol. Technol.*, 1137 (October, 1963).
31. Thomas, G. W., *J. Petrol. Technol.*, 1145 (October, 1963).
32. Kunii, D., and Smith, J. M., *Soc. Petrol. Engrs. J.*, 37 (March, 1961).
33. Somerton, W. H., and Boozer, G. D., *Trans. AIME* **219**, 418 (1960).
34. Asaad, Yousri, "A Study of the Thermal Conductivity of Fluid Bearing Porous Rocks," Ph.D. Thesis, U. of Calif. (June, 1955).
35. Somerton, W. H., *Trans. AIME* **213**, 375 (1958).
36. Kelley, K. K., *U. S. Bur. Mines Bull.* 476 (1949).
37. Lange, N. A., "Handbook of Chemistry," 9th Ed., p. 874 and 1540, Sandusky, Ohio, Handbook Publishers, Inc., 1956.
38. Adivarahan, P., Kunii, D., and Smith, J. M., *Soc. Petrol. Engrs. J.*, 290 (September, 1962).
39. Bailly, F. H., and Mangold, G. B., paper no. SPE 455, presented 37th annual fall mtg., *Soc. Petrol. Engrs.* of **AIME**, Los Angeles, October, 1962.
40. Barshad, I., *Amer. Mineral* **37**, 667 (1952).
41. Brindley, G. W., and Nakahira, M., *J. Am. Ceram. Soc.* **40**, 10, 346 (1957).
42. Somerton, W. H., and Selim, M. A., *Soc. Petrol. Engrs. J.*, 249 (December, 1961).

43. Owens, W. D., and Suter, V. E., *Oil Gas J.*, B-2 (April 26, 1965).
44. Dew, J. N., and Martin, W. L., *Petrol. Engr.*; *Part I*, 82 (December, 1964); *Part II*, 82 (January, 1965).
45. Smith, R. L., and Watson, K. M., U. S. Patent 2,642,943 (June 23, 1953).
46. Grant, B. F., and Szasz, S. E., *Trans. AIME* **201**, 108 (1954).
47. Walther, H., *J. Petrol. Technol.*, 16 (February, 1957).
48. Langnes, G. L., and Beeson, C. M., *Petr. Engr.*; *Part I*, 92 (July, 1965); *Part II*, 98 (August, 1965).
49. Morse, R. A., U. S. Patent 2,793,696 (June 23, 1957).
50. Berry, V. J., Jr., and Parrish, D. R., *Trans. AIME* **219**, 124 (1960).
51. Marx, J. W., and Trantham, J. C., *Oil Gas J.*, **123** (May 17, 1965).
52. Warren, J. E., Reed, R. L., and Price, H. S., *Trans. AIME* **219**, 109 (1960).
53. Reed, R. L., Reed, D. W., and Tracht, J. H., *Trans. AIME* **219**, 99 (1960).

PROBLEMS

1. Sketch a typical isotherm and radial and vertical temperature profile as you would expect it to appear where the formation thickness is 80 feet, the radial distance to the combustion front is 50 feet, and where the formation is assumed to be homogeneous as to fluid and reservoir properties over the entire vertical and areal interval. (The reader is referred to references (31) and (30) for additional helpful discussion of this problem.)

2. With increased radial distances of the combustion front from the well-bore, the vertical coverage tends to become less than 100 per cent. What is the reason for this behavior, and what may be done to offset such an occurrence?

3. As long as flame extinction does not occur, how do you visualize the recovery mechanism by which some of the heated oil above and below the burned zone can be produced?

4. The following data from a laboratory tube burn and from a viscous oil reservoir are available:

Combustion tube internal diameter	0.333 ft
Length of burned pack	6.0 ft
Porosity of pack	35 per cent
Volume of produced gas (dry basis)	190 SCF

Composition of injected air:

Component	Volume per cent
N_2	79.0
O_2	21.0

Composition of produced gas:

Component	Volume per cent
N_2	84.2
O_2	1.1
CO_2	11.7
CO	3.0

Field data:

Pattern area	5 acres
Formation thickness	80 ft
Reservoir temperature	70°F
Production well bottom hole pressure	55 psia
Porosity	18 per cent
Absolute permeability	440 md
Oil saturation	79.9 per cent
Water saturation	20.1 per cent
Production and injection well radius	0.375 ft
Volumetric sweep of burned zone	40 per cent

Calculate the following:

(a) Volume of nitrogen and oxygen injected in the laboratory run.

(b) Volume of carbon dioxide, carbon monoxide and unreacted oxygen produced in the laboratory run.

(c) Carbon in fuel burned, hydrogen in fuel burned and weight of water formed in pounds in the laboratory run.

(d) Weight of total fuel consumed in the laboratory run.

(e) Fuel consumed per cubic foot of the sand pack burned and per acre-ft of the reservoir burned.

(f) Air injected per acre-ft of reservoir burned and per acre-ft of pattern.

(g) Total air required for a five-spot pattern.

(h) Air flux for a minimum burning front advance of 0.125 ft/day.

(i) Maximum daily air injection rate for the field pattern, and the time required to reach this maximum rate.

(j) Volume of air injected while reaching maximum air injection rate, air injected during constant rate period, and air injected during declining rate period.

(k) Time for each of the three rate intervals, plus the time to complete the entire burning operation.

(l) Maximum air injection pressure required for an assumed air viscosity of 0.0186 cp and a permeability to injection air of 25 md.

(m) Oil displaced from the reservoir burned, oil displaced from the unburned reservoir, total oil recovery, and overall oil recovery efficiency.

(n) Barrels of water produced per acre-ft.

(o) Oil recovery per MMSCF of air injected.

(p) Maximum oil-producing rate from the pattern.

(q) Barrels of water produced per MMSCF of air injected and the maximum water-producing rate.

5. Prepare a graph of fuel requirements in barrels per acre-ft for forward combustion as a function of temperature rise in the formation, using heat capacity of the reservoir rock as a parameter. Prepare four curves for heat capacities of 0.22, 0.24, 0.26, and 0.28 Btu/lb-°F. The heat loss to the system is 35 per cent, 80 per cent of the oxygen is utilized, 100 Btu/scf of heat is generated by the injected air, sand grain specific gravity is 2.65, and system porosity is 27 per cent. The heating value of coke is 18,000 Btu/lb.

CHAPTER 12

Heat Injection Methods
of Secondary Recovery

Introduction

The immediately preceding chapter dealt with the recovery of oil by in situ combustion means, in which a heat front, moving as a cylindrical heat source, passed from the injection well(s) to the producing well(s). In the heat injection methods of secondary recovery treated in this present chapter, heat is generated at the surface and injected into the oil-bearing reservoir through the injection well(s). The injection of heat bearing fluids for the recovery of oil have their most effective application to the low-gravity or viscous crudes. This is illustrated in Figure 11-11 of Chapter 11, which shows the significant effect of temperature on oil viscosity. In general, the more viscous the crude is, the greater will be the proportionate reduction in viscosity for a given temperature increase. It is this particular feature which makes the thermal recovery methods attractive, since they are the only presently known techniques which can be applied effectively to viscous crude recoveries.

Viscous crude reservoirs are known to exist in most parts of the United States[1] and in widely scattered locations throughout the world. Until recently, most of the tar sand or heavy oil deposits were not included in oil reserves records, since they were not considered to be of commercial significance. A preliminary estimate[2] indicates that 40 billion barrels of the heavy oil deposits may now be producible in the United States by the thermal oil recovery methods—including in situ combustion methods. If true, this will result in a more than doubling of the 32 billion barrels of proven reserves of the United States (see Chapter 1).

Heating Oil Producing Wells

The application of heat for the purposes of increasing the producing rate and recovery of oil has been used for a number of years. A patent describing an electrical oil well heater was issued in 1865 to Perry and Warner[3]—just six years after the drilling of the Drake well. At least 100 patents have been issued on the subject of oil well heating since that time. Snelling[4] in 1914, described a method of heating wells by using a slow-burning explosive such as gunpowder, nitrocellulose, or thermite to generate hot gases for the purpose of melting paraffin and other plugging agents at the sandface. Mills[5] has described the use of electrical heaters, hot water, steam, and downhole combustion for the removal of paraffin deposits from the producing sandface of wells. In 1925, the Hope Gas Company used a natural gas burner at the sandface for heating purposes[5]. Temperatures of 2300 to 2700°F at the burner were attained using 200 cu ft of gas for each 3000 cu ft of air injected[6].

The benefits of using heat for the removal of paraffin and other sandface deposits were recognized very early. It has been only in recent years[7] that heating techniques have been used for the express purpose of reducing the viscosity of the oil in the vicinity of the wellbore where paraffin problems did not necessarily exist. Heat transfer considerations will show the heat release that must be effected to permit heat entry into the formation counter-current to the flow of the produced fluids. A theory for such conduction of heat into a formation, and the convection of the heat back to the wellbore by the movement of the produced fluids, has been presented by Melby[7] and by Schild[8]. There are three types of heating equipment currently being used. These are the electrical, natural gas burner and hot fluid circulating types. Hot fluid types are commonly used where the oil sand is shallow, since the wellbore heat losses to such depths can be tolerated, and fluid heating costs are low in surface heaters[9]. Electrical downhole heaters have application where heat requirements and well depths are moderate, since line losses become significant with greater depths. Down-the-hole gas burners have application where large quantities of heat are required, but have the disadvantage of requiring a sizeable investment in injection equipment.

Strictly speaking, the above-outlined heat methods for the recovery of viscous oil cannot be termed secondary recovery techniques, but are procedures to increase the oil production at the
producing wells, by means of improving oil mobility in the vicinity
of the wellbore. Nelson and McNiel[10] have presented cost and
statistical data for electrical and hot fluid circulating types of
heating equipment.

Injection of Hot Fluids

In those secondary recovery operations using heat injection
methods to aid in oil recovery, hot water, hot inert gases, or saturated or superheated steam are injected to improve the mobility of
the oil, and to physically displace the viscous oil toward producing
wells. If the temperature of the fluid is high when reaching the
reservoir fluid displacement front, additional oil recovery beyond
that due to increased oil mobility may be attained by condensation
and vaporization mechanisms, acting in the high temperature
region.

The idea of hot fluid injection is not new. Field applications
of either steam, or hot water injection for the recovery of oil, have
been reported by Stovall[11], Breston and Pearman[12] and Tucker[13].
More recently, hot water flooding to increase oil recovery in the
Pembina Field of Alberta, Canada, has been proposed by Lewis[14].
Nelson and McNiel[10] report that very large heat losses percentage-
wise occur in wellbores, especially if the formation of interest is
at considerable depth where hot fluids are used. This loss is
particularly severe for the case of hot gas injection, since gases
have a low heat-carrying capacity. High injection rates of the heat-
carrying fluid are usually necessary, if the heat losses to the formation from the wellbore are to be minimized. As the injection
rate increases, the percentage heat losses to the formation will
decrease. It is the high heat losses encountered in the transporta-
tion of hot fluids to the displacement front which have restricted
the use of hot water or gases as an injection media for most
projects.

Water and Steam as Heat Carriers

The primary advantage of water and/or steam, as a medium for
moving heat to a displacement front, is due to its relatively high
heat-carrying capacity, plus the large amount of available heat

which may be transferred to the formation as heat of condensation (in the case of steam). Figure 12-1 presents a volume-temperature phase diagram for water and steam. At atmospheric pressure the boiling point is 212°F while at 135 psia the boiling point is increased to 350.2°F. The constant temperature line across the two-phase region separating the liquid and vapor regions represents a vaporization of the liquid as heat is applied at constant pressure. When the saturated-vapor is reached, the vaporization is complete, and temperature may again increase as volume increases with constant pressure.

During the vaporization of the water to steam, the internal energy of the system is increased due to the addition of heat. Work is

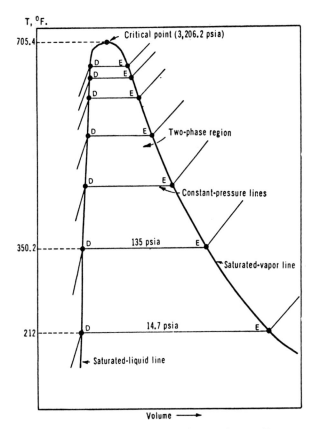

Figure 12-1. Temperature-volume phase diagram for water and steam (after Bleakley[15]).

also being accomplished against the system, as evidenced by either an increase in pressure or volume, as temperature is increased due to heating. The term "enthalpy" defines this change in energy of the fluids involved[15], and is expressed by the following equation:

$$\underline{H} = u + pv/J \tag{1}$$

where

$\underline{H}$ = enthalpy, Btu/lb$_m$
u = internal energy, Btu/lb$_m$
p = pressure, psfa
v = volume, cu ft
J = mechanical equivalent of heat, 778 lb$_f$-ft/Btu.

Since it is the heat which a fluid can transmit to its surroundings that is of interest in thermal recovery techniques, equation (1) can be rewritten to reflect this change in enthalpy:

$$\Delta \underline{H} = \Delta u + \Delta(pv)/J \tag{2}$$

Tables 12-1 and 12-2 provide enthalpy and specific volume data on saturated and superheated steam, respectively. Figures 12-2 and 12-3 are equivalent charts for saturated steam, showing lines of constant steam quality. A more complete set of steam properties may be found in Keenan and Keyes[16]. The enthalpy given in the tables arbitrarily uses a zero for saturated water at 32°F. All listed enthalpy values represent the increase in enthalpy from that initial condition. For reasons that will be discussed at the end of the chapter, most steam generated for injection to an oil-bearing formation will be of approximately 80 per cent quality, or that is, will contain 80 weight per cent saturated vapor, with the remainder being liquid which contains the concentrated salts carried into the steam generator by the feedwater. The determination of the amount of heat which a saturated steam can carry is best illustrated by an example calculation:

EXAMPLE 1:
    Find the amount of heat required to generate steam of 80 per cent quality at 500 psia if the feedwater temperature is 50°F. Use 1 lb as a basis. From Keenan and Keyes[16], the enthalpy of saturated water at 50°F is 18.07 Btu/lb$_m$. From Table 12-1 the enthalpy of saturated liquid at 500 psia, $\underline{H}_f$, is 449.4 Btu/lb$_m$, while the enthalpy of the saturated vapor is 1204.4 Btu/lb$_m$. The amount of heat required would be:

$$\Delta \underline{H} = 0.2(449.4) + 0.8(1204.4) - 18.07 = 1035.3 \text{ Btu/lb}_m$$

TABLE 12-1. Abridged Table of Saturated Steam Properties
(from Keenan & Keyes[16]).

p, psia	T, °F	Specific Volume, cu ft/lb$_m$		Enthalpy, Btu/lb$_m$		
		Sat. Liquid, v_f	Sat. Vapor, v_g	Sat. Liquid, H_f	Evap., H_{fg}	Sat. Vapor, H_g
100	327.81	0.01774	4.432	298.40	888.8	1187.2
150	358.42	0.01809	3.015	330.51	863.6	1194.1
200	381.79	0.01839	2.288	355.36	843.0	1198.4
300	417.33	0.01890	1.5433	393.84	809.0	1202.8
400	444.59	0.0193	1.1613	424.0	780.5	1204.5
500	467.01	0.0197	0.9278	449.4	755.0	1204.4
600	486.21	0.0201	0.7698	471.6	731.6	1203.2
700	503.10	0.0205	0.6554	491.5	709.7	1201.2
800	518.23	0.0209	0.5687	509.7	688.9	1198.6
900	531.98	0.0212	0.5006	526.6	668.8	1195.4
1000	544.61	0.0216	0.4456	542.4	649.4	1191.8
1100	556.31	0.0220	0.4001	557.4	630.4	1187.8
1200	567.22	0.0223	0.3619	571.7	611.7	1183.4
1300	577.46	0.0227	0.3293	585.4	593.2	1178.6
1400	587.10	0.0231	0.3012	598.7	574.7	1173.4
1500	596.23	0.0235	0.2765	611.6	556.3	1167.9
2000	635.82	0.0257	0.1878	671.7	463.4	1135.1
2500	668.13	0.0287	0.1307	730.6	360.6	1091.1
3000	695.36	0.0346	0.0858	802.5	217.8	1020.3
3206.2	705.40	0.0503	0.0503	902.7	0	902.7

or alternately:

$$\Delta H = 0.8(1204.4 - 499.4) + 449.4 - 18.07 = 1035.3 \text{ Btu/lb}_m$$

If the formation conditions were 150 psia and 80°F, the maximum heat which the above steam could release would be less than that required to generate the steam, i.e.,

From Keenan and Keyes and Table 12-1, the release of heat would be:

$$\Delta H = 0.2(449.4) + 0.8(1204.4) - 48.02 = 1005.4 \text{ Btu/lb}_m$$

Figure 12-2 would have yielded approximately the same answer.

The specific volume of an 80 per cent quality steam may also be readily calculated from the steam tables.

Although superheated steam is not presently used for carrying heat to a formation, Table 12-2 has been included for completeness. The injection of superheated steam would be possible by removing the hot water in a separator, and then increasing the heat of the steam at constant pressure. The reason for not taking this approach at the present time is due to the heat loss which would re-

TABLE 12-2.　Abridged Table of Superheated Steam Properties, Btu/lb$_m$ (from Keenan & Keyes [16]).

p psia (sat. temp)	Symbol*	Temperature, °F										
		450	500	550	600	700	800	900	1000	1200	1400	1600
300 (417.33)	v	1.6364	1.7675	1.8891	2.005	2.227	2.442	2.652	2.859	3.269	3.674	4.078
	$\underline{H}$	1225.8	1257.6	1286.8	1314.7	1368.3	1420.6	1472.8	1525.2	1631.7	1741.0	1853.7
350 (431.72)	v	1.3734	1.4923	1.6010	1.7036	1.8980	2.084	2.266	2.445	2.798	3.147	3.493
	$\underline{H}$	1217.7	1251.5	1282.1	1310.9	1365.5	1418.5	1471.1	1523.8	1630.7	1740.3	1853.1
400 (444.59)	v	1.1744	1.2851	1.3843	1.4770	1.6508	1.8161	1.9767	2.134	2.445	2.751	3.055
	$\underline{H}$	1208.8	1245.1	1277.2	1306.9	1362.7	1416.4	1469.4	1522.4	1629.6	1739.5	1852.5
450 (456.28)	v	—	1.1231	1.2155	1.3005	1.4584	1.6074	1.7516	1.8928	2.170	2.443	2.714
	$\underline{H}$	—	1238.4	1272.0	1302.8	1359.9	1414.3	1467.7	1521.0	1628.6	1738.7	1851.9
500 (467.01)	v	—	0.9927	1.0800	1.1591	1.3044	1.4405	1.5715	1.6996	1.9504	2.197	2.442
	$\underline{H}$	—	1231.3	1266.8	1298.6	1357.0	1412.1	1466.0	1519.6	1627.6	1737.9	1851.3
600 (486.21)	v	—	0.7947	0.8753	0.9463	1.0732	1.1899	1.3013	1.4096	1.6208	1.8279	2.033
	$\underline{H}$	—	1215.7	1255.5	1289.9	1351.1	1407.7	1462.5	1516.7	1625.5	1736.3	1850.0
700 (503.10)	v	—	—	0.7277	0.7934	0.9077	1.0108	1.1082	1.2024	1.3853	1.5641	1.7405
	$\underline{H}$	—	—	1243.2	1280.6	1345.0	1403.2	1459.0	1513.9	1623.5	1734.8	1848.8
800 (518.23)	v	—	—	0.6154	0.6779	0.7833	0.8763	0.9633	1.0470	1.2088	1.3662	1.5214
	$\underline{H}$	—	—	1229.8	1270.7	1338.6	1398.6	1455.4	1511.0	1621.4	1733.2	1847.5
900 (531.98)	v	—	—	0.5264	0.5873	0.6863	0.7716	0.8506	0.9262	1.0714	1.2124	1.3509
	$\underline{H}$	—	—	1215.0	1260.1	1332.1	1393.9	1451.8	1508.1	1619.3	1731.6	1846.3
1000 (544.61)	v	—	—	0.4533	0.5140	0.6084	0.6878	0.7604	0.8294	0.9615	1.0893	1.2146
	$\underline{H}$	—	—	1198.3	1248.8	1325.3	1389.2	1448.2	1505.1	1617.3	1730.0	1845.0
1100 (556.31)	v	—	—	—	0.4532	0.5445	0.6191	0.6866	0.7503	0.8716	0.9885	1.1031
	$\underline{H}$	—	—	—	1236.7	1318.3	1384.3	1444.5	1502.2	1615.2	1728.4	1843.8
1200 (567.22)	v	—	—	—	0.4016	0.4909	0.5617	0.6250	0.6843	0.7967	0.9046	1.0101
	$\underline{H}$	—	—	—	1223.5	1311.0	1379.3	1440.7	1499.2	1613.1	1726.9	1842.5

Pressure (Temp)												
1400 (587.10)	v	—	—	—	0.3174	0.4062	0.4714	0.5281	0.5805	0.6789	0.7727	0.8640
	H	—	—	—	1193.0	1295.5	1369.1	1433.1	1493.2	1608.9	1723.7	1840.0
1600 (604.90)	v	—	—	—	—	0.3417	0.4034	0.4553	0.5027	0.5906	0.6738	0.7545
	H	—	—	—	—	1278.7	1358.4	1425.3	1487.0	1604.6	1720.5	1837.5
1800 (621.03)	v	—	—	—	—	0.2907	0.3502	0.3986	0.4421	0.5218	0.5968	0.6693
	H	—	—	—	—	1260.3	1347.2	1417.4	1480.8	1600.4	1717.3	1835.0
2000 (635.82)	v	—	—	—	—	0.2489	0.3074	0.3532	0.3935	0.4668	0.5352	0.6011
	H	—	—	—	—	1240.0	1335.5	1409.2	1474.5	1596.1	1714.1	1832.5
2500 (668.13)	v	—	—	—	—	0.1686	0.2294	0.2710	0.3061	0.3678	0.4244	0.4784
	H	—	—	—	—	1176.8	1303.6	1387.8	1458.4	1585.3	1706.1	1826.2
3000 (695.36)	v	—	—	—	—	0.0984	0.1760	0.2159	0.2476	0.3018	0.3505	0.3966
	H	—	—	—	—	1060.7	1267.2	1365.0	1441.8	1574.3	1698.0	1819.0
3206.2 (705.40)	v	—	—	—	—	—	0.1583	0.1981	0.2288	0.2806	0.3267	0.3703
	H	—	—	—	—	—	1250.5	1355.2	1434.7	1569.8	1694.6	1817.2
3500	v	—	—	—	—	0.0306	0.1364	0.1762	0.2058	0.2546	0.2977	0.3381
	H	—	—	—	—	780.5	1224.9	1340.7	1424.5	1563.3	1689.8	1813.6
4000	v	—	—	—	—	0.0287	0.1052	0.1462	0.1743	0.2192	0.2581	0.2943
	H	—	—	—	—	763.8	1174.8	1314.4	1406.8	1552.1	1681.7	1807.2
4500	v	—	—	—	—	0.0276	0.0798	0.1226	0.1500	0.1917	0.2273	0.2602
	H	—	—	—	—	753.5	1113.9	1286.5	1388.4	1540.8	1673.5	1800.9
5000	v	—	—	—	—	0.0268	0.0593	0.1036	0.1303	0.1696	0.2027	0.2329
	H	—	—	—	—	746.4	1047.1	1256.5	1369.5	1529.5	1665.3	1794.5
5500	v	—	—	—	—	0.0262	0.0463	0.0880	0.1143	0.1516	0.1825	0.2106
	H	—	—	—	—	741.3	985.0	1224.1	1349.3	1518.2	1657.0	1788.1

*where: v = specific volume, cu ft/lb$_m$
$\underline{H}$ = enthalpy, Btu/lb$_m$

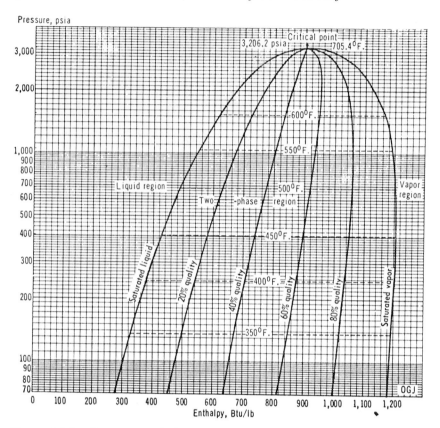

Figure 12-2. Pressure-enthalpy chart for saturated steam showing steam quality lines (after Bleakley[15]).

sult in the disposing of the hot water that has been removed. The use of the hot water to preheat the entering feedwater would partially eliminate this source of heat loss. The advantage of injecting superheated steam is the carrying of a larger amount of heat to the formation for a given injected volume of fluid.

Heat Losses in Hot Fluid Injection

The following sections will treat the theory and methods for determining temperatures and heat transmission to the formation which can be attained by means of hot fluid injection. As Ramey[17] has pointed out, the heat losses in a system begin at the thermal unit or heat source, with subsequent heat losses occurring in the

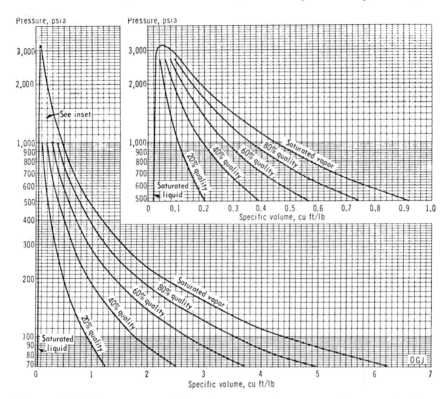

Figure 12-3. Pressure-specific volume chart for saturated steam showing steam quality lines (from Bleakley[15]).

surface injection lines, the injection wellhead, the wellbore and finally the formation itself and the adjacent strata. Heat losses of such a system are illustrated in Figure 12-4. These heat losses are due to heat transmission away from the system by conduction, by convection, by radiation, or by combinations of all three means.

(1) *Heat Transmission by Conduction.* When heat passes from one part of a body to another part of the same body, or to another body with which it is in physical contact without a physical displacement of the particles of the body.

(2) *Heat Transmission by Convection.* When heat is transferred from one place to another within a fluid by the mixing of one part of the fluid with the other.

(3) *Heat Transmission by Radiation.* When heat passes from one body to another body at a different temperature by means of a flow of massless particles, which are not a part of the two bodies. The

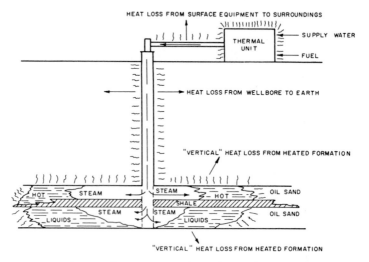

Figure 12-4.   Illustration of heat losses which occur in a heat injection system (after Ramey[17]).

best example of radiant heat transfer is solar energy which passes through interstellar space by means of radiation.

The various sources of heat loss in a heat injection system are best treated in separate sections.

Loss of Heat in Surface Facilities. While a certain loss of heat occurs in the thermal unit itself, this is usually taken into account in thermal efficiencies. Then too, the temperature of the fluid and its quality (if steam) can be measured at the outlet of the thermal unit. Detailed theory and design information can be found in the literature from which heat losses can be estimated for the surface injection lines[8,9,20,21].

Loss of heat from fluid in a pipe at a higher temperature than its surrounding media is due to a combination of the several heat transfer mechanisms. There must be heat transfer through the fluid by convection to the vicinity of the pipe wall. All fluids are bounded at the retaining walls by a film of stagnant fluid. Heat will be transferred through these films by conduction. In turn, the heat will be transferred through the pipe wall by conduction. If the pipe is insulated, the heat will in turn flow by conduction through this media. In fluid injection projects, it is common practice to leave the insulated steam lines in contact with air at the surface of the ground. Heat transfer to the air will be primarily by radiation and convection.

In a line carrying a hot fluid in oil field usage, it will usually suffice for engineering purposes to assume that the outside wall temperature of the surface injection line will be at the temperature of the fluid inside, if no external insulating material is being used. Where insulating material is being used, conduction through the pipe should be considered. Conduction through solids is the simplest method of heat transfer. Fourier's law for conduction of heat may be written in the following form:

$$Q = \frac{kA(T_2 - T_1)}{L} \qquad (3)$$

where

Q = heat transferred, Btu/hr
k = thermal conductivity, Btu/°F-sq ft-hr for 1 ft thickness, or Btu/°F-ft-hr
A = conducting surface area, sq ft
T_2 = temperature of hot surface, °F
T_1 = temperature of cold surface, °F
L = solid thickness, ft

Care should be taken in the selection of thermal conductivity values, since engineers and scientists have reached no general agreement on the proper units. The following units have been used: (1) Btu/°F-sq ft-hr/ft or Btu/°F-ft-hr, (2) Btu/°F-sq ft-hr/in, (3) Joules/°C-sq cm-sec/cm, (4) Gm cal/°C-sq cm-sec/cm, (5) Kg cal/°C-sq meter-hr/meter, (6) Btu/°F-sq in-hr/in, and (7) Watts/°C-sq meter/meter.

Table 12-3 presents handy conversion factors for k. If, for instance, a foreign manufacturer has an insulating material with a thermal or heat conductivity of 0.00025 gm cal/°C-sq cm-sec/cm, the conversion to units of Btu/°F-sq ft-hr/ft would be achieved by

TABLE 12-3. Conversion Factors for k.

Btu/°F-sq ft-hr/ft.	Btu/°F-sq ft-hr/in.	Joules/°C sq cm-sec/cm.	Gm cal/°C-sq cm-sec/cm.	Kg cal/°C-sq meter-hr/meter	Btu/°F-sq in-hr/in.	Watts/°C sq meter/meter
57.7	693	1	0.239	86.1	4.81	100
242	2903	4.185	1	360	20.2	419
0.672	8.06	0.0116	0.00278	1	0.0560	1.16
0.0833	1	0.00144	0.000345	0.124	0.00695	0.144
1	12.0	0.0173	0.00414	1.49	0.0835	1.73
12.0	144	0.208	0.0497	17.9	1	20.8
0.577	6.93	0.0100	0.00239	0.861	0.0481	1

coming down the column of the initial unit to unity, and preceding horizontally to the column bearing the desired unit for the proper multiplying factor of 242. This would result in a heat conductivity value of 0.06050 Btu/°F-sq ft-hr/ft.

Since several materials make up the wall of an insulated pipe-line, it will often be necessary to modify equation (3). Due to continuity considerations, the quantity of heat passing through each material will be the same, although the temperature drop across each material may be different. Assuming three materials in series have conductivities of k_1, k_2, k_3, thicknesses of L_1, L_2, L_3, and temperature differences of ΔT_1, ΔT_2, and ΔT_3, then in view of equation (3):

$$\frac{Q}{A_1} = \frac{\Delta T_1}{\dfrac{L_1}{k_1} \dfrac{A_1}{A_1}} = \frac{\Delta T_2}{\dfrac{L_2}{k_2} \dfrac{A_1}{A_2}} = \frac{\Delta T_3}{\dfrac{L_3}{k_3} \dfrac{A_1}{A_3}}$$

The total temperature drop across the materials, ΔT, may then be written as:

$$\Delta T = \Delta T_1 + \Delta T_2 + \Delta T_3 = \frac{Q}{A_1}\left(\frac{L_1}{k_1}\frac{A_1}{A_1} + \frac{L_2}{k_2}\frac{A_1}{A_2} + \frac{L_3}{k_3}\frac{A_1}{A_3}\right) \quad (4)$$

Equation (4) may be simplified if the conducting surface is flat, or if the pipe diameter is large compared to the thickness of the pipe insulation, since the areas A_1, A_2, A_3, etc., would be approximately equal. In the insulation of pipe of small diameter, the logarithmic mean area must be used:

$$\text{Logarithmic mean area} = \frac{\text{larger area} - \text{smaller area}}{\ln\left(\dfrac{\text{larger area}}{\text{smaller area}}\right)} \quad (5)$$

Tables 12-9-A through 12-9-F provide a brief listing of the properties of interest in the determination of thermal behavior of fluids and of solids. This information coupled with information on the pipe and its insulating material will permit a calculation of either the temperature drop, ΔT, or the heat transferred, Q, across the pipe, depending upon which one of these factors is known or can be estimated. Actually, heat transfer calculations are not quite this simple, since all fluids are bounded by a film of stagnant fluid through which the heat must be transferred by conduction. These films, both on the inside and outside of the pipe, are very thin, and are not easily measured. Nelson[18] reports that the apparent film thickness is about 0.1 inch for a gas (such as air), and about 0.0001 inches for condensing steam. If h_o and h_i are taken to be the outside and inside film coefficients of heat transfer in Btu/hr-sq ft-°F, equation (2) for heat conduction can be rewritten in the following form[18]:

$$Q = \frac{A_o (\Delta T)}{\dfrac{1}{h_o} \dfrac{A_o}{A_o} + \dfrac{L_w}{k_w} \dfrac{A_o}{A_w} + \dfrac{1}{h_i} \dfrac{A_o}{A_i} + R_o \dfrac{A_o}{A_o} + R_i \dfrac{A_o}{A_i}} \qquad (6)$$

for a pipe with no insulation. If a layer of insulation is present, an additional term may be added to the denominator. The nomenclature used is as follows

A_o, A_i, and A_w = wall area at the outside, inside, and mean points of the wall, sq ft

L_w = pipe wall thickness, ft

k_w = thermal conductivity of the wall, Btu/°F-ft-hr

R_o and R_i = fouling resistances at the outside and inside surface of the pipe due to dirt, corrosion, and roughness, ft/Btu/°F-sq ft-hr

Unfortunately, very little meaningful data have been published by which the heat transfer rate across a pipe to the outside air can be reliably calculated. Nelson[18] has published information on the approximate heat losses from plant equipment and pipes, a summary of which is given in Table 12-4. Notice that the equipment is

TABLE 12-4. Approximate Heat Losses from Equipment (after Nelson[18]). Btu/hr

Surface	Conditions	Inside Temperature			
		200°F	400°F	600°F	800°F
Bare metal	Still air, 0°F	540	1560	3120	
(loss per sq ft)	Still air, 100°F	210	990	2250	
	10 mph wind, 0°F	1010	2540	4680	
	10 mph wind, 100°F	440	1710	3500	
	40 mph wind, 0°F	1620	4120	7440	
	40 mph wind, 100°F	700	2760	5650	
Magnesia pipe insulation	Standard on 3-in. pipe	50	150	270	440
(loss per ft of length)	Standard on 6-in. pipe	77	232	417	620
(80°F air temperature)	1½-in. on 3-in. pipe	40	115	207	330
	1½-in. on 6-in. pipe	64	186	335	497
	3-in. on 3-in. pipe	24	75	135	200
	3-in. on 6-in. pipe	40	116	207	322

listed merely as bare metal, since the film coefficient for air on
the outside of the pipe controls the heat loss, as is evident when
a wind is present or the ambient temperature is low.

EXAMPLE 2.
   Calculate the heat loss per day from a 3 in. line 1000 ft in length operating at 600°F and insulated with 1½ in. of magnesia pipe insulation
where ambient temperature is 80°F.

$$Q = (207 \text{ Btu/hr-ft})(1000 \text{ ft})(24 \text{ hr/day}) = 4.96 \times 10^6 \text{ Btu/day}$$

Information in Table 12-4 may also be used for the estimation of
heat losses from the wellhead, or any other equipment used to
move the hot fluid to the wellbore.

Loss of Heat in the Wellbore—Injection of Single Phase Fluids.
As fluids are displaced down the wellbore to the formation being
flooded, heat is transferred from or to the surrounding beds due to
the difference in geothermal and injected fluid temperatures. In
the case of hot fluid injection, the magnitude of this heat transfer
is important in the calculation of the effective amount of heat carried by the fluid, and its temperature when it reaches the sandface.
The importance of heat transfer in a wellbore is illustrated in Figure 12-5. Here the injection of hot natural gas has failed to increase the wellbore temperature at a depth of 1300 feet after 19
months of injection. This behavior is due to the low heat capacity and low injection rate of natural gas. In fact, it is the behavior
illustrated by Figure 12-5 which has inhibited the use of heated
fluids in secondary recovery until very recent times. Hot fluid injection had been limited primarily to the injection of hot fluid at
high rates as a stimulation technique for the removal of wellbore
deposits.
   A number of authors have treated the problem of heat transmission in the wellbore[23,25,26]. The method presented here will parallel very closely that published by Ramey[22], whose approach is
similar to that of Moss and White[24]. Figure 12-6 is a schematic
representation of the wellbore heat loss problem. Let us assume that
the hot fluid is injected at known rates and at a known surface
temperature down the tubing in a well which is cased to the top of
the sandface. It will be necessary to develop a relationship to
yield the temperature of the injected fluid as a function of time
and depth. The first law of thermodynamics may be written as:

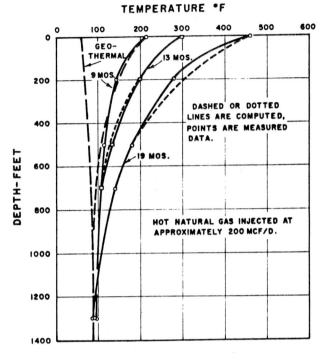

Figure 12-5. Measured and computed temperatures in a wellbore down which hot gas is being injected (after Ramey[22]).

$$\left(\underline{H} + \frac{v^2}{2\,g_c\,J} + \frac{gZ}{g_c\,J} \right) \delta m_{\text{in}} - \left(\underline{H} + \frac{v^2}{2\,g_c\,J} + \frac{gZ}{g_c\,J} \right) \delta m_{\text{out}} +$$

$$\delta Q - \frac{\delta w}{J} = d \left[\left(E + \frac{v^2}{2\,g_c\,J} + \frac{gZ}{g_c\,J} \right)_m \right]_{\text{system}} \quad (7)$$

where

$\underline{H}$ = enthalpy, Btu/lb$_m$
v = velocity, ft/sec
Z = elevation, ft
g = acceleration due to gravity, 32.17 ft/sec^2
g_c = ma/F, lb$_m$-ft/sec^2-lb$_f$
J = 778, a conversion factor, lb$_f$-ft/Btu
m = lb$_m$

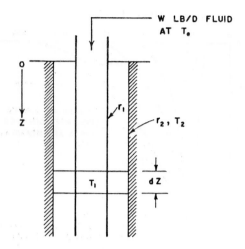

Figure 12-6. Schematic of the well-bore heat problem (after Ramey[22]).

$+Q$ = heat flux out, Btu/sec
w = shaft work
δ = indicates small increment flowing
d = indicates differential change

where an underline means unit mass. For the flow system being considered, there is no accumulation of energy (steady-state) and no shaft work (no compressor in the system) so that the right hand side and δw of equation (7) become zero:

$$d\underline{H} + \frac{dv^2}{2Jg_c} + \frac{g\,(dZ)}{Jg_c} = dq \tag{8}$$

where

$dq = \delta Q/\delta m$, Btu/lb$_m$-sec

$d\underline{H} = TdS + \dfrac{V\,(dp)}{J}$, Btu/lb$_m$

T = Temperature, °R
S = Entropy, Btu/°R

In equation (8), the kinetic and potential energy terms can be neglected since, practically speaking, the change in enthalpy will be the controlling term. Then, in view of the above definition of change in enthalpy, and the fact that TdS is equivalent to $c_p\,dT$

when the system is at constant pressure, equation (8) may be written as:

$$d\underline{H} = c_p(dT) + \frac{V(dp)}{J} \tag{9}$$

where

c_p = specific heat at constant pressure of the fluid, Btu/lb$_m$-°F
V = specific volume, ft^3/lb$_m$

Equation (18) of this chapter will result if the potential energy term is retained. For the incompressible fluid case, $V(dp)/J$ is approximately zero. The first law of thermodynamics has been reduced to the following expression:

$$dq \approx c_p(dT) \tag{10}$$

for an incompressible fluid flowing vertically in tubing of constant diameter.

Reference to Figure 12-6 shows that an energy balance may be written over the depth interval dZ:

$$dq = \text{heat lost by liquid} = \text{heat transferred through casing} \tag{11}$$
$$= -W c_p dT_1 = 2\pi r_1 U(T_1 - T_2) dZ$$

where the nomenclature is that illustrated by Figure 12-6, and the units are Btu, ft, °F, and lb$_m$/day, and U is the overall heat transfer coefficient with units of Btu/day-sq ft-°F. Ramey[22] gives the following equation to express the rate of heat conduction from the casing to the formation:

$$dq = \frac{2\pi k (T_2 - T_e) dZ}{f(t_D)} \tag{12}$$

where

T_e = geothermal temperature = $aZ + b$, °F
a = geothermal gradient, °F/ft
b = geothermal surface temperature
$f(t_D)$ = dimensionless transient heat conduction time function

The geothermal surface temperature is usually taken to be roughly 60°F in the nothern United States or southern Canada, and 80°F in the southern United States. Equation (12) assumes that heat conduction, from the casing to the formation above the injection interval, is radially away from the wellbore.

In view of equation (12), equation (11) can be rewritten as:

$$\frac{\partial T_1}{\partial Z} + \frac{T_1}{A} - \frac{(aZ + b)}{A} = 0 \tag{13}$$

where

$$A = \frac{W c_p [k + r_1 U f(t_D)]}{2 \pi r_1 U k} \tag{14}$$

W = fluid injection rate, lb_m/day

Here, A is termed the time function. The differential equation, equation (13), can be integrated using the integrating factor, $e^{Z/A}$:

$$T_1(Z, t) = e^{-Z/A} \int \frac{(aZ + b) e^{Z/A}}{A} \, dZ + C(t) e^{-Z/A}$$

$$= aZ - aA + b + C(t) e^{-Z/A} \tag{15}$$

The function, $C(t)$, may be evaluated from the boundary condition, $T_1 = T_o$, for $Z = 0$. Here, T_o is the temperature at the surface of the injected fluid. Then:

$$C(t) = T_o + aA - b \tag{16}$$

Equation (15) can now be written in final form to yield an expression for temperature in an injected *liquid* as a function of depth and time:

$$T_1(Z, t) = aZ - aA + b + [T_o + aA - b] e^{-Z/A} \tag{17}$$

where A is defined by equation (14).

Since a number of the hot fluids which can be injected for secondary recovery, or well stimulation are not liquids but are gases, such as steam, an expression is needed for temperature in the injected *gas* as a function of depth and time. If the fluid is a perfect gas, then equation (11) can be altered, if desired, to include a potential energy term:

$$dq = W c_p dT_1 \pm \frac{W \, dZ}{778} = 2 \pi r_1 U (T_1 - T_2) dZ \tag{18}$$

A positive sign on the potential energy term indicates flow downward while a negative sign is for upward flow.

The combining of equations (12) and (18) with an integration as performed in equation (15) results in a final expression for temper-ture of an injected gas as a function of depth and time:

$$T_1(Z, t) = aZ - A\left(a \pm \frac{1}{778\,c_p}\right) + b +$$

$$\left[T_o + A\left(a \pm \frac{1}{778\,c_p}\right) - b\right] e^{-Z/A} \quad (19)$$

Equation (19) may also be used for gas flow up a wellbore. In this case, the negative sign on the potential energy term is used, and depth is taken to be positive and increasing upward from the producing interval. Geothermal temperatures must also be taken, with respect to depth increasing in an upward direction.

In order that equations (17) or (19) may be used to determine the temperature of the injected fluid as it enters the formation, it is necessary that the transient heat conduction time function, $f(t_D)$, be evaluated. The function is defined by the rearranged form of equation (12):

$$f(t_D) = \frac{2\pi k(T_2 - T_e)}{dq/dZ} \quad (20)$$

A rigorous treatment of the transient heat conduction in a wellbore for increasing injection times would be difficult, since both the temperature and heat flux are continuously changing. An exception would be for the case when saturated steam is injected, since the temperature would be essentially constant, so long as no long fluid columns and phase change due to increased pressure is involved. Then too, a time-consuming calculation using short intervals in combination with a superposition problem would make such an approach of questionable engineering value if desk calculations are attempted, but relatively simple if computer facilities are available. A number of solutions have been published which permit an estimation for the value of $f(t_D)$. Carslaw and Jaeger[27] have presented a solution from which $f(t_D)$ can be determined for the case where the heated tubing and casing are considered to be a line source losing heat at a constant rate by convective transport to the formation, which is considered to be an infinite radial system. Figure 12-7 presents the dimensionless relationships for three cases: (1) for a cylinder with constant heat flow from a line source, (2) for a constant temperature at the external radius of the well casing, and (3) for a casing losing heat at a constant rate by convective transport to the formation. It can be observed that all three

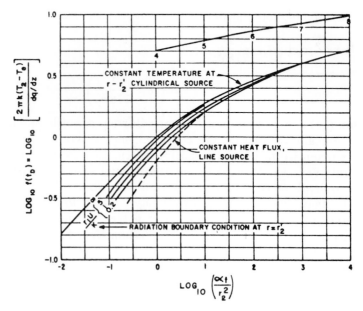

Figure 12-7. Transient heat conduction in a wellbore and surrounding formation (after Ramey[17]).

cases yield identical results when the log $(\alpha t/r_2'^2)$ term of the abscissa has a value in excess of 2.5. Ramey points out that this condition is often satisfied if the injection time exceeds one week. Nomenclature appearing in Figure 12-7 not previously defined are:

$\alpha = k/\rho\, c_f$, the thermal diffusivity of the earth, sq ft/day
c_f = specific heat of the earth, Btu/lb$_m$-°F
r_1 = inside tubing radius, ft
r_2' = outside casing radius, ft
t = time from the initiation of fluid injection, days
k = thermal conductivity of the earth, Btu/day-ft-°F

When injection times are for less than one week, it has been found that the radiation boundary condition gives reasonable values for $f(t_D)$, according to the following equation[22]:

$$-k \left(\frac{\partial T}{\partial r} \right)_{r=r_2'} = U_2 (T_1 - T_2) \tag{21}$$

where

$U_2 = r_1 U_1/r_2'$, the overall heat transfer coefficient based upon the outside of the casing, Btu/day-sq ft-°F

U_1 = the overall heat transfer coefficient between the outside of the tubing and the outside of the casing, Btu/day-sq ft-°F and where area is based on r_1

Figure 12-7 provides values for the radiation boundary condition at values of $r_1 U/k$ of 0.2, 1, 5, and infinity.

In order to use equations (17) and (19), a method must be found through which the overall heat transfer coefficient from the injected fluid to the formation, U, can be evaluated. A number of references[19,20,21,28] contain treatments of the heat transfer coefficient. In the problem of wellbore heat transmission, the coefficient must contain the resistance to heat flow offered by: (1) the fluid in the tubing, (2) all scale or deposits on the inner or outer wall of the tubing and casing, (3) solids or fluids in the annular space, and (4) the casing and tubing wall. Ignoring the effect of scale or deposits on the casing and tubing, and the small contribution of heat radiation in the annulus, the following equation applies[22,28]:

$$\frac{1}{U} = \frac{dA_1}{h_1 \, dA_1} + \frac{x_t \, dA_1}{k_t \, dA_t} + \frac{dA_1}{h_2 \, dA_1} + \frac{dA_1}{h_2 \, dA_2} + \frac{x_c \, dA_1}{k_c \, dA_c} \quad (22)$$

where

h_1, h_2 = film coefficient of heat transfer for fluid within the tubing or casing, respectively, Btu/day-sq ft-°F

A_t, A_c = log mean cross-sectional area of the tubing and casing, respectively, sq ft

x_t, x_c = tubing and casing well thickness, respectively, ft

A_1, A_2 = inside area of the tubing and of the casing, respectively, sq ft

A_1' = outside area of the tubing, sq ft

k_t, k_c = thermal conductivity of the tubing and casing material, respectively, Btu/day-ft-°F

The logarithmic mean area of the casing and the tubing may be calculated using equation (5).

Should the tubing-casing annulus be filled with an insulating material, the third and fourth terms on the right hand side of equation (22) would not be applicable, since in this case, the predominant part of the resistance to heat flow is provided by the insulating material. The film resistance would still be present however. The form of the equation would become[22]:

$$\frac{1}{U} = \frac{dA_1}{h_1 \, dA_1} + \frac{x_t \, dA_1}{k_t \, dA_t} + \frac{x_a \, dA_1}{k_a \, dA_a} + \frac{x_c \, dA_1}{k_c \, dA_c} \qquad (23)$$

where the subscript a refers to a property of the annular space. Again, logarithmic mean areas should be used.

The use of equations (22) and (23) can be simplified somewhat where an overall heat transfer coefficient must be calculated through the use of the following "rules of thumb," which have been presented by Ramey[22]:

(1) In view of the form of equations (22) and (23), the terms representing the thermal resistance of the casing and the tubing may often be neglected due to the relatively high thermal conductivity of steel.

(2) Since heat film coefficients for liquid water or condensing steam are in the range of 200 to 2000 Btu/hr-sq ft-°F, the thermal resistance of the water or condensing steam can usually be neglected with only minor error resulting.

(3) Gas film coefficients in the annulus in conjunction with insulating materials, if present, influence the overall heat transfer coefficient to the largest degree. For turbulent flow, gas film cc-efficients are in the range from 2 to 5 Btu/hr-sq ft-°F.

Table 12-9-A does provide thermal conductivity values for several grades of steel and for aluminum, from which values for k_t and k_c may be estimated for use in equations (22) and (23). Jakob[29] has presented information on the thermal conductivity of dry building materials, which are shown in Table 12-5a.

In view of the data from Table 12-5a, the thermal conductivity of a dry sandstone of density 2.3 gm/cm³ would approximate 0.65 Btu/hr-sq ft-°F/ft. Jakob[29] states that the thermal conductivity of massive rocks such as granite, gneiss, and marble, having densities of 2.4 to 3.1 gm/cm³, lies in the range of 1.2 to 2.4 Btu/hr-sq ft-°F/ft. Where a material contains moisture, the thermal conductivity will be larger than for the dry condition. Table 12-5b provides correction factors (multipliers) which may be applied to obtain approximate values for the thermal conductivity of a material containing a range of moisture in volume per cents.

As only limited data has been published on the thermal conductivity of different formations at the conditions of temperature, pressure and fluid saturations prevailing (see Chapter 11), Tables 12-5a and 12-5b provide a reasonable basis upon which approximate values may be estimated. In example calculations, several authors

TABLE 12-5a. Thermal Conductivity of
Dry Building Materials
(after Jakob[29]).

Density, gm/cm^3	k, Btu/hr-sq ft-$^\circ$F/ft
0.2	0.035
0.4	0.045
0.6	0.065
0.8	0.090
1.0	0.120
1.2	0.155
1.4	0.195
1.6	0.240
1.8	0.300
2.0	0.400
2.2	0.540
2.4	0.765

have used 1.0 Btu/hr-sq ft-$^\circ$F/ft as the thermal conductivity of the formation. Tables 12-5a and 12-5b may also be used to approximate the thermal conductivity of the cement sheath between the casing and the formation. In some instances, it will often suffice to assume that the cement sheath and the formation have the same thermal conductivity, if their densities and moisture content can be assumed to be nearly equal.

The problem of heat transfer between a fluid within a pipe and the pipe wall has been treated by a large number of investigators[18,28,30]. Essentially, the problem is one of heat transfer by either natural or forced convection. In natural convection, the

TABLE 12-5b. Factors to Correct
Thermal Conductivity of a
Material for Moisture
Content (after
Jakob[29]).

Water Content, Vol. Per cent	Correction Factor
1	1.30
2.5	1.55
5	1.75
10	2.10
15	2.35
20	2.55
25	2.75

convective movements in the body of the fluid are brought about by differences in densities due to differences in temperature; in forced convection, the transfer of heat due to contact of the fluid at a given temperature, with a surface of different temperature, is due to velocity of the fluid because of applied pressures, and is independent of heat flow.   If a fluid is being injected down the tubing, heat transfer to the tubing walls is by forced convection, while the heat transfer from the tubing wall to the casing wall across an annulus filled with stagnant fluid, is by natural convection.   Calculations of heat transfer by convection are complicated by the large number of variables which are involved.   By means of the Pi theorem, it is possible to arrange the variables into a select number of dimensionless groups to which the heat transfer, h, is proportional.   These groups are:

$$ h \approx \frac{k}{L} \left(\frac{v \rho L}{\mu} \right)^{x_1} \left(\frac{c_p \mu}{k} \right)^{x_2} \left(\frac{ag(\Delta T)L^3 \rho^2}{\mu^2} \right)^{x_3} \tag{24} $$

where

ΔT = temperature difference, °F
L = length, ft
v = velocity, ft/hr
ρ = fluid density, lb_m/ft^3
μ = fluid viscosity, $\text{lb}_m/\text{ft-hr}$ (note that 1 cp = 2.42 $\text{lb}_m/\text{ft-hr}$)
c_p = specific heat of fluid at constant pressure, $\text{Btu}/\text{lb}_m\text{-°F}$
a = volumetric expansion coefficient, vol/vol, equal to $1/T$, where T is in °R
g = acceleration due to gravity, 4.17×10^8 ft/hr^2

and the exponents, x_1, x_2, and x_3, are numerical constants which normally have to be determined by experimentation for a fluid in a system of a given geometry.   The four dimensionless groups of expression (24) have been given names, hL/k, is called the Nusselt number, N_{Nu}, and is considered the dependent variable.   The dependent dimensionless variables are $v\rho L/\mu$, the Reynolds number, N_{Re}; $c_p \mu/k$, the Prandtl number, N_{Pr}; and $ag(\Delta T)L^3 \rho^2/\mu^2$, the Grashof number, N_{Gr}.

The Nusselt number, N_{Nu}, represents the ratio of the actual heat transfer term, h, to the terms, k/L, which represents the heat transfer due to conduction in a fluid at rest.   The Reynolds number, N_{Re}, can be considered as the ratio of the inertia forces, ρv^2, to the viscous forces, $\mu v/L$.   The Prandtl number can be visualized as the ratio of the kinematic viscosity, μ/ρ, to the thermal diffusivity, $k/c_p\rho$. In forced convection, the temperature distribution, and the resulting heat transfer which occurs, are specified by the Reynolds and Prandtl numbers. The Grashof number, N_{Gr}, is not particularly amenable to a physical interpretation. Essentially, the number represents a ratio between buoyancy forces per unit area, $ag(\Delta T)\rho L$, and viscous forces, $\mu v/L$. This form would result in an expression for the Grashof number of $ag(\Delta T)\rho L^2/\mu v$, which includes the dependent variable, v. However, in view of the Reynolds number, v is proportional to $\mu/\rho L$, so that the Grashof number takes the final form, $ag(\Delta T)L^3\rho^2/\mu^2$, which contains only independent variables.

When the fluid moving through a pipe has a velocity due to applied pressure, the flow regime resulting can be expressed by means of the Reynolds number. If the Reynolds number calculation for a given fluid flowing in a pipe of given size exceeds 10,000, turbulent flow conditions exist. The following equation describes the system:

$$\frac{hD}{k} = a_1 \left(\frac{DG}{\mu}\right)^{0.8} \left(\frac{c_p\mu}{k}\right)^p \tag{25}$$

where

h = film coefficient, Btu/hr-sq ft-°F
D = pipe diameter, ft
k = thermal conductivity, Btu/hr-ft-°F
p = experimentally determined constant
a_1 = experimentally determined constant

Equation (25) is developed from equation (22) by replacing the proportionality by the constant a_1. The exponent x_1 has been defined by experimentation as 0.8, for a wide range of fluids flowing through pipes under many different temperature difference conditions. Study of the systems flowing under forced convection has further shown that the Grashof number can be neglected. There is no general agreement as to the proper values for the constants a_1 and p. Table 12-6 summarizes values for the constants, as presented by a number of authors. In the usage of the table, μ is the bulk viscosity of fluid, μ_f is the viscosity of fluid in the film at the average temperature of the film, and μ_w is the viscosity of the fluid at the wall. The differences found by the investigators for the values of the constants a_1 and p are largely due to the differences in the

TABLE 12-6. Values of Constants a_1 and p in Equation (25)
(from McAdams[28]).

Author	Date	a_1	p	μ or μ_f	Heating, H Cooling, C
Hinton[31]	1928	0.0281	0.355	μ	H and C
Dittus and Boelter[32]	1930	0.0243	0.4	μ	H
Dittus and Boelter[32]	1930	0.0265	0.3	μ	C
Sherwood and Petrie[33]	1932	0.024	0.4	μ	H
Colburn[34]	1933	0.023	0.333	μ_f	H and C
Sieder and Tate[35]	1936	$0.027(\mu/\mu_w)^{0.14}$	0.333	μ, μ_w	H and C

fluid viscosity used, and whether the fluid in the pipe was at either a higher or lower temperature than the pipe wall. For gases and liquids of low viscosity, it is recommended that equation (25) be used with values of 0.023 and 0.4 for a_1 and p, respectively. In this instance, h, the heat transfer coefficient, is for the film. Care should be taken to be consistent in data used, since to calculate the film coefficient, h, for the inside of a pipe, the internal diameter, D, must be used and the viscosity determined at the bulk temperature of the fluid. Film coefficients for the outside of the pipe may be calculated in a similar manner. The film thickness on the inside and the outside of the pipe would only rarely be the same.

If the fluid temperature is considerably higher or lower than the temperature of the wall, the constants proposed by Sieder and Tate[35] should be used with the additional multiplying factor of $(\mu/\mu_w)^{0.14}$ applied to equation (25). The equation apparently is valid for petroleum oils, organic liquids, aqueous solutions, and gases[28]. Serious errors may result for liquids of high viscosity.

In the transitional flow regime between viscous and turbulent flow, where Reynolds numbers range from 2,100 to 10,000, equation (25) may be used for gases unless the mass velocity, G, is less than $1200 P^{2/3}$ where P is pressure in atmospheres[28]. For viscous or laminar flow, i.e. Reynolds number, $Dv\rho/\mu$, between 100 and 2,100, the applicable equation is[30]:

$$\frac{hD}{k} = 1.86 \left(\frac{DG}{\mu} \frac{c_p \mu}{k} \frac{D}{L}\right)^{0.33} \left(\frac{\mu}{\mu_w}\right)^{0.14} \qquad (26)$$

where

L = pipe length, ft
v = flow velocity, ft/sec

An additional dimensionless group has been added, D/L, to characterize the length of the pipe relative to the pipe diameter. A

study of equations (25) and (26) reveals that for the transition between turbulent and viscous flow, the value of Nusselt number
(contains the film heat transfer coefficient) depends upon the exponent on the Reynolds number decreasing from 0.8 to 0.333, and
upon the exponent on D/L increasing from zero to 0.333. At very
low Reynolds numbers corresponding to very low injection fluid
velocities, it may not be possible to neglect the Grashof number as
given in the proportionality of equation (24), since buoyancy forces,
due to temperature differences, could become significant. At zero
fluid velocity, transfer of heat is by free convection.

The following example problem will show the application of the
equations in the calculation of heat transfer by forced convection.

EXAMPLE 3.

The following field data are available for hot water injection down the
casing to a formation: injection rate, 500 B/D; water temperature at the
surface, 200°F; casing length, 5000-ft; casing size, 7 in. (6.366 in ID,
23 lb/ft); geothermal gradient, 1°F/100 ft of depth; geothermal surface
temperature, 70°F; with injection for a period of 30 days. Assuming an
average flowing temperature in the casing of 150°F, the viscosity of
water, according to Table 12-9-F, is 1.047 lb/ft-hr. The Reynolds number
is:

$$N_{Re} = \frac{DG}{\mu} = (0.53 \text{ ft}) \left(\frac{500 \text{ B/D} \times 350 \text{ lb/bbl}}{24 \text{ hr/day} \times \pi/4 \times (0.53)^2 \text{ ft}^2} \right) \left(\frac{1}{1.047 \text{ lb/ft-hr}} \right)$$

$$= 16,770$$

To calculate the film coefficient of heat transfer, h_1, for the inside of
the casing, use equation (25) in the following form:

$$h_1 = \frac{0.023 \, k}{D} \left(\frac{DG}{\mu} \right)^{0.8} \left(\frac{c_p \, \mu}{k} \right)^{0.4}$$

The specific heat for water is 1.0 Btu/lb$_m$-°F (see Table 12-9-F), while
the thermal conductivity at bulk or average flowing temperature is 0.381
Btu/lb$_m$-ft-°F. Thus,

$$h_1 = \frac{0.023 \, (0.381)}{0.53} \left[\frac{0.53 \, (33,100)}{1.047} \right]^{0.8} \left[\frac{1.0 \, (1.047)}{0.381} \right]^{0.4} = 59.4 \text{ Btu/hr-ft}^2\text{-°F}$$

Equation (22) can be modified to the following form for the calculation of
the overall thermal conductivity, U:

$$\frac{1}{U} = \frac{1}{h_1} + \frac{x_c}{k_c}$$

Table 12-9-A gives a k_c for steel of 26 Btu/hr-ft-°F.

$$U = 1 \bigg/ \left[\frac{1}{59.4} + \frac{7 - 6.366}{2\,(12)\,(26)} \right] = 56.0 \text{ Btu/hr-ft}^2\text{-}^{\circ}\text{F}$$

The transient aspect of heat transfer can be taken into account by evaluating log $(\alpha t/r_2'^2)$, where α is the thermal diffusivity of the earth, $k/\rho\,c_f$. Assuming that the average formation density is 2.3 gm/cm^3, Table 12-5a yields a k value of 0.65 Btu/hr-ft-$^{\circ}$F.

The k value will be arbitrarily increased by a factor of 2.75 for 25 per cent moisture content to yield an effective k of 1.79 Btu/hr-ft-$^{\circ}$F. The specific heat should approximate 0.619 Btu/lb$_m$-$^{\circ}$F.

$$\log (\alpha t/r_2'^2) = \log \left[\frac{1.79\,(30)\,(24)\,(12)^2}{143.5\,(0.619)\,(3.5)^2} \right] = 2.233$$

From Figure 12-7, log $f(t_D) = 0.48$, or $f(t_D) = 3.02$. Equation (14) for the unsteady-state term may now be evaluated:

$$
\begin{aligned}
A &= \frac{W\,c_p\,[k + r_1\,U\,f(t_D)]}{2\,\pi\,r_1\,U\,k} \\[2mm]
&= \frac{500\,(350)\,(1)\,[1.79 + (0.53/2)\,(56.0)\,(3.02)]}{2\,\pi\,(0.53/2)\,(56.0)\,(1.79)\,(24)} \\[2mm]
&= 2{,}036 \text{ ft}
\end{aligned}
$$

The temperature of the hot injection water at the bottom of the casing can be calculated using equation (17):

$$
\begin{aligned}
T &= aZ + b - Aa + (T_0 + Aa - b)\,e^{-Z/A} \\[2mm]
&= 0.01\,(5000) + 70 - 0.01\,(2036) + [200 + 0.01\,(2036) - 70]\,e^{\frac{-5000}{2036}} \\[2mm]
&= 112.5^{\circ}\text{F}
\end{aligned}
$$

For a more accurate solution to the problem as outlined above, the average temperature in the system should be reassumed as the average between the 200°F surface temperature and the 112.5°F bottomhole temperature and the calculation repeated. Only a minor change in the bottom hole temperature of the water would result. If the calculation were repeated for an injection period of 365 days, the indicated bottomhole temperature would still be relatively low. For this particular example, it is evident that little benefit has been derived from injecting hot water, since the geothermal bottomhole temperature was already 120°F. On the other hand, the injection of water at ambient temperatures would have resulted in an even larger cooling effect—a fact which could be pointed out by repeating the above calculation for such a case. Physically, the reason for the large drop in the temperature of the injection water is the presence of a very large heat sink, the formation, in direct contact with the casing carrying the hot water. A considerable reduction in heat losses and accompanying temperature drop could be achieved by injecting the hot water through tubing with a packer set just above the injection sandface. The casing could be filled with a granular insulating material, vented so that air at atmospheric pressure was present, or filled with a gas of low thermal conductivity (e.g., carbon dioxide) at low pressure. As pressure increases, the thermal conductivity of a gas also increases. Tables 12-9-D and 12-9-E provide thermal conductivities for some of the solid-type insulating materials. Table 12-9-B presents the thermal properties of a variety of gases at atmospheric pressure.

For the case where a hot or cold fluid is being pumped down tubing, and where a fluid is present in the annular space, the calculation of the heat losses and fluid temperature at the sandface would require a determination of the overall heat transfer coefficient through the film coefficient on the inside of the tubing, through the tubing wall, through the stagnant fluid film on the outside of the tubing, through the fluid in the annulus by free convection, through the fluid film on the inside of the casing, through the casing and through the cement (if the thermal conductivity of the cement is markedly different from that of the formation). The overall heat transfer coefficient can be calculated using equations (22) or (23), with addition or deletion of terms as needed. It should be apparent to the reader that the calculation of an applicable overall heat transfer coefficient in a wellbore can be complicated by the fact that the casing is seldom cemented from surface to the sandface, that a film coefficient may exist between the casing, the cement and the formation, that the formation thermal conductivity is unlikely to remain constant over the length of the wellbore, and that any insulating material or gas of low thermal conductivity in the annular space may have altered thermal properties due to moisture or contamination with other gases. Yet for all those uncertainties which are common to all engineering problems, reasonably accurate values for the overall heat transfer coefficient can be determined, and a good prediction of the heat losses and temperature drop in the injected fluid obtained.

Tables 12-9-A through 12-9-F provide a summary of the thermal properties of a variety of materials which might be encountered in the engineering design where thermal phenomena are encountered. Additional information can be found in the International Critical

Tables[36], in handbooks[37,38] or in thermodynamics or heat transfer text and reference books[21,27,39,40].

Loss of Heat in the Wellbore—Injection of a Condensing Fluid. The preceding section describes analytical and empirical methods by which the loss or gain of heat, with consequential drop or rise in temperature, can be predicted for the case when a single phase fluid is injected. The method has application for most of the hot fluids which would normally be injected into or produced from subsurface reservoirs. The method should be used for superheated steam and for hot water injection temperature behavior predictions. During the condensation of saturated steam, the preceding technique does not apply, although the principles certainly do, since no way is provided for estimating the quality of the steam or for allowing for the constant temperature which results until condensation is complete. Satter[41] has developed an analytical method by which the quality of the steam may be determined at set depth intervals from the top of the injection well. The assumptions which are necessary for the calculation technique developed are as follows[41]:

(1) The quality, temperature, and pressure of the steam is known at the wellhead.

(2) The annulus between the casing and tubing is isolated by a packer set above the injection interval and is filled with low pressure air.

(3) Steady-state heat transfer conditions prevail in the wellbore.

(4) Heat transfer from the casing to the surrounding formation is by unsteady-state radial heat conduction.

(5) There are no changes in kinetic energy in the system.

(6) Steam pressure in the tubing is constant from the surface to the sandface.

(7) Thermal properties of the formations do not change with depth.

Figure 12-8 shows a schematic diagram of the wellbore heat problem and illustrates the nomenclature. The necessary equation may be developed by applying the law of conservation of energy to fluid element of height dx[41]:

$$-W\,d\underline{H} + \frac{Wg}{778\,g_c}\,dx + dq = 0 \tag{27}$$

where

$\qquad W$ = steam injection rate, lb_m/hr
$\qquad q$ = heat-transfer rate, Btu/hr
$\qquad dx$ = differential height, ft

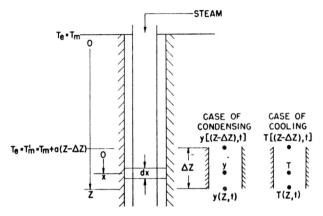

Figure 12-8.   Schematic diagram of wellbore heat problem where steam is condensing (after Satter[41]).

The enthalpy per lb_m of mixture of vapor and liquid can be expressed as:

$$\underline{H} = y \underline{H}_{fg} + \underline{H}_w \qquad (28)$$

where

> y = quality of the steam, mass fraction of vapor in the mixture
> $\underline{H}_{fg}$ = latent heat of steam at the prevailing pressure, Btu/lb_m
> $\underline{H}_w$ = enthalpy of saturated water at prevailing pressure, Btu/lb_m

If the change in pressure in the column of fluid due to friction and gravity is neglected, then:

$$d\underline{H} = \underline{H}_{fg} dy \qquad (29)$$

since the change in enthalpy would be due to the condensation taking place. Combining equations (27) and (29) results in:

$$-W \underline{H}_{fg} dy + \frac{Wg}{778 \, g_c} dx = dq \qquad (30)$$

Equations (11) and (12) may be rewritten in the following modified forms:

$$dq = 2 \pi r_1 U \left(T_s - T_c\right) dx \qquad (31)$$

$$dq = \frac{2 \pi k_{hf}(T_c - T_e) dx}{f(t_D)} \qquad (32)$$

where

T_s = saturation temperature of steam at prevailing pressure, °F
T_c = casing temperature, °F
T_e = earth temperature, °F

Equating equations (30) and (32) results in an equation for the casing temperature:

$$T_c = \frac{k_{hf} T_e + T_s r_1 U f(t_D)}{r_1 U f(t_D) + k_{hf}} \tag{33}$$

where k_{hf} = thermal conductivity of the earth, Btu/hr-ft-°F.

It can be reasonably assumed that the geothermal temperature varies linearly with depth from the earth's surface:

$$T_e = T_m' + ax; \quad T_m' = T_m + a(Z - \Delta Z) \tag{34}$$

where T_m = mean geothermal surface temperature of the earth, °F.

The following differential equation results from the combining of equations (30), (31), (33), and (34):

$$\frac{\partial y}{\partial x} + \frac{T_s}{A'} + \frac{A'B' + T_m' + ax}{A'} = 0 \tag{35}$$

where

$$A' = \frac{W \underline{H}_{fg} [k_{hf} + r_1 U f(t_D)]}{2 \pi r_1 U k_{hf}} \tag{36}$$

$$B' = \frac{g}{778 \, g_c \underline{H}_{fg}} \tag{37}$$

y = quality of the steam, weight fraction

Since the pressure drop in the column has been neglected, the temperature of the steam may be considered to be constant and the following solution to equation (35) obtained[41]:

$$y = \frac{ax^2}{2A'} + \left[\frac{A'B' + T_m' - T_s}{A'} \right] x + c(t) \tag{38}$$

The constant of integration, $c(t)$, can be evaluated from the boundary condition that $y = y[(Z - \Delta Z), t]$ at the top of the interval of interest or when $x = 0$:

$$y[Z, t] = y[(Z - \Delta Z), t] + \frac{a(\Delta Z)^2}{2 A'} +$$

$$\frac{[A'B' + T_m + a(Z - \Delta Z) - T_s] \Delta Z}{A'} \quad (39)$$

Equation (39) provides an important analytical means by which the quality of the steam at the bottom of an interval, ΔZ, may be estimated from the steam quality at the top of the interval. No field method is presently available for determining the quality of an injected steam at a given depth from the surface.

To determine the loss of heat in a wellbore when a condensing or saturated steam is being injected, it is necessary to provide a means by which the overall heat transfer coefficient may be calculated. Equation (22) or a modification of it may be used. Where the loss of heat to the formation above the oil zone is minimized by the use of a packer, the annulus is filled with a low pressure gas (such as air), and the steam is injected down the tubing, equation (22) may be written in the following specialized form[41]:

$$\frac{1}{U} = \frac{1}{h_1} + \frac{(r_1' - r_1)}{k_t} + \frac{1}{h_{tc} + h_r} \quad (40)$$

where

h_1 = film coefficient between injected fluid and tubing, Btu/ hr-sq ft-°F

r_1 = inside radius of the tubing, ft

r_1' = outside radius of the tubing, ft

k_t = thermal conductivity of the tubing, Btu/hr-ft-°F

h_{tc} = convective heat transfer coefficient between tubing and casing in a stagnant gas, Btu/hr-sq ft-°F

h_r = radiation heat transfer coefficient between tubing and casing, Btu/hr-sq ft-°F

In this instance, the overall heat transfer coefficient includes the convective heat transfer coefficient between the injected fluid and the tubing, the natural convection heat transfer coefficient between the tubing and the casing; the thermal resistance of the tubing wall and the radiation heat transfer coefficient between the tubing and the casing. Figure 12-9 presents a plot of the log of the Nusselt number versus the log of the product of the Prandtl and Grashof numbers for a range of casing internal diameter to tubing external

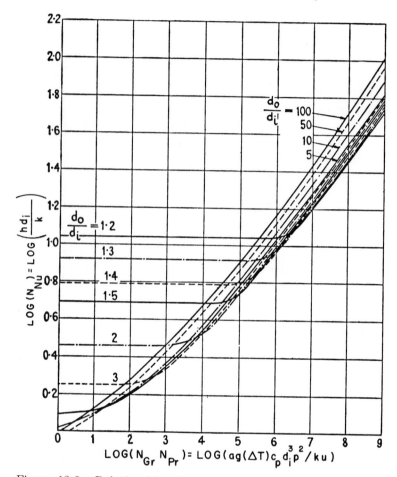

Figure 12-9. Relationships for natural convection across an annular space between two concentric cylinders (after Fishenden and Saunders[30]).

diameter ratios. Figure 12-9 is used by calculating the log $(N_{Gr} N_{Pr})$ where the data are obtained from Tables 12-9-B and 12-9-F. The difference in temperature between the casing and tubing, ΔT, will not be known. As a first approximation, the temperature of the casing may be taken as the average of the formation temperature and the injected fluid temperature at the particular level of interest in the wellbore. Leutwyler[42] has presented a technique for the calculation of casing temperatures in steam injection wells. With the ratio of the casing internal diameter to the tubing external diameter, d_o/d_i, and the log $(N_{Gr} N_{Pr})$ known, Figure 12-9 yields a value for log (N_{Nu}). A value for the overall thermal conductivity between the tubing and the casing, h, or h_{tc}, in equation (40), may then be determined. The accuracy of h_{tc} will depend upon the value used for ΔT. Equations (25) or (26) can be used to determine h_1 in equation (40).

The radiation heat transfer coefficient in equation (40) remains to be evaluated. The propagation of heat by radiation is governed by the same laws that apply to visible radiation of light. The fourth power law for radiation from black bodies, observed experimentally by Stefan in 1879, and analytically by Boltzmann in 1884, states that:

$$q_r = \sigma T^4 \tag{41}$$

where

q_r = radiant heat transfer, Btu/sq ft-hr
σ = Stefan-Boltzmann constant, 0.173×10^{-8}
T = temperature of a perfect radiator or black body, °R

Since no known material is a perfect radiator, the net effect is for a surface to both receive radiant heat from its surroundings, and at the same time radiate heat to the surroundings. The net transfer of heat which results is:

$$q_r = 0.173 \times 10^{-8} A \, \varepsilon_{eff} (T_1^4 - T_2^4) \tag{42}$$

where

ε_{eff} = effective emissivity of a material, or the ratio of the heat radiated by the material to that of a black body under the same conditions
A = area of radiating surface, sq ft

It can be shown[30] that for the concentric cylinder case where the inner surface of the outer cylinder reflects heat diffusely, the net radiation per unit area from the outer surface of the inner cylinder will be:

$$q_r = 0.173 \times 10^{-8} A \; \frac{\varepsilon_1 \varepsilon_2}{\varepsilon_2 + \varepsilon_1 (1 - \varepsilon_2) \dfrac{r_1'}{r_2}} (T_1^4 - T_2^4) \tag{43}$$

where r_1' is the outer radius of the inner cylinder (the tubing) and r_2 is the inner radius of the outer cylinder (the casing). The subscripts 1 and 2 refer to properties of the tubing and casing, respectively. Table 12-7 provides a summary of emissivity data on surfaces which might be encountered in thermal oil recovery projects. Equation (43) will yield information on the radiant heat transfer between the tubing and the casing in Btu/sq ft-hr, which can readily be converted to a value for h_r in Btu/sq ft-hr-°F by dividing by the difference in the casing and tubing temperatures at the level of interest in the well.

TABLE 12-7. Emissivities of Various Surfaces

Surface	0°F-100°F	250°F	500°F	1000°F	2000°F
Pure polished metals	0.04	0.05	0.06	0.07	0.14
White surfaces	0.95	0.94	0.88	0.70	0.45
Dark painted surfaces	0.95	0.94	0.90	0.85	0.80
Various aluminum paints	0.30-070				
Various lacquers & oils	0.50-0.90				
Rough cast iron	0.82				
Rough cast iron, oxidized	0.98				
Oxide on steel	0.80	0.80	0.80	0.80	
Water	0.95				
Polished aluminum	0.04		0.95	0.92	0.83
Aluminum foil (clean)	0.05				
Oxidized aluminum	0.11		0.12	0.18	

Manipulation of equations (40) and (43) will point out the decrease in heat loss to the casing and the formation, which can be effected by decreasing the emissivity of the outer surface of the tubing by means of aluminum jackets, or less effectively by the use of aluminum paint. Leutwyler[42] reports that the emissivity of aluminum-coated tubing centered in black casing was 53 per cent of that obtained for uncoated tubing by laboratory tests. Figure 12-10 reports the results of laboratory studies where the overall heat transfer coefficient, U, between 2.375-inch O.D. tubing centered in 7-inch O.D. casing was measured.

Where a condensing fluid is being injected or where a fluid initially superheated begins condensing before reaching the injection well sandface, a calculation of the per cent heat loss is required. Satter[41] has presented applicable equations where the change in enthalpy and potential energy have been taken into account for the case where superheated steam is injected at the surface.

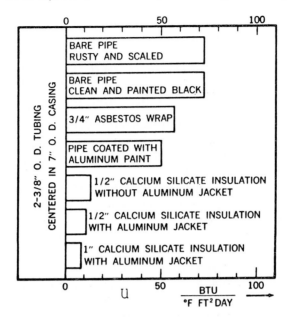

Figure 12-10.   Overall heat transfer coefficient for the tubing, casing, and tubing-casing annulus (from Leutwyler[42]).

(1) For the loss of heat through fluid cooling:

$$\text{Per cent heat loss} = \frac{\left[\underline{H}_i - \underline{H}[Z, t] + \dfrac{gZ}{778 g_c}\right] \times 100}{\underline{H}_i - (\underline{H}_w)_{T_m}} \qquad (44)$$

(2) For the loss of heat through fluid condensation:

$$\text{Per cent heat loss} = \frac{\left[\underline{H}_i - \{\underline{H}_w + y[Z, t]\underline{H}_{fg}\} + \dfrac{gZ}{778 g_c}\right] \times 100}{\underline{H}_i - (\underline{H}_w)_{T_m}} \qquad (45)$$

where

$\underline{H}_i$ = specific enthalpy of saturated water at initial pressure and temperature, Btu/lb$_m$

$\underline{H}[Z, t]$ = enthalpy at a given depth and injection time, Btu/lb$_m$

y_i = quality of the steam at the well head, fraction

$(\underline{H}_w)_{T_m}$ = specific enthalpy of water at T_m, the mean surface temperature, Btu/lb$_m$

$y[Z, t]$ = time and depth dependent steam quality, fraction

For the case where the steam is saturated or undersaturated, the loss of heat is due to condensation alone:

$$\text{Per cent heat loss} = \frac{\left[\{y_i - y[Z, t]\}\, \underline{H}_{fg} + \dfrac{gZ}{778 g_c}\right] \times 100}{\underline{H}_w + y_i \underline{H}_{fg} - (\underline{H}_w)_{T_m}} \quad (46)$$

A calculation procedure which could be used to determine the heat loss, and the quality of the steam entering the formation would be as follows (temperature, pressure and quality of the steam known at the surface):

(1) If the steam is superheated at the surface, divide the depth into several intervals, assume the temperature of the steam at the bottom of the first interval and the average temperature of the casing at the midpoint of the first interval, then estimate the average temperature of the tubing at the midpoint of the first interval (tubing and steam temperatures will be essentially equal and temperature profile with depth should be approximately linear).

(2) The average gas temperature in the annulus (in a boiled-off annulus, air would be present at atmospheric pressure) over the interval is taken as the average of the casing and tubing temperatures at the midpoint of the interval. The overall heat transfer coefficient may now be calculated from equation (40).

(3) Figure 12-7 is used to calculate the transient heat transfer in the wellbore by evaluating the term in the abscissa and obtaining the $\log_{10} f(t_D)$ from the ordinate. Equation (19), written in a more convenient form, can be used to calculate the steam temperature at the bottom of the interval, and is as follows[41]:

$$T[Z, t] = aZ + T_m - aA - AB + \{T\,[(Z - \Delta Z), t] - T_m - a(Z - \Delta Z) + aA + AB\}\, e^{-\Delta Z/A} \quad (47)$$

where

$$A = \frac{W c_g \,[k_{hf} + r_1 U f(t_D)]}{2\pi r_1 U k_{hf}}$$

$$B = \frac{g}{778 g_c c_g}$$

g = acceleration due to gravity, 4.17×10^8 ft/hr^2

g_c = conversion factor, 4.17×10^8 lb$_m$-ft/lb$_f$-hr^2

c_g = specific heat of gas at constant pressure (c_p), Btu/lb$_m$-°F

Figure 12-11 presents values for the specific heat of steam for a wide range of pressures and temperatures. Equation (33) can now be used to calculate the temperature of the casing.

(4) The calculated temperatures are compared to those assumed for the steam temperature at the bottom of the interval and for the casing. If a discrepancy exists, the calculated values are assumed to be the more-nearly correct values and the calculation procedure repeated until satisfactory agreement between assumed and calculated temperatures is obtained. The temperature at the bottom of the interval will reveal whether the steam is superheated or saturated. If the steam appears to be saturated or undersaturated, equation (39) should be tried for the calculation of part (3).

(5) Equations (44) and (45) or (46) are used to calculate the heat loss in the interval.

(6) The calculation procedure is repeated for each interval until the total depth of the injection well has been reached.

If the steam is saturated at the wellhead, the calculation will be somewhat simplified, since the equations describing the behavior of the superheated steam will not be needed. If the condensation of the steam is completed before reaching the bottom of the injec-

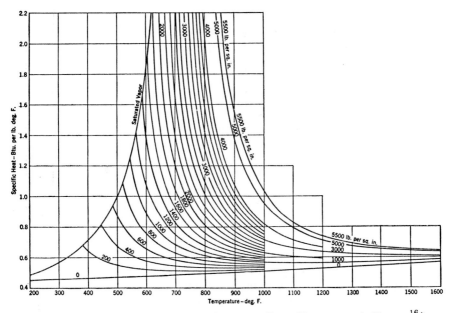

Figure 12-11.   Specific heats of steam (from Keenan and Keyes[16]).

tion well, the above-outlined calculation procedure is still valid. Equation (47) would be used to calculate the temperature at the bottom of the interval and all subsequent intervals. In this case, A would be defined by equation (14), and B would be modified by replacing c_g, the specific heat of the steam, by c_p, the specific heat of the hot water.

Satter[41] has presented the results of calculations where superheated steam was injected down tubing and where the casing contains low pressure air. The steam injection rate was 5,000 lb/hr at a wellhead pressure of 500 psia and an initial temperature of 1000°F. The following formation and well conditions were used:

Mean surface temperature	75 °F
Geothermal gradient	0.011 °F/ft
Thermal conductivity of the earth	1.0 Btu/hr-ft-°F
Thermal diffusivity of the earth	0.046 sq ft/hr
Casing inner diameter	5.989 inches
Casing outer diameter	6.625 inches
Tubing inner diameter	2.441 inches
Tubing outer diameter	2.875 inches

Figure 12-12 shows the temperature and steam quality with depth for total injection times of 0.001, 0.01, 0.1, 1 and 10 years. It is evident that for very short injection times, the steam will begin to condense at relatively shallow depths, 1100 feet in the case of 0.001 years injection time. At the end of 10 years, steam would still begin to condense at 3100 feet. Since temperature is the primary factor controlling heat losses where the physical system is not changed, an increase in the steam injection rate would reduce the percentage heat loss incurred in the wellbore. Figure 12-13 illustrates the effect of well depth upon the percentage of total input heat lost for a range of total injection times. Injection of steam to deep sands will be limited, due to the high heat losses which occur, especially for short term projects which the "huff-and-puff" techniques represent. Where boiler fuel is expensive, such high heat losses are serious. Care should be taken to design the wellbore equipment in a manner to minimize these losses. Packers to restrict the steam to the tubing, the use of boiled-off annulii containing low pressure air (or another gas of low-thermal conductivity) the use of high injection rates and elevated steam temperatures will minimize heat losses for a well of given depth. The use of a low pressure gas in the annulus also has the benefi-

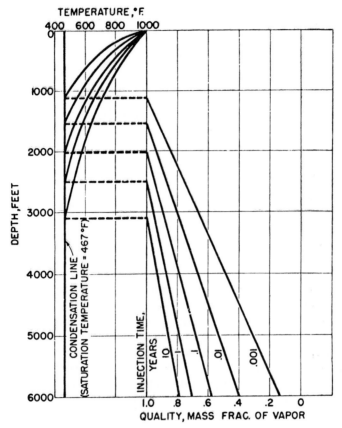

Figure 12-12. Temperature and quality of steam as function of depth and injection time for the stated conditions (from Satter[41]).

cial effect of decreasing the casing temperature, thereby aiding in maintaining the structural integrity of this part of the system under the operating pressures and temperatures imposed. Satter[41] has found that the reduction of casing temperature using a low pressure annulus may be 100°F or more. Leutwyler[42] advises against the use of insulating cements between the casing and the formation, since this tends to increase the casing temperature and increase the possibility of casing failure. The use of an emissivity-reducing coating on the tubing will also lower the casing temperatures as would insulating material in the tubing-casing annular space above a packer.

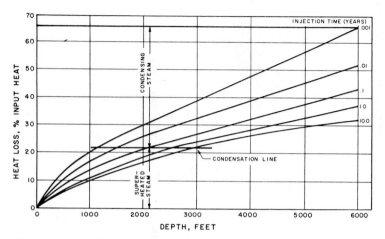

Figure 12-13. Heat loss as a function of depth and time (after Satter[41]).

Satter[41] suggests the use of Figures 12-14 and 12-15 for an approximate correlation of heat losses per foot of depth, which is based upon the observation that steam heat losses are a function of injection rate and injection pressure. The correlation is strictly applicable for the reservoir and wellbore properties previously stated, where the steam is not superheated, and where injection is for a one-year period. The answers obtained apply at the end of the one-year period. The use of the figures is best illustrated by an example calculation.

EXAMPLE 4.
A saturated steam is to be injected through tubing at 4000 lb/hr and 750 psia to a sand at 1500 feet. The wellbore and formation physical properties are those previously outlined. Calculate the percentage heat loss in the wellbore.

Figure 12-15 must be used to determine the location of the point from the surface where the steam would be totally condensed, since Figure 12-14 applies only to the constant-temperature saturated steam case. The hot water point is at approximately 7500 feet, therefore, the steam will not be completely condensed. Figure 12-14 yields a heat loss value of 0.9 per cent of input heat per 100 feet of depth. The total heat loss in the wellbore will be approximately 13.5 per cent.

Ramey[17] has published useful equations for the determination of the total heat loss rate from the wellbore when steam is injected down tubing:

$$q = \frac{2\pi r_1 U k_{hf}}{[k_{hf} + r_1 U f(t_D)]} \left[(T_s - T_m) Z - \frac{aZ^2}{2} \right] \qquad (48)$$

If the thermal resistance is low, as would be the case with injection down the casing, the overall thermal conductivity, U, would be very large. In this instance, equation (48) can be written as:

$$q = \frac{2\pi r_1 k_{hf}}{f(t_D)} \left[(T_s - T_m) Z - \frac{aZ^2}{2} \right] \qquad (49)$$

The time function, $f(t_D)$, is defined by equation (20), and can be evaluated from Figure 12-7 if the thermal diffusivity of the earth, the outside radius of the casing, and the injection time is known. Equations (48) and (49) require that saturated steam exist over the entire length of the wellbore of the injection well.

Loss of Injected Heat in the Formation. A number of papers have appeared in the past ten years which provide an analytical means for predicting heat loss from a heated formation. Lauwerier[43]

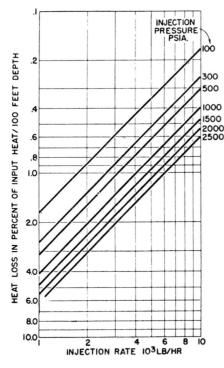

Figure 12-14. Heat loss as a function of injection rate and pressure for cases where saturated steam is injected for one year (from Scatter[41]).

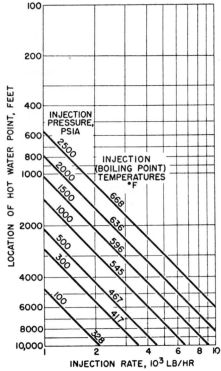

Figure 12-15. Location of the hot water point in tubing based on injection rate and pressure where saturated steam is injected for one year (from Satter[41]).

has presented a mathematical heat model, which should apply to the case where a hot liquid, such as water, is injected at the sandface. Marx and Langenheim[44] have presented a calculation technique which is especially applicable to the case where steam is injected at prevailing bottomhole conditions. In addition, Rubinshtein[45] and Willman *et al.*[46] have presented papers dealing with the problem of heat transfer and heat loss in the formation.

Ramey[17] has made a graphical comparison of the predicted fraction of the total heat lost to adjacent formations due to conduction, W_o^*, versus the logarithm of the dimensionless time function, t_D, controlling heat loss to the formations yielded by the published analytical techniques. Figure 12-16 presents the results of this comparison. The Rubinshtein solution[45] is the most acceptable one to date. The dimensionless time function, t_D, is defined as $4\alpha t/h^2$, where α is the thermal diffusivity in sq ft/hr, t is the total injection time in hours, and h is the thickness of the forma-

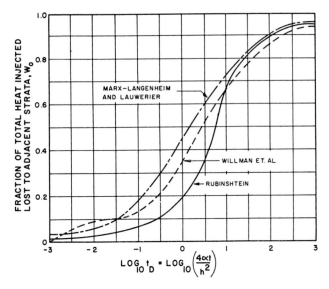

Figure 12-16. Vertical heat loss to adjacent strata as a function of the dimensionless time function (from Ramey[17]).

tion accepting hot fluid in feet. It is interesting to note the rather good agreement between the four separate approaches to determining the extent of the vertical heat loss. Table 12-9-D presents information on thermal diffusivity for various materials. Present information indicates that a value of 0.04 to 0.05 sq ft/hr is approximately correct for formations, both for those containing oil and for those serving as the capping and underlying formations.

The following heat-loss equation derived by Ramey[17], from the results obtained by Lauwerier[43] and by Marx and Langenheim[44], is useful:

$$W_o^*(t_D) = 1 - \frac{1}{t_D}\left[e^{t_D}\ \text{erfc}\ \sqrt{t_D} + 2\sqrt{\frac{t_D}{\pi}} - 1\right] \qquad (50)$$

where

erfc(x) = complementary error function, dimensionless

$$= 1 - \text{erf}(x) = 1 - \frac{2}{\sqrt{\pi}}\int_0^x e^{-\beta^2}\,d\beta$$

The form of the Rubinshtein equation is

$$W_o^*(t_D) = \frac{2}{3}\sqrt{\frac{t_D}{\pi}}\left[1 - \left(1 + \frac{1}{t_D}\right)e^{-t_D^{-1}}\right] + \left(1 + \frac{2}{3t_D}\right)\text{erfc}\frac{1}{\sqrt{t_D}} \quad (51)$$

The equation developed by Ramey, from the results of Willman *et al.* is as follows:

$$W_o^*(t_D) = 1 - \frac{1}{t_D}\left[\sqrt{\pi t_D} - \frac{\pi}{2}\ln\left(2\sqrt{\frac{t_D}{\pi}} + 1\right)\right] \quad (52)$$

Rubinshtein has also published an equation which can be used to determine the vertical heat loss for the case where a hot fluid is injected, then followed by "cold" fluid[17]:

$$W_o(t_D) = \frac{t_D}{t_{Do}}\left[W_o^*(t_D) - \left(\frac{t_D - t_{Do}}{t_D}\right)W_o^*(t_D - t_{Do})\right] \quad (53)$$

where

W_o = fraction of the injected heat which is lost

W_o^* = fraction of the injected heat lost to adjacent strata

t_{Do} = dimensionless time at which hot fluid injection was terminated

t_D = total dimensionless injection time for the period of hot and "cold" fluid injection

The mathematical arguments needed for the development of equations (50) through (53) will not be presented here. The utility of Figure 12-16 in solving the equations for the fraction of the heat which is lost, to bounding formations and sections of the sand not accepting hot fluids, is best shown by an example.

EXAMPLE 5.

Determine the percentage heat loss to adjacent strata for the injection of hot water to a 100 ft thick sand for a period of one year. The formation thermal diffusivity is 0.045 sq ft/hr. Determine the heat lost if cold water is injected for the following year. The value required for the abscissa of Figure 12-16 is:

$\text{Log}_{10}(4\alpha t/h^2) = \text{Log}_{10}(4 \times 0.045 \times 365 \times 24/100^2) =$

$\text{Log}_{10}(0.1417) = -0.849$

from which fractional heat loss, using the curve attributed to Rubinshtein, is 14.5 per cent. Equation (53) may be used to determine the total injected heat, which is injected at the sandface, lost at the end of the two year period:

$$W_o(t_D) = \frac{0.1417 \times 2}{0.1417}[0.209 - 0.5 \times 0.145] = 0.273 \quad \text{or} \quad 27.3\%$$

The reader is referred to Chapter 11 on oil recovery by in situ combustion, where the role of pressure, temperature and saturating liquid on the thermal conductivity of the reservoir and the overlaying and underlaying rocks is discussed. It should be noted that the presence of clays which undergo an endothermic reaction will increase the amount of heat lost to the formation, beyond that indicated by the above unsteady-state heat loss analysis. In this instance, the temperature of the heated fluid in the reservoir, during the oil displacement process, and the composition of the reservoir rock will determine whether such a source of heat loss could exist.

Equations (50) through (53) are for the case where the rate of heat injection, Btu/hr, is constant during the life of the project. Eventually, a radial distance will be reached where the heat loss rate will equal the heat injection rate—the limiting point at which the project would normally be terminated. The analytical development of Marx and Langenheim[44] permits an estimation of the limiting area which can be heated at a specified heat injection rate. The equation is:

$$[e^{x^2} \text{ erfc } x]_1 = (5.618 \times 10^{-6})\left[\frac{\$_h M(\Delta T)}{\$_o \phi(S_o - S_{or})}\right] \tag{54}$$

where

$$x = \frac{2k_{ob}t^{\frac{1}{2}}}{Mh\sqrt{\alpha_D}}, \text{ dimensionless time.} \tag{54a}$$

$$M = (1 - \phi)\rho_f c_f + S_w \phi \rho_w c_w + S_o \phi \rho_o c_o$$

c_f, c_w, c_o = dry formation, water, and oil specific heat, respectively, Btu/lb$_m$-°F

ρ_f, ρ_w, ρ_o = formation grain, water, and oil density, lb$_m$/ft^3

S_o, S_w = the initial reservoir oil and water saturations, respectively

ΔT = difference between the formation and steam zone temperature, °F

$\$_h$ = cost of steam energy, dollars/1,000,000 Btu

$\$_o$ = net value of the displaced oil to the producer with royalty and lifting costs, etc., deducted. An ultimate recovery factor should be applied to this value—perhaps 80 per cent or less.

α_D = overburden and underburden thermal diffusivity, sq ft/hr.

h = pay thickness contacted by steam injected, ft

k_{ob} = thermal conductivity of the over and underlaying formations and/or the heated oil formation which is not contacted directly by the steam, Btu/hr-ft-°F

ϕ = porosity of formation, fraction

t = total steam injection time, hr

Of the factors appearing in equation (54), S_h and S_o would be developed by engineering analysis, and the S_{or} from data on nearby floods, by laboratory study, or by estimation based on a Buckley-Leverett frontal displacement analytical treatment, with a correction applied for steam distillation and thermal expansion (treated in a later section of this chapter). When a value for the term e^{x^2} erfc x has been obtained in equation (54), Table 12-8 may be used to determine x. Marx and Langenheim provide an equation for the cumulative heated area in square feet at time t:

$$A(t) = \frac{H_o M h \alpha_D}{4 k_{ob}^2 (\Delta T)} \left[e^{x^2} \text{ erfc } x + \frac{2x}{\sqrt{\pi}} - 1 \right] \qquad (55)$$

where H_o = constant heat injection rate, Btu/hr

The terms in the brackets on the right hand side of equation (55) may be easily evaluated from Table 12-8.

TABLE 12-8. Values for terms comprising the complementary error function.

x	e^{x^2} erfc x	e^{x^2} erfc $x + \dfrac{2x}{\sqrt{\pi}} - 1$	x	e^{x^2} erfc x	e^{x^2} erfc $x + \dfrac{2x}{\sqrt{\pi}} - 1$
0.00	1.00000	0.00000	0.20	0.80902	0.03470
.02	.97783	.00039	.22	.79318	.04142
.04	.95642	.00155	.24	.77784	.04865
.06	.93574	.00344	.26	.76297	.05635
.08	.91576	.00603	.28	.74857	.06451
0.10	0.89646	0.00929	0.30	0.73460	0.07311
.12	.87779	.01320	.32	.72106	.08214
.14	.85974	.01771	.34	.70792	.09157
.16	.84228	.02282	.36	.69517	.10139
.18	.82538	.02849	.38	.68280	.11158

TABLE 12-8. Cont.

x	e^{x^2} erfc x	e^{x^2} erfc $x + \dfrac{2x}{\sqrt{\pi}} - 1$	x	e^{x^2} erfc x	e^{x^2} erfc $x + \dfrac{2x}{\sqrt{\pi}} - 1$
0.40	0.67079	0.12214	1.50	0.32159	1.01415
.42	.65912	.13304	.55	.31359	.06258
.44	.64779	.14428	.60	.30595	.11136
.46	.63679	.15584	.65	.29865	.16048
.48	.62609	.16771	.70	.29166	.20991
0.50	0.61569	0.17988	1.75	0.28497	1.25964
.52	.60588	.19234	.80	.27856	.30964
.54	.59574	.20507	.85	.27241	.35991
.56	.58618	.21807	.90	.26651	.41043
.58	.57687	.23133	.95	.26084	.46118
0.60	0.56780	0.24483	2.00	0.25540	1.51215
.62	.55898	.25858	.05	.25016	.56334
.64	.55039	.27256	.10	.24512	.61472
.66	.54203	.28676	.15	.24027	.66628
.68	.53387	.30117	.20	.23559	.71803
0.70	0.52593	0.31580	2.25	0.23109	1.76994
.72	.51819	.33062	.30	.22674	.82201
.74	.51064	.34564	.35	.22255	.87424
.76	.50328	.36085	.40	.21850	.92661
.78	.49610	.37624	.45	.21459	1.97912
0.80	0.48910	0.39180	2.50	0.21081	2.03175
.82	.48227	.40754	.60	.20361	.13740
.84	.47560	.42344	.70	.19687	.24350
.86	.46909	.43950	.80	.19055	.35001
.88	.46274	.45571	.90	.18460	.45690
0.90	0.45653	0.47207	3.00	0.17900	2.56414
.92	.45047	.48858	.10	.17372	.67169
.94	.44455	.50523	.20	.16873	.77954
.96	.43876	.52201	.30	.16401	.88766
.98	.43311	.53892	.40	.15954	2.99602
1.00	0.42758	0.55596	3.50	0.15529	3.10462
.05	.41430	.59910	.60	.15127	.21343
.10	.40173	.64295	.70	.14743	.32244
.15	.38983	.68746	.80	.14379	.43163
.20	.37854	.73259	.90	.14031	.54099
1.25	0.36782	0.77830	4.00	0.13700	3.65052
.30	.35764	.82454	.10	.13383	.76019
.35	.34796	.87127	.20	.13081	.87000
.40	.33874	.91847	.30	.12791	3.97994
.45	.32996	0.96611	.40	.12514	4.09001

TABLE 12-8. Cont.

x	$e^{x^2}\text{ erfc } x$	$e^{x^2}\text{ erfc } x + \dfrac{2x}{\sqrt{\pi}} - 1$	x	$e^{x^2}\text{ erfc } x$	$e^{x^2}\text{ erfc } x + \dfrac{2x}{\sqrt{\pi}} - 1$
4.50	0.12248	4.20019	7.00	0.07980	6.97845
.60	.11994	.31048	.20	.07762	7.20195
.70	.11749	.42087	.40	.07556	.42557
.80	.11514	.53136	.60	.07361	.64929
.90	.11288	.64194	.80	.07175	7.87311
			8.00	0.06999	8,09702
5.00	0.11070	4.75260	.20	.06830	.32101
.20	.10659	4.97417	.40	.06670	.54508
.40	.10277	5.19602	.60	.06517	.76923
.60	.09921	.41814	.80	.06371	8.99344
.80	.09589	.64049	9.00	0.06231	9.21772
6.00	0.09278	5.86305	.20	.06097	.44206
.20	.08986	6.08581	.40	.05969	.66645
.40	.08712	.30874	.60	.05846	9.89090
.60	.08453	.53184	.80	.05727	10.11539
.80	.08210	.75508	10.00	0.05614	10.33993

The area of the cumulative heated area may be converted to acres by the factor 43,560 sq ft/acre. Even though the area swept in a pattern flood would be substantially different from the radial system which equations (54) and (55) model, the resulting area value permits an approximate sizing of the patterns and a more-or-less optimum spacing of the wells.

It should be noted that equation (54a) would permit a calculation of the total steam injection time, t, in hours for the radial system where heat is injected at a constant rate, H_o, if the limiting value of x just calculated is used. Again, the time value obtained would not apply directly to a pattern injection system, but would result in order-of-magnitude answers.

Waldorf[47] has reported the effect of steam on the permeabilities of water-sensitive formations. The montmorillionite in the samples tested were dehydrated when exposed to superheated steam with an increase in permeability to flow resulting. However, the water sensitivity remained, since the permeability again decreased when exposed again to fresh water. In those sands containing water-sensitive clays, which have retained higher than original oil producing rates after a "huff-and-puff" steam treatment and temperature has returned to original levels, it is generally believed that the cause is due to an improvement in the permeability in the vicinity of the wellbore. This could well be due to the semi-permanent shrinking of clays, and to the removal of asphaltic and/or paraffinic materials that previously were present.

Estimation of Steam-Drive Performance

Several authors [44,46,48] have presented approaches to the problem of calculating the anticipated performance of a steam-drive project. Willman *et al.*[46] have identified the principal mechanisms responsible for additional oil beyond that which would be obtained by "cold" water flooding as that due to:

(1) thermal expansion of the oil so that the residual oil saturaration is reduced when the reservoir is cooled.

(2) viscosity reduction which permits a more efficient immiscible fluid displacement due to the improved mobility ratio of the displaced and displacing fluids.

(3) steam distillation if the reservoir oil contains distillable light components.

Figure 12-17 presents the results of a cold water, hot water and steam flood upon a 91.4 cm long linear Torpedo sandstone core, having 26.2 per cent porosity, 856 md permeability to water, 26.0 per cent connate water saturation, and containing a 138 cp (at

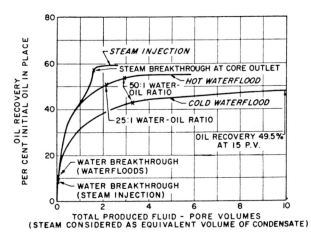

Figure 12-17. Per cent oil recovery by cold water, hot water (212°F), and steam flooding (after Willman *et al.*[46]).

80°F) non-distillable oil. In this case, the advantage of using steam over hot water for flooding purposes is not large. Had the oil been distillable, oil recovery would have been increased from approximately 60 per cent to 82 per cent.

The displacement mechanism during steam flooding is quite complex. In a radial or pattern system containing a partially-distillable oil, the heat-carrying steam will be cooled so that at the front of the displacement, a condensed water zone will form. As the displacement moves forward, the hot condensed water will cool to the reservoir temperature. If the oil is steam-distillable at the temperature of the steam, a vaporization of the residual oil left behind by the hot condensate and water flood will occur. The vaporized oil and steam will condense, thereby supporting the hot condensate and water flood. Essentially then, the initial oil displacement is by a cold water flood, followed closely by a hot water and hot condensate flood, with a final displacement by the steam front. A number of discrete displacements are taking place simultaneously—a cold water-oil displacement, a hot water and condensate oil displacement and finally a gas (steam)-condensate and hot water displacement (partially miscible).

Willman *et al.*[46] have developed a technique for predicting steam-drive performance where the three distinct and separate floods are included.

Position of the Steam and Hot Water Fronts. During the early stages of steam injection to an oil-bearing sand body, the displacement will be radial. The following equation permits a calculation of the radial positions of the steam front[46]:

$$R_{st}^2 = \frac{14.6\, i_{st} \underline{H}_{fg}}{k(T_{st} - T_f)} \sqrt{\frac{\alpha}{\pi}} \left[\frac{\sqrt{t}}{2} - \frac{h}{8} \sqrt{\frac{\pi}{\alpha}} \frac{(\rho c_p)_f}{(\rho c_p)_{OB}} \times \right.$$
$$\left. \ln\left(\frac{4}{h} \sqrt{\frac{\alpha}{\pi} \frac{(\rho c_p)_{OB}}{(\rho c_p)_f}} \sqrt{t} + 1 \right) \right] \quad (56)$$

where

t = time, hr

st, f, OB = subscripts meaning steam, formation, overburden, respectively

$\underline{H}_{fg}$ = latent heat of steam at sandface injection pressure, Btu/lb$_m$

Figure 12-18 presents a schematic drawing for the injection of steam into a formation.

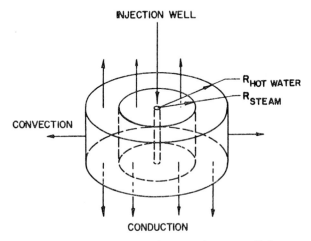

Figure 12-18. Injection of steam into a radial system (from Willman et al.[46]).

Since heat losses vertically increase as the radius of the steam zone increases, the steam injection rate could be increased to maintain steam front temperatures. The following equation permits a calculation of the steam injection rate required:

$$i_{st} = \frac{\pi R_{st}^2 (T_{st} - T_f)}{14.6 \underline{H}_{fg} t} \left[h(\rho c_p)_f + 4k \sqrt{\frac{t}{\pi \alpha}} \right] \qquad (57)$$

Equation (56) does not take into account the sensible heat which would make the answer for R_{st} conservative, or the radial conduction of heat, which would reduce the value of R_{st}. A study of the variables in equation (56) indicates that the radius of the steam zone will be maximized when the thickness of the oil-bearing formation is large, when formation permeability is high, since then the injection pressure would be lower and the steam injection rate higher.

If it is assumed that the average temperature of the hot water zone is the same as that of the steam zone, the radius of the hot water zone may be calculated from a slightly modified form of equation (56). In this instance, the latent heat of steam term, $\underline{H}_{fg}$, would be replaced by $[(\underline{H}_g)_{T_{st}} - (\underline{H}_f)_{T_f}]$. An average (ρc_p) term would be used for the steam *and* the hot water zones.

Prediction for Oil Recovery by Steam Injection. An accepted method for determining the saturation distribution during an immiscible fluid-fluid displacement is that due to Buckley and

Leverett (see Chapter 5). The applicable equation for the radial system is[46] :

$$\left[\frac{\partial(R^2)}{\partial t}\right]_{S=\text{const}} = \frac{i_w}{13.43}\left(\frac{\partial f_w}{\partial S_w}\right)_{S=\text{const}} \tag{58}$$

where

i_w = effective water injection rate (condensed steam), bbl/day

$$f_w = \frac{1}{1 + \dfrac{k_o}{k_w}\dfrac{\mu_w}{\mu_o}}$$

Since both i_w and the change of water fraction flowing, f_w, with respect to water saturation, are functions of temperature, the solution of equation (58) for the radial position of a given constant saturation would be:

$$(R^2)_{S=\text{const}} = \frac{1}{13.43}\int_0^t i_w\left(\frac{\partial f_w}{\partial S_w}\right)_{S=\text{const}} dt \tag{59}$$

The graphical solution of equation (59) has been outlined in Chapter 5 for immiscible fluid-fluid displacement in porous media. The only modification will be to assume isothermal steps, for which the water and oil viscosities can be specified (so that f_w may be determined), and the effective water injection rate, i_w, calculated from the volume of steam injected and the prevailing temperature. At least several isothermal steps should be made between the steam zone temperature and the reservoir temperature.

The effective water injection rate should normally be corrected for the rate at which the steam is displacing the water front due to steam expansion. The following equation will correct the value for Q upward:

$$i_T = \frac{(\rho_w)_{T_f}}{(\rho_w)_T}\left[i_{st} + 13.43\,\phi h(S_{st} - \Delta S_o)\frac{\partial R_{st}^2}{\partial t}\right] \tag{60}$$

where

$\Delta S_o = (S_{or})_{HWF} - (S_{or})_{st}$
HWF = subscript for hot water flood

In this instance, i_T would replace the i_w in equation (59), if this correction is to be used. A graphical integration of equation (59) may now be readily accomplished since i_w will no longer be under the integration sign.

One complication in the calculation procedure remains—that of "cylinders" of constant saturation, which initially flow faster than the leading edge of the hot water zone. Willman *et al.*[46] pointed out that in each given isothermal region between the steam zone and the reservoir which is at the initial formation temperature, the "cylinders" of constant saturation must necessarily have a constant speed since i_w or i_T and $\partial f_w/\partial S_w$ are assumed to be constant. This problem can be overcome by preparing a plot such as Figure 12-19 through the use of equation (60), where the temperatures correspond to the isothermal steps taken in dropping from the

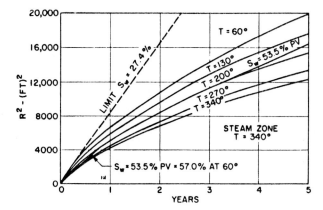

Figure 12-19. Reservoir temperature distribution with radial distance versus time (from Willman *et al.*[42]).

steam temperature to the reservoir temperature. A plot on the same figure of the radial location of a given saturation from equation (59) versus the injection time, as for the illustrated water saturation of 53.5 per cent, results in an intersection with the constant temperature lines. This intersection occurs because the tempera-discontinuities are slowing down. When the intersection is obtained, the values of $(\partial f_w/\partial S_w)_{S=\text{const}}$ and i_w corresponding to the new and lower temperature are used. Figure 12-20 is the type of performance that would be obtained in a radial homogeneous system, using this type of steam flood performance calculation.

The primary difficulty with the above-outlined method for predicting the performance of a steam flooding project is that no system is strictly radial. Many floods will be made on a pattern basis, perhaps a five-spot or a line-drive. Chapter 7 outlines a

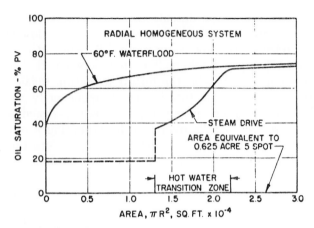

Figure 12-20. Reservoir saturation distributions in a radial system after a period of injection (from Willman *et al.*[46]).

procedure which can be used with some modification, proposed by Higgins and Leighton, using the concept of shape factors and the Buckley-Leverett type fluid-fluid displacement. The approach could be readily modified for the case of steam injection, where the isothermal displacements would be treated in the manner outlined above, but for the linear case with the shape factors applied. Then too, Chapter 4 outlined techniques for correction for the observation that the producing wells may not be true pressure sinks due to the fact that many reservoirs containing a viscous crude have been, or are, pressure depleted. In this instance, oil production at the producing wells cannot begin until the "pressure wave," or moving oil front has moved to the producing wellbore(s). If a pilot test of the steam flooding technique is to be attempted, methods discussed in Chapter 4 will also permit an estimate to be made of performance to be expected on an enclosed pattern basis, even though the pilot flood well pattern is not "enclosed."

Both Willman *et al.*[46] and Fournier[49] have presented calculation methods by means of which a prediction of oil recovery by hot water injection may be estimated. The Fournier method takes into account thermal expansion of the oil, heat losses to adjacent strata and the effect of viscosity ratio reduction. The analytical development is for a radial system, and would require modification to properly reflect the reservoir performance which would be expected for a given producing and injection well pattern. Then too, reservoir heterogeneities and formation stratification have not been treated, but could be incorporated by methods presented in Chapter 7. The Willman method has the advantage of simplicity and has the potential, when carefully used, of providing good engineering answers to the problem of oil recovery with hot water injection.

The reader is referred to the original work of Willman *et al.* and of Fournier for calculation details, as well as for the reference list which these technical papers provide.

Estimation of Hot-Water Drive Performance

The preceding section on heat loss in a formation will permit an estimate to be made of the prevailing temperature in a system at various points and at various times of injection. The procedures of Chapter 7, for the prediction of water flooding performance, may then be used for the proper oil-water viscosity ratio corresponding to the various temperatures prevailing in the reservoir. Since the temperature of the hot water flood will not be constant in the system, either an average temperature should be used or a temperature and resulting performance applied to sections of the reservoir. Then, a weighted average performance could be calculated, based upon relative reservoir areas or relative reservoir volumes contacted by the water, at various temperatures. To this recovery should be added the additional recovery due to the thermal swelling of the water-flood residual oil. The coefficient for expansion may be obtained from a PVT analysis from that section of the oil curve where the temperature is dropped from the reservoir value to laboratory temperature in the determination of the oil formation volume factor curve. The most accurate value for the thermal expansion of a given oil would be that determined in a laboratory where the oil contains the proper amount of gas in solution, at the approximate operating pressure of the steam flood. An estimate of the remaining or residual oil saturation would then be that bypassed by the hot water flood, as predicted by the Buckley-Leverett theory, less the contraction which would occur as the reservoir is cooled down to the initial geothermal temperature.

Estimation of Oil Recoveries by Cyclic Steam Injection

Considerable success has been reported[50,51,52] for the cyclic steam injection, "huff-and-puff" or "steam soak" methods for oil recovery where steam is used. Steam has also been circulated around a packer set high in thick oil-saturated sand sections. Figure 12-21 illustrates these two methods for the recovery of a viscous oil. The "huff-and-puff" presently enjoys the greatest popularity among the steam using methods for oil recovery, since no great amount of engineering talent is required for the pilot testing of the technique on a given heavy oil producing lease. The steam generator and water treating equipment may be rented on a day-to-day basis, and only minor modifications at most[51] made to the wells before steaming is begun. The "huff-and-puff" well steaming method benefits the oil producing rate of the reservoir in two main ways: (1) removes accumulated asphaltic and/or paraffinic deposits which have decreased the permeability in the vicinity of the sandface, and (2) radically decreases the oil viscosity in those parts of the reservoir which receive significant quantities of heat.

It is instructive to consider the productivity increase which could be attributed to the decrease in oil viscosity due to reservoir heating by the "huff-and-puff" well steaming technique. If for simplicity, it may be assumed that the reservoir has been

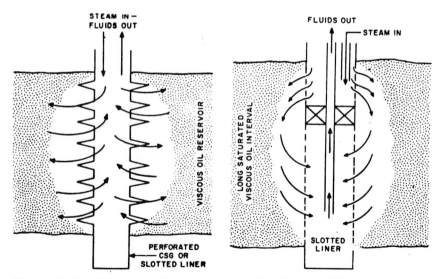

Figure 12-21. Schematic diagrams for the "huff-and-puff" and steam circulation around a packer technique for the recovery of heavy oils (from Kastrop[53]).

heated out to a radius r to a uniform temperature, then the heated oil viscosity out to this radius could be represented by $\mu_{o_{hot}}$. The pressure drop between the drainage radius, r_e, and the wellbore radius, r_w, could be written as:

$$(\Delta p)_{hot} = (p_e - p) + (p - p_w) \tag{61}$$

where p corresponds to the pressure at radius r. The Darcy's Law expression for pressure drop in a radial system in the unheated reservoir is:

$$(\Delta p)_{cold} = \frac{q_{o_{cold}} \, \mu_{o_{cold}} \, \ln (r_e/r_w)}{7.07 \, k_o \, h} \tag{62}$$

where the units are pressure in psi, viscosity in cp, permeability in darcys, and pay thickness in feet. The Darcy's Law expression for the heated reservoir case is, in view of the statement of equation (61):

$$(\Delta p)_{hot} = \frac{q_{o_{cold}} \, \mu_{o_{cold}} \, \ln (r_e/r)}{7.07 \, k_o \, h} + \frac{q_{o_{hot}} \, \mu_{o_{hot}} \, \ln (r/r_w)}{7.07 \, k_o \, h} \tag{63}$$

Assuming that the pressure drop across the radial system will be the same for either the hot or cold reservoir case, equations (61) and (62) can be rearranged to yield the following relationship:

$$\frac{q_{o_{hot}}}{q_{o_{cold}}} = \frac{\mu_{o_{cold}} \, \ln (r_e/r_w)}{\mu_{o_{hot}} \, \ln (r/r_w) + \mu_{o_{cold}} \, \ln (r_e/r)} \tag{64}$$

The substitution of reasonable values for hot and cold viscosities, for the wellbore radius, the heated zone radius and the drainage radius, into equation (64), will show that a productivity increase in excess of 3 is unlikely to occur through the mechanism of oil viscosity reduction in "huff-and-puff" well steaming methods. The larger productivity increases, which a number of field projects have realized, are due to the improvement of the permeability of the oil-producing sand in the vicinity of the wellbore.

Since the primary function of the injected steam is to decrease the viscosity of the oil, it is evident that results will be best where oil saturation and porosity is high, water saturation is low, where no endothermic reactions take place with the rock, where pay sections are relatively thick and where depths are sufficiently shallow so that heat losses in the wellbore are minimized. The depth to the heavy oil reservoir must be great enough so that an adequate pressure can be applied to inject sufficient daily volumes of steam. Adequate reservoir pressure must exist, or gravity drainage be possible, so that a return flow of condensed steam and oil to the wellbore will occur. It is well to note that the "huff-and-puff" steaming method is not, strictly speaking, a secondary oil recovery technique, since no physical displacement of oil by an injected fluid from one well to another is occurring.

In the interest of confining most of the injected heat to the oil-bearing zone, a reservoir containing highly permeable sand sections in the middle of the section would be the best candidate for the "huff-and-puff" steaming technique. The injected steam would then only displace a small fraction of the oil away from the wellbore, and would transfer heat vertically to oil-bearing strata rather than to the over and underlaying formations, as would occur if the sand was homogeneous with constant permeability throughout the section. The amount of steam injected will depend upon the stratification present, the pay thickness and the number of wells which a steam generator of a given size must service. If a reasonable estimate of the section accepting steam can be made, the maximum radius of the injected steam would be:

$$R = \sqrt{\frac{5.615\ V_{st}}{\pi\,h\,\phi\,(1 - S_w - S_{or})}} \qquad (65)$$

Equation (65) assumes a radial distribution of the steam with an interval accepting steam of h feet, with an oil saturation remaining in the steam-displaced zone of S_{or}, and with no condensation of the steam occurring. Of course, a decrease in the pressure of the steam would result in an increase in the radius above that calculated; condensation at the leading edge would result in a hot water zone, decreasing the calculated radius.

The following heat balance equation, based upon adiabatic heating, may also be used for estimating the radius to which steam can be injected:

$$R = \sqrt{\frac{\text{Heat Input}}{\pi\,h\,M\,(\Delta T)}} \qquad (66)$$

where the heat input is in Btu's injected at the sandface of the input well and where:

$M = (1 - \phi)\,\rho_f\,c_f + S_w\,\phi\,\rho_w\,c_w + S_o\,\phi\,\rho_o\,c_o$, Btu/ft^3-°F

ΔT = difference between steam and formation temperatures, °F

The pay thickness, h, is the same as that defined in equation (65). It is suggested that the radius to which steam can be injected be calculated by both equations (65) and (66). Careful consideration of the relative accuracy of the data used in the generation of the radius values will indicate the proper weight which should be attached to each calculated value.

Where the oil-bearing section is thick and the reservoir pressure low, the producing mechanism is predominately that due to gravity. The thermal reduction of the oil viscosity then permits a more rapid drainage of the oil into the wellbore.

The preceding sections permit an estimate to be made of the time for heat to be either partly or totally-dissipated in the formation. Ideally, either superheated or saturated steam would be injected, sufficient steam "soaking" time allowed for all the latent heat of the steam to be recovered, and then oil production begun after only hot water at the temperature of the saturated steam remains. In this way, the maximum available heat of the steam would be at the lowest possible value. This would maximize the oil producing rate from the treated well.

In the thick, oil-saturated sections, the steam-soak method should be successful for a number of cyclic steam-injection operations even if pressure is very low, since oil production by gravity drainage would be attractive. Where the treated sand is thin and the reservoir pressure low, the number of cycles which are possible would be less since radial distances over which the oil would have to move the wellbore after displacement by the steam would become excessively large. No theoretical analysis for the number of steam cycles possible has yet appeared in the literature. All operations are presently tested by pilots in a section of the field of possible application.

Obviously, any successful "huff-and-puff" steaming operation will be affected by the value received for the treated oil, by the cost to make the recovered oil marketable and by the cost per barrel for injected steam. Both natural gas and lease crude have been used for fueling the steam generators. A source of good inexpensive water for steam generation is also important if a project is to be economically attractive. Figures 12-22 and 12-23 provide a convenient means for estimating the approximate cost of generating 10,000,000 Btu/hr of heat through burning of natural gas and crude oil, respectively. The actual heat output of a steam generator would be determined by the amount of fuel consumed, and the per cent thermal efficiency of the generating system.

COST OF LIBERATING 10,000,000 BTU PER HOUR BY
BURNING NATURAL GAS

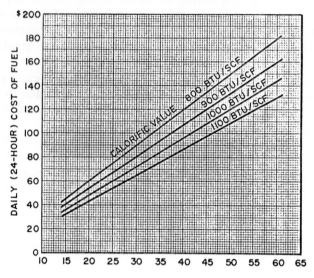

FUEL GAS PRICE, Cents Per MSCF

Figure 12-22. Natural gas fuel costs in steam generation (from *The Petr. Engr.*[54]).

COST OF LIBERATING 10,000,000 BTU PER HOUR BY
BURNING 10° TO 25° API GRAVITY CALIFORNIA CRUDE OILS

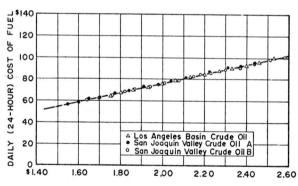

CRUDE OIL PRICE, ¢ PER BBL.

Figure 12-23. Crude oil fuel costs in steam generation (from *The Petr. Engr.*[54]).

Water Quality for the Generation of Steam

The treatment of water to be used for steam generation in the "once-through" steam generators[55,56,57] is somewhat different from the treatment required to make a suitable injection water as discussed in Chapter 8. In conventional steam systems, the system water is re-circulated time and again, with only small additions of outside water added to replace that consumed. Any solids in the water are periodically removed by blowing down the boiler. In the oil field, the only requirement of a steam generator is that it provide heat for injection into an oil reservoir. In actual practice, it is customary to produce steam of approximately 80 per cent quality, that is, 80 per cent by weight vapor. The remaining 20 weight per cent is hot liquid, which carries hopefully all the feedwater constituents consisting of soluble salts, now concentrated 5 times, out of the boiler.

In general, boiler feedwater for steam generators used in steaming operations must meet the following requirements[55]:

(1) Be essentially free of non-ionic suspended matter, having less than 5 ppm turbidity, 0.4 ppm iron, 0.1 ppm manganese, 0.1 ppm sulphides or hydrogen sulfide gas, zero organic matter, and zero oil.

(2) Have zero hardness so that scales formed by calcium and magnesium ions will not form in the steam generator.

(3) Be free of oxygen and carbon dioxide to avoid corrosion in the vaporizers.

The most common method for the removal of hardness in the feedwater is by sodium zeolite cation exchange, which replaces the objectionable calcium and magnesium ions by sodium ions. The system is regenerated with sodium chloride, and the magnesium and calcium ions are flushed to waste. To avoid the possibility of exceeding the capacity of the water softening unit, it is common practice to place a second softening unit in series with the first which is termed the "polishing" unit. Figure 12-24 shows a schematic layout of a typical water treating plant employing automatic devices for checking treated water hardness and oxygen concentration, and for switching and regenerating the water softening units. Obviously, such a water softening system would have difficulty processing a brackish water, due to the tendency of the sodium in the water to "back regenerate" the zeolite resin bed

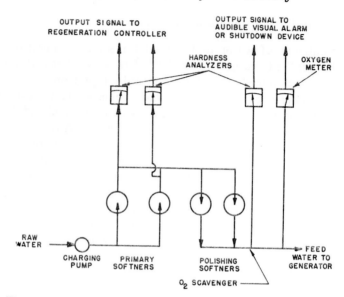

Figure 12-24. Schematic diagram of a typical water softening plant (from Fincher *et al.*[58]).

and permit the hardness to pass through. If the total $CaCO_3$ hardness exceeds 3000 ppm, it will usually be necessary to resort to a lime-soda softening process[57]. Oxygen in small concentrations is removed by chemical "oxygen scavengers," such as hydrazine (N_2H_4) or sodium sulfite (Na_2SO_3). The removal of other foreign materials is usually accomplished upstream from the softening units by methods such as are treated in Chapter 8.

Treatment of the Produced Oil

Many of the steam and in situ combustion projects have produced tight oil and water emulsions. The two main requirements for the formation of a stable emulsion are[59]:

(1) Mixing energy to produce water-in-oil or oil-in-water emulsions where the internal phase has a diameter not exceeding 25 microns.

(2) Surface active agents to prevent the coalescence of the internal phase.

Surface active agents such as asphaltenes, resins, porphyrin-metal compounds and finely dispersed solids[60] are present in most crudes. Sufficient mixing energy is applied during the production

(cont'd on p. 488)

TABLE 12-9A. Properties of Metals (Eckert and Drake[61]).

Metals	Properties at 68°F				k, thermal conductivity, Btu/hr-ft°F						
	ρ lb$_m$/ft³	$\frac{c_p}{\text{lb}_m\,°F}$ Btu	$\frac{k}{\text{hr-ft°F}}$ Btu	$\frac{\alpha}{\text{hr}}$ ft²	32F 0C	212F 100C	392F 200C	572F 300C	752F 400C	1112F 600C	1472F 800C
Aluminum:											
Pure	169	0.214	118	3.665	117	119	124	132	144		
Al-Cu (Duralumin) 94-96 Al, 3-5 Cu, trace Mg	174	0.211	95	2.580	92	105	112				
Al-Mg (Hydronalium) 91-95 Al, 5-9 Mg	163	0.216	65	1.860	63	73	82				
Al-Si (Silumin) 87 Al, 13 Si	166	0.208	95	2.773	94	101	107				
Al-Si (Silumin, copper bearing) 86.5 Al, 12.5 Si, 1 Cu.	166	0.207	79	2.311	79	83	88	93			
Al-Si (Alusil) 78-80 Al, 20-22 Si	164	0.204	93	2.762	91	97	101	103			
Al-Mg-Si 97 Al, 1 Mg, 1 Si, 1 Mn	169	0.213	102	2.859	101	109	118				
Lead	710	0.031	20	0.924	20.3	19.3	18.2	17.2			
Iron:											
Pure	493	0.108	42	0.785	42	39	36	32	28	23	21
Wrought iron (C < 0.5%)	490	0.11	34	0.634	34	33	30	28	26	21	19
Cast iron (C ≈ 4%)	454	0.10	30	0.666							
Steel (C max ≈ 1.5%)											
Carbon steel C ≈ 0.5%	489	0.111	31	0.570	32	30	28	26	24	20	18
1.0%	487	0.113	25	0.452	25	25	24	23	21	19	17
1.5%	484	0.116	21	0.376	21	21	21	20	19	18	16
Nickel steel Ni ≈ 0%	493	0.108	42	0.785							
10%	496	0.11	15	0.279							
20%	499	0.11	11	0.204							
30%	504	0.11	7	0.118							
40%	510	0.11	6	0.108							
50%	516	0.11	8	0.140							
60%	523	0.11	11	0.182							
70%	531	0.11	15	0.258							
80%	538	0.11	20	0.344							
90%	547	0.11	27	0.452							
100%	556	0.106	52	0.892							

TABLE 12-9A. Cont.

Metals	Properties at 68°F				k, thermal conductivity, Btu/hr-ft°F						
	ρ lb$_m$/ft³	c_p Btu/lb$_m$°F	k Btu/hr-ft°F	α ft²/hr	32F 0C	212F 100C	392F 200C	572F 300C	752F 400C	1112F 600C	1472F 800C
Invar Ni = 36%	508	0.11	6.2	0.108							
Chrome Steel Cr = 0%	493	0.108	42	0.785	42	39	36	32	28	23	21
1%	491	0.11	35	0.645	36	32	30	27	24	21	19
2%	491	0.11	30	0.559	31	28	26	24	22	19	18
5%	489	0.11	23	0.430	23	22	21	21	19	17	17
10%	486	0.11	18	0.344	18	18	18	17	17	16	16
20%	480	0.11	13	0.258	13	13	13	13	14	14	15
30%	476	0.11	11	0.204							
Cr-Ni (chrome-nickel): 15 Cr, 10 Ni	491	0.11	11	0.204							
18 Cr, 8 Ni(V2A)	488	0.11	9.4	0.172	9.4	10	10	11	11	13	15
20 Cr, 15 Ni	489	0.11	8.7	0.161							
25 Cr, 20 Ni	491	0.11	7.4	0.140							
Ni-Cr (nickel-chrome): 80 Ni, 15 Cr	532	0.11	10	0.172							
60 Ni, 15 Cr	516	0.11	7.4	0.129							
40 Ni, 15 Cr	504	0.11	6.7	0.118							
20 Ni, 15 Cr	491	0.11	8.1	0.151	8.1	8.7	8.7	9.4	10	11	13
Cr-Ni-Al: 6 Cr, 1.5 Al, 0.5 Si (Sicromal 8)	482	0.117	13	0.237							
24 Cr, 2.5 Al, 0.5 Si (Sicromal 12)	479	0.118	11	0.194							
Manganese steel Mn = 0%	493	0.118	42	0.784	22	21	21	21	20	19	
1%	491	0.11	29	0.538							
2%	491	0.11	22	0.376							
3%	490	0.11	13	0.247							
10%	487	0.11	10	0.194							
Tungsten steel W = 0%	493	0.108	42	0.785	36	34	31	28	26	21	
1%	494	0.107	38	0.720							
2%	497	0.106	36	0.677							
5%	504	0.104	31	0.591							
10%	519	0.100	28	0.527							
20%	551	0.093	25	0.484							

Material				223	219	216	213	210	204	36
Silicon steel Si = 0%	493	0.108	42	0.785						
1%	485	0.11	24	0.451						
2%	479	0.11	18	0.344						
5%	463	0.11	11	0.215						
Copper:										
Pure	559	0.0915	223	4.353		41				
Aluminum bronze 95 Cu, 5 Al	541	0.098	48	0.903	34	74	83	85	85	
Bronze 75 Cu, 25 Sn	541	0.082	15	0.333		18	23	26	28	
Red brass 85 Cu, 9 Sn, 6 Zn	544	0.092	35	0.699			15			
Brass 70 Cu, 30 Zn	532	0.092	64	1.322						
German Silver 62 Cu, 15 Ni, 22 Zn	538	0.094	14.4	0.290						
Constantan 60 Cu, 40 Ni	557	0.098	13.1	0.237		12.8				
Nickel:										
Pure (99.9%)	556	0.1065	52	0.882	54	48	42	37	34	32	36
Impure (99.2%)	556	0.106	40	0.677	40	37	34	32	30	
Ni-Cr 90 Ni, 10 Cr	541	0.106	10	0.172	9.9	10.9	12.1	13.2	14.2	13.0
80 Ni, 20 Cr	519	0.106	7.3	0.129	7.1	8.0	9.0	9.9	10.9	

TABLE 12-9B.   Properties of Gases at Atmospheric Pressure
(Eckert and Drake[61]).

T_F	ρ, lb_m/ft^3	c_p, $Btu/lb_m\,°F$	μ, $lb_m/sec\ ft$	k, $Btu/hr\text{-}ft\,°F$	α, ft^2/hr	N_{Pr}	$a \times 10^3$
			Air				
80	0.0735	0.2402	1.241×10^{-5}	0.01516	0.8587	0.708	1.85
170	0.0623	0.2410	1.394	0.01735	1.156	0.697	1.59
260	0.0551	0.2422	1.536	0.01944	1.457	0.689	1.39
350	0.0489	0.2438	1.669	0.02142	1.636	0.683	1.23
440	0.0440	0.2459	1.795	0.02333	2.156	0.680	1.11
530	0.0401	0.2482	1.914	0.02519	2.531	0.680	1.01
620	0.0367	0.2520	2.028	0.02692	2.911	0.680	0.96
710	0.0339	0.2540	2.135	0.02862	3.324	0.682	0.85
800	0.0314	0.2568	2.239	0.03022	3.748	0.684	0.79
890	0.0294	0.2593	2.339	0.03183	4.175	0.686	
980	0.0275	0.2622	2.436	0.03339	4.631	0.689	
1070	0.0259	0.2650	2.530	0.03483	5.075	0.692	
1160	0.0245	0.2678	2.620	0.03628	5.530	0.696	
1250	0.0232	0.2704	2.703	0.03770	6.010	0.699	
1340	0.0220	0.2727	2.790	0.03901	6.502	0.702	
1520	0.0200	0.2772	2.955	0.04178	7.536	0.706	
			Nitrogen				
-280	0.2173	0.2561	4.611×10^{-6}	0.005460	0.09811	0.786	
-100	0.1068	0.2491	8.700	0.01054	0.3962	0.747	
80	0.0713	0.2486	11.99	0.01514	0.8542	0.713	
260	0.0533	0.2498	14.77	0.01927	1.447	0.691	
440	0.0426	0.2521	17.27	0.02302	2.143	0.684	
620	0.0355	0.2569	19.56	0.02646	2.901	0.686	
800	0.0308	0.2620	21.59	0.02960	3.668	0.691	
980	0.0267	0.2681	23.41	0.03241	4.528	0.700	
1160	0.0237	0.2738	25.19	0.03507	5.404	0.711	
1340	0.0213	0.2789	26.88	0.03741	6.297	0.724	
1520	0.0194	0.2832	28.41	0.03958	7.204	0.736	
1700	0.0178	0.2875	29.90	0.04151	8.111	0.748	
			Carbon dioxide				
-64	0.1544	0.187	7.462×10^{-6}	0.006243	0.2294	0.818	
-10	0.1352	0.192	8.460	0.007444	0.2868	0.793	
80	0.1122	0.208	10.051	0.009575	0.4103	0.770	
170	0.0959	0.215	11.561	0.01183	0.5738	0.755	
260	0.0838	0.225	12.98	0.01422	0.7542	0.738	
350	0.0744	0.234	14.34	0.01674	0.9615	0.721	
440	0.0670	0.242	15.63	0.01937	1.195	0.702	
530	0.0608	0.250	16.85	0.02208	1.453	0.685	
620	0.0558	0.257	18.03	0.02491	1.737	0.668	
			Carbon monoxide				
-64	0.09699	0.2491	9.295×10^{-6}	0.01101	0.4557	0.758	
-10	0.0525	0.2490	10.35	0.01239	0.5837	0.750	
80	0.07109	0.2489	11.990	0.01459	0.8246	0.737	
170	0.06082	0.2492	13.50	0.01666	1.099	0.728	
260	0.05329	0.2504	14.91	0.01864	1.397	0.722	
350	0.04735	0.2520	16.25	0.0252	1.720	0.718	

TABLE 12-9B. Cont.

T, °F	ρ lb_m/ft^3	c_p, Btu/lb_m °F	μ $lb_m/sec\ ft$	k, $Btu/hr\text{-}ft°F$	α, ft^2/hr	N_{Pr}	$a \times 10^3$
440	0.04259	0.2540	17.51	0.02232	2.063	0.718	
530	0.03872	0.2569	18.74	0.02405	2.418	0.721	
620	0.03549	0.2598	19.89	0.02569	2.786	0.724	
			Steam (H_2O vapor)				
224	0.0366	0.492	8.54×10^{-6}	0.0142	0.789	1.060	
260	0.0346	0.481	9.03	0.0151	0.906	1.040	
350	0.0306	0.473	10.25	0.0173	1.19	1.010	
440	0.0275	0.474	11.45	0.0196	1.50	0.996	
530	0.0250	0.477	12.66	0.0219	1.84	0.991	
620	0.0228	0.484	13.89	0.0244	2.22	0.986	
710	0.0211	0.491	15.10	0.0268	2.58	0.995	
800	0.0196	0.498	16.30	0.0292	2.99	1.000	
890	0.0183	0.506	17.50	0.0317	3.42	1.005	
980	0.0171	0.514	18.72	0.0342	3.88	1.010	
1070	0.0161	0.522	19.95	0.0368	4.38	1.019	

TABLE 12-9C. Thermal Conductivities of Some Hydrocarbon Gases (Perry[37]).

Constituent	Measuring Temperature, °F	k, Btu/hr-ft-° F
Air	32	0.0140
	212	0.0183
	392	0.0226
	572	0.0265
n-Butane	32	0.0078
	212	0.0135
iso-Butane	32	0.0080
	212	0.0139
Carbon Dioxide	32	0.0085
	212	0.0133
	392	0.0181
	572	0.0228
Ethane	32	0.0106
	212	0.0175
Methane	32	0.0175
	122	0.0215
Methylene Chloride	32	0.0039
	115	0.0049
	212	0.0063
	413	0.0095

TABLE 12-9C. Cont.

Constituent	Measuring Temperature, °F	k, Btu/hr-ft-°F
Nitrogen	32	0.0140
	122	0.0160
	212	0.0180
Oxygen	32	0.0142
	122	0.0164
	212	0.0185
n-Pentane	32	0.0074
	68	0.0083
iso-Pentane	32	0.0072
	212	0.0127
Propane	32	0.0087
	212	0.0151
Water Vapor, zero pressure	32	0.0132
	200	0.0159
	400	0.0199
	600	0.0256
	800	0.0306
	1000	0.0495

TABLE 12-9D. Properties of Selected Non-Metallic Solids (Marks[38] and Nelson[18]).

Material	T, °F	ρ, lb_m/ft^3	c_p, $\dfrac{Btu}{lb_m\,°F}$	k, $\dfrac{Btu}{hr\text{-}ft\,°F}$	α, $\dfrac{ft^2}{hr}$
Aerogel, silica	248	8.5		0.013	
Asbestos	−328	29.3		0.043	
Asbestos	32	29.3		0.090	
Asbestos	32	36.0	0.195	0.087	
Asbestos	212	36.0	0.195	0.111	
Asbestos	392	36.0		0.120	
Asbestos	752	36.0		0.129	
Asbestos	−328	43.5		0.090	
Asbestos	32	43.5		0.135	
Brick, dry	68	110–113	0.20	0.22–0.30	0.011–0.013
Bakelite	68	79.5	0.38	0.134	0.0044
Cardboard, corrugated				0.037	
Clay	68	91.0	0.21	0.739	0.039
Concrete	68	119–144	0.21	0.47–0.81	0.019–0.027
Coal, anthracite	68	75–94	0.30	0.15	0.005–0.006
Coal, powdered	86	46.0	0.31	0.067	0.005
Cotton	68	5.0	0.31	0.034	0.075
Cork, board	86	10.0		0.025	
Cork, expanded scrap	68	2.8–7.4	0.45	0.021	0.006–0.017
Cork, ground	86	9.4		0.025	
Diatomaceous earth	100	20.0		0.036	

TABLE 12-9D. Cont.

Material	T, °F	ρ, lb_m/ft^3	c_p, $\dfrac{Btu}{lb_m\ °F}$	k, $\dfrac{Btu}{hr\text{-}ft\ °F}$	α, $\dfrac{ft^2}{hr}$
Diatomaceous earth	1600	20.0		0.082	
Earth, coarse gravelly	68	128.0	0.44	0.30	0.0054
Felt, wool	86	20.6		0.03	
Fiber, insulating board	70	14.8		0.028	
Fiber, red	68	80.5		0.27	
Glass plate	68	169.0	0.2	0.44	0.013
Glass, borosilicate	86	139.0		0.63	
Glass, wool	68	12.5	0.16	0.023	0.011
Granite				1.0–2.3	
Ice	32	57.0	0.46	1.28	0.048
Kapok				0.023	
Limestone (dry)				0.3–0.75	
Marble	68	156–169	0.193	1.6	0.054
Rubber, hard	32	74.8		0.087	
Sandstone	68	135–144	0.17	0.94–1.2	0.041–0.049
Sawdust (dry)				0.042	
Silk	68	3.6	0.33	0.021	0.017
Soil, sandy, dry, 24-in cover				0.24–0.4	
Soil, sandy, moist, 24-in cover				0.5–0.6	
Soil, sandy, soaked, 24-in cover				1.1–1.3	
Soil, sandy, dry, 8-in cover				0.6–0.7	
Soil, sandy, moist, 8-in cover				1.2–2.4	
Soil, clay, dry, 24-in cover				0.2–0.3	
Soil, clay, moist, 24-in cover				0.4–0.6	
Soil, clay, wet, 24-in cover				0.6–0.9	
Wood, oak radial	68	38–50	0.57	0.10–0.12	0.0043–0.0047
Wood, fir (20% moisture) radial	68	26.0–26.3	0.65	0.08	0.0048

TABLE 12-9E.   Thermal Conductivities of Solids for Varied Temperatures Btu/hr-ft-°F (Marks[38] and Nelson[16]).

Material	Mean temp, F										Limiting use temp, F
	100	200	300	400	500	600	800	1000	1500	2000	
Asbestos (36 lb/ft³) laminated asbestos felt	0.097	0.110	0.117	0.121	0.123	0.125	0.130				
Approx 40 laminations/in	0.033	0.037	0.040	0.044	0.048						700
Approx 20 laminations/in	0.045	0.050	0.055	0.060	0.065						500
Corrugated asbestos (4 plies/in)	0.050	0.058	0.069								300
85% magnesia (13 lb/ft³)	0.034	0.036	0.038	0.040							600
Diatomaceous earth, asbestos and bonder	0.045	0.047	0.049	0.050	0.053	0.055	0.060	0.065			1600
Diatomaceous earth brick	0.054	0.056	0.058	0.060	0.063	0.065	0.069	0.073			1600
Diatomaceous earth brick	0.127	0.130	0.133	0.137	0.140	0.143	0.150	0.158	0.176		2000
Diatomaceous earth brick	0.128	0.131	0.135	0.139	0.143	0.148	0.155	0.163	0.183	0.203	2500
Diatomaceous earth powder (density, 18 lb/ft³)	0.039	0.042	0.044	0.048	0.051	0.054	0.061	0.068			
Magnesium 85%, insulation		0.037		0.043							
Mineral wool, insulating cement		0.058		0.067		0.075	0.083				
Rock wool	0.030	0.034	0.039	0.044	0.050	0.057					
Rock wool blanket		0.034		0.044		0.054					
Superex, Johns-Mansville		0.054		0.057			0.064				
Unibestos, Standard, Union Asbestoes Co.		0.038		0.04		0.049					
Unibestos, Super, Union Asbestoes Co.							0.059	0.068			
Vermiculite		0.044		0.052			0.068				

TABLE 12-9F. Physical Properties of Water at Atmospheric or
Saturation Pressures (after Gröber *et al.*[39]).

T	p	ρ	c_p	H_{fg}	k	μ	$α × 10^2$	N_{Pr}	$a × 10^3$
32	14.697	62.42	1.0074	1075.8*	0.319	3.335	0.508	13.67	−0.039
40		62.42	1.0048	1071.3*	0.325	3.713	0.519	11.46	+0.010
60		62.35	0.9999	1059.9*	0.341	2.760	0.546	8.11	0.084
80		62.20	0.9983	1048.6*	0.352	2.057	0.566	5.84	0.151
100		61.99	0.9979	1037.2*	0.361	1.647	0.584	4.55	0.203
120		61.70	0.9985	1025.8*	0.370	1.349	0.600	3.64	0.250
140		61.37	0.9994	1014.1*	0.376	1.130	0.614	3.00	0.291
160		61.00	1.0009	1002.3*	0.382	0.963	0.625	2.53	0.328
180		60.57	1.0028	990.2*	0.387	0.831	0.637	2.15	0.363
200		60.11	1.0056	977.9*	0.391	0.733	0.648	1.88	0.398
212	↓	59.83	1.0070	970.3	0.394	0.684	0.653	1.75	0.418
240	24.97	59.10	1.013	952.2	0.395	0.586	0.661	1.50	0.467
280	49.20	57.94	1.023	924.7	0.395	0.487	0.667	1.26	0.534
320	89.66	56.66	1.037	894.9	0.394	0.415	0.671	1.09	0.610
360	153.0	55.22	1.055	862.2	0.389	0.362	0.669	0.98	0.694
400	247.3	53.65	1.080	826.0	0.383	0.322	0.661	0.91	0.797
440	381.6	51.92	1.113	785.4	0.373	0.293	0.647	0.87	0.932
480	566.1	50.00	1.157	739.4	0.361	0.270	0.624	0.87	1.10
520	812.4	47.85	1.225	686.4	0.344	0.249	0.584	0.89	1.35
560	1133	45.25	1.343	624.2	0.320	0.231	0.524	0.97	1.76
600	1543	42.37	1.51	548.5	0.291	0.212	0.452	1.11	2.36
640	2060	38.46	1.94	452.0	0.252	0.189	0.338	1.45	4.14
680	2708	32.79	3.20	309.9	0.202	0.165	0.194	2.59	
705.4	3206.2	19.9	∞	0	0.121	0.122	0	∞	∞

*At saturation temperature
T = temperature, °F
p = pressure, psia.
$ρ$ = density, lb_m/cu ft.
c_p = specific heat at constant pressure, Btu/lb_m-°F.
H_{fg} = latent heat of vaporization, Btu/lb_m
k = thermal conductivity, Btu/hr-ft-°F
$μ$ = viscosity, lb_m/ft-hr.
$α$ = thermal diffusivity, sq ft/hr.
N_{Pr} = Prandtl number.
a = coefficient of thermal expansion, vol/vol-°F.

process to form the emulsion. Emulsions formed at the high temperatures encountered in thermal recovery processes are unusually stable, and can present considerable problems in preparing a crude for sale at low treating cost. Usually the common wash tanks, heater-treaters and electric dehydrators will be inadequate to treat out stable emulsified water without chemical aid. The selection of the proper chemical is usually accomplished by bottle testing techniques, as is common to the oil-producing industry.

References

1. Directory of Thermal Oil Recovery Projects, *Petrol. Engr.*, 86 (May, 1965).
2. Journal Staff Study, Thermal Recovery, *Oil Gas J.*, vol. 62, no. 42, 75 (October 19, 1964).
3. Perry, G. T., and Warner, W. S., U. S. Patent No. 48,584, July 4, 1865.
4. Snelling, W. O., U. S. Patent No. 1,104,011, July 21, 1914.
5. Mills, R., *Natl. Petrol. News* **15**, No. 56, 58 (1923).
6. Smith, E. E., *Natl. Petrol. News* **17**, No. 4, 55 (1925).
7. Melby, T. S., **AIME** Tech. paper 575-G, presented at fall mtg., So. California Petr. Sec., Los Angeles, Oct. 20–21, 1955.
8. Schild, A., *Petrol. Trans. AIME* **210**, 1 (1957).
9. Walker, E. W., *Petrol. Engr.* **26**, No. 13, B-36 (December, 1954).
10. Nelson, T. W., and McNiel, J. S., Jr., "Fundamentals of Thermal Oil Recovery," p. 26, The Petroleum Engineer Publishing Co., 1965.
11. Stovall, S. L., *Oil Weekly* **74**, No. 9, 17 (August 3, 1934).
12. Breston, J. N., and Pearman, B. R., *Producers Monthly* **18**, No. 1, 15 (November, 1953).
13. Tucker, M., *Oil Gas J.* **35**, No. 37, 172 (January 28, 1937).
14. Anon., *Oil Gas J.* **43**, No. 58, 80 (June 11, 1956).
15. Bleakley, W. B., *Oil Gas J.*, **75** (January 11, 1965), 121 (February 15, 1965).
16. Keenan, J. H., and Keyes, F. G., "Thermodynamic Properties of Steam," New York, John Wiley & Sons, Inc., 1936.
17. Ramey, H. J., Jr., "Fundamentals of Thermal Oil Recovery," p. 165, Dallas, The Petroleum Engineer Publishing Co., 1965.
18. Nelson, W. L., "Petroleum Refinery Engineering," chap. 17 & 18, 4th Ed., New York, McGraw-Hill Book Co., Inc., 1958.
19. Spalding, D. B., and Cole, E. H., "Engineering Thermodynamics," chap. 8, New York, McGraw-Hill Book Co., Inc., 1959.
20. Short, B. E., Kent, H. L., Jr., and Treat, B. F., "Engineering Thermodynamics," chap. 9, New York, Harper & Brothers, 1953.
21. Hutchinson, F. W., "Industrial Heat Transfer," New York, The Industrial Press, 1952.
22. Ramey, H. J., Jr., *Petr. Trans. AIME* **225**, 427 (1962).
23. Nowak, T. J., *Petr. Trans. AIME* **198**, 203 (1953).

24. Moss, J. T., and White, P. D., *Oil Gas J.* **57**, No. 11, 174 (March 9, 1959).
25. Lesem, L. B., Greytok, F., Marotta, F., and McKetta, J. J., *Petr. Trans. AIME* **210**, 169 (1957).
26. Squier, D. P., Smith, D. D., and Dougherty, E. L., *Petr. Trans. AIME* **225**, 436 (1962).
27. Carslaw, H. S., and Jaeger, J. C., "Conduction of Heat in Solids," chap. 13, 2nd Ed., London, Oxford Univ. Press, 1959.
28. McAdams, W. H., "Heat Transmission," 2nd Ed., New York, McGraw-Hill Book Co., Inc., 1942.
29. Jakob, Max, "Heat Transfer," vol. 1, p. 94, New York, John Wiley & Sons, Inc., 1949.
30. Fishenden, M., and Saunders, O. A., "An Introduction to Heat Transfer," London, Oxford Univ. Press, 1961.
31. Hinton, A. G., quoted by Squiers, H. M., "World Power Conference, London, 1928, Technical Data on Fuel," p. 101, World Power Conference, 1928.
32. Dittus, F. W., and Boelter, L. M. K., *Univ. Calif. Pub. in Engrg.* **2**, 443 (1930).
33. Sherwood, T. K., and Petrie, J. M., *Ind. Eng. Chem.* **24**, 736 (1932).
34. Colburn, A. P., *Trans. Am. Inst. Chem. Engrs.* **29**, 174 (1933).
35. Sieder, E. N., and Tate, G. E., *Ind. Eng. Chem.* **28**, 1429 (1936).
36. "International Critical Tables," vol. 2, p. 55, vol. 5, p. 84, New York, McGraw-Hill Book Co., Inc., 1929.
37. Perry, J. H., "Chemical Engineers' Handbook," 3rd Ed., p. 235, New York, McGraw-Hill Book Co., Inc., 1950.
38. Marks, L. W., "Mechanical Engineers' Handbook," p. 273, New York, McGraw-Hill Book Co., Inc., 1951.
39. Grober, H., Erk, S., and Grigull, U., "Fundamentals of Heat Transfer," p. 497, New York, McGraw-Hill Book Co., Inc., translated by J. R. Moszynski, 1961.
40. Wilkes, G. B., "Heat Insulation," New York, John Wiley & Sons, Inc., 1950.
41. Satter, Abdus, *J. Petrol. Technol.*, 845 (July, 1965).
42. Leutwyler, K., paper no. SPE 1264, presented Denver, fall mtg. of Petr. Br. of **AIME**, 1965.
43. Lauwerier, H. A., *Applied Science Research, Sec. A.* **5**, 145 (1955).
44. Marx, J. W., and Langenheim, R. H., *Petr. Trans. AIME* **216**, 312 (1959).
45. Rubinshtein, L. I., *Neft I Gaz* **2**, No. 9, 41 (1959).
46. Willman, B. T., Valleroy, V. V., Runberg, G . W., Cornelius, A. J., and Powers, L. W., *Petr. Trans. AIME* **222**, 681 (1961).
47. Waldorf, D. M., paper no. SPE 1118, presented fall mtg., Denver, *Soc. of Petrol. Engrs. of AIME* (October, 1965).
48. Landrum, B. L., Smith, J. E., and Crawford, P. B., paper no. 1389-G, presented fall mtg., Corpus Christi, *Soc. of Pet. Engrs. of AIME* (October, 1959).
49. Fournier, K. P., *Soc. of Petrol. Engrs., AIME*, 131 (June, 1965).
50. Owens, W. D., and Suter, V. E., "Fundamentals of Thermal Oil Recovery," p. 61, Dallas, Petroleum Engineer Publishing Co., 1965.

51. Long, R. J., paper no. SPE 1168, presented Denver, fall mtg. of Petr. Br. of AIME, 1965.
52. Payne, R. W., and Zambrano, G., *Oil Gas J.*, 78 (May 24, 1965).
53. Kastrop, J. E., "Fundamentals of Thermal Oil Recovery," p. 11, Dallas, Petroleum Engineer Publishing Co., 1965.
54. Anon., "Fundamentals of Thermal Oil Recovery," Dallas, Petroleum Engineer Publishing Co., 1965.
55. Burns, W. C., paper no. SPE 1000, presented fall mtg., Houston, *Soc. of Petrol. Engrs. of AIME* (October, 1964).
56. Hagist, F. C., and Fincher, D. R., *Oil Gas J.*, 64 (January 11, 1965).
57. C. E. Fieber, Reference 50, p. 127.
58. Fincher, D. R., Hagist, F. C., and Gallaher, D. L., paper no. SPE 1265, presented fall mtg., Denver, *Soc. of Petrol. Engrs. of AIME* (October, 1965).
59. Becher, Paul, "Emulsions: Theory and Practice," Amer. Chem. Soc. Monograph Series, New York, Reinhold Publ. Corp., 1965.
60. Bertness, T. A., paper no. SPE 1266, presented fall mtg., Denver, *Soc. of Petrol. Engrs. of AIME* (October, 1965).
61. Eckert, E. R. G., and Drake, R. M., Jr., "Heat and Mass Transfer," p. 496, New York, McGraw-Hill Publ. Co., Inc., 1959.

PROBLEMS

1. A steam generator has an output of 15 MMBtu/hr of 80 per cent quality steam.

(a) Calculate the heat loss from the steam generator to an injection well 300 feet away. The line has an outside diameter of 2.875 inches and a wall thickness of 0.217 inches, and is steel. Assume the line is bare metal, the air is still, and ambient temperature is $0°F$.

(b) Calculate the temperature and steam quality at the injection well of the steam, assuming no pressure drop in the line. Repeat the calculation for a friction pressure drop of 10 psi.

(c) Calculate the heat available at the injection well for the conditions of Part (a) and (b).

(d) The pipe of Part (a) is to be insulated with 1½ inch of magnesia pipe insulation. Repeat the calculations of Parts (a), (b), and (c) for this case.

(e) What would be the thermal expansion of the line if the mean coefficient of thermal expansion of steel is 6.5×10^{-6} ft/ft-$°F$?

2. Calculate the wellbore heat loss in Btu/day which would result when injecting a high quality steam at 514.7 psia down 2 inch nominal diameter tubing inside 7 inch nominal diameter casing, after 10 days of injection. The data applicable to the well are as follows:

Depth	950 ft
Surface temperature (geothermal)	70°F
Thermal conductivity of formation	2.3 Btu/hr-ft-°F
Rock density	167 lb/ft³
Specific heat of rock	0.226 Btu/lb$_m$-°F
Overall thermal conductivity	70 Btu/ft²-°F-day

What would be the heat loss after 100 days? After 1 year?

3. Use dimensional analysis (Pi theorem) to derive equation (25); which is commonly called the Dittus-Boelter equation.

4. Derive equation (15) from equations (11) and (12). Note that the heat flow is in series, and that the resistance will be proportional to the difference in temperature between each boundary. Solve equation (11) for $(T_1 - T_2)$, and equation (12) for $(T_2 - T_e)$, eliminating T_2.

5. When a high quality steam is injected down tubing set on a packer and the annulus vented, there is a negligible change in steam temperature with depth. The total heat loss may be calculated from equation (48). Derive this equation using equation (11) and the answer from Problem 3.

6. Calculate the total heat loss in the wellbore when high quality steam is injected down tubing set on a packer, and where the annulus is vented, after injection times of 1, 10, and 100 days, for the following conditions:

Well depth	950 ft	r_1	0.0833 ft
Injection pressure	666 psia	r_2	0.29 ft
Steam temperature	497°F	U	70 Btu/ft^2-°F-day
k	33.6 Btu/day-ft^2-°F	Surface temperature	60°F
Rock density	125 lb$_m$/ft^3	Geothermal Gradient	0.0105 °F/ft
c_p	0.20 Btu/lb$_m$-°F		

Repeat the calculation for the case where steam is injected down the casing.

7. Calculate the bottom hole temperature for hot water injection after injection times of 1, 10, and 100 days for the following conditions:

Depth	950 ft	r_1	0.0833 ft
Injection rate	500 bbl/day	r_2	0.29 ft
Surface water temperature	400°F	U	70 Btu/ft^2-°F-day
k	33.6 Btu/day-ft^2-°F	Geothermal surface temperature	60°F
c_p	0.20 Btu/lb$_m$-°F		

8. Six thousand lb/hr of 80 per cent quality steam is injected at 450 psig into a well. The available data follow:

Porosity	25 per cent
Irreducible water saturation	25 per cent
Gas saturation (initial)	6 per cent
Residual oil saturation	15 per cent
Formation temperature	85°F
Oil pay thickness	22 ft
Rock specific heat	0.20 Btu/lb$_m$-°F
Oil specific heat	0.48 Btu/lb$_m$-°F
Rock grain density	165 lb$_m$/ft

Oil gravity	12° API
Overburden thermal conductivity	1.38 Btu/hr-ft-°F
Overburden thermal diffusivity	0.048 ft^2/hr

If the cost to generate steam is $0.52 per MMBtu, and the net value of the oil to the working interest is $1.90 per STB after royalty and overhead expenses, calculate:

(a) The swept area after an injection period of 30 days.

(b) The oil displacement rate at the end of 30 days.

(c) The time when the economic limit is reached.

(d) An approximate answer for the well spacing which could be used for a five-spot pattern steam injection program. Discuss the method which you have used to obtain such an answer.

(e) Use the method of Willman *et al.* to calculate the radial position of the steam and water fronts at the end of 30 days of steam injection. Compare the resulting answer with that calculated in Part (a).

Author Index

Subject Index